T0142838

Lecture Notes in Computational Science and Engineering

115

More information about this series at http://www.springer.com/series/3527

Michael Griebel • Marc Alexander Schweitzer
Editors

Meshfree Methods for Partial Differential Equations VIII

Editors
Michael Griebel
Institut für Numerische Simulation
Universität Bonn
Bonn, Germany

Fraunhofer Institute for Algorithms
and Scientific Computing SCAI
Sankt Augustin, Germany

Marc Alexander Schweitzer
Institut für Numerische Simulation
Universität Bonn
Bonn, Germany

Fraunhofer Institute for Algorithms
and Scientific Computing SCAI
Sankt Augustin, Germany

ISSN 1439-7358 ISSN 2197-7100 (electronic)
Lecture Notes in Computational Science and Engineering
ISBN 978-3-319-84787-0 ISBN 978-3-319-51954-8 (eBook)
DOI 10.1007/978-3-319-51954-8

Mathematics Subject Classification (2010): 65N30, 65N35, 65N75, 65M60, 65M70, 65M75

Printed on acid-free paper

This Springer imprint is published by Springer Nature
The registered company is Springer International Publishing AG
The registered company address is: Gewerbestrasse 11, 6330 Cham, Switzerland

Preface

The Eighth International Workshop on *Meshfree Methods for Partial Differential Equations* was held from September 7 to September 9, 2015, in Bonn, Germany. It was dedicated to the memory of Ted Belytschko, who passed away in September 2014. Ted Belytschko was one of the leading experts in meshfree methods and co-organized the workshop series over many years. He is dearly missed.

This workshop series was installed in 2001 to bring together European, American and Asian researchers working in this exciting field of interdisciplinary research on a regular basis. To this end, Ivo Babuška, Jiun-Shyan Chen, Michael Griebel, Antonio Huerta, Wing Kam Liu, Marc Alexander Schweitzer and Harry Yserentant invited scientist from all over the world to Bonn to strengthen the mathematical understanding and analysis of meshfree discretizations but also to promote the exchange of ideas on their implementation and application.

The workshop was again hosted by the Institut für Numerische Simulation at the Rheinische Friedrich-Wilhelms-Universität Bonn with the financial support of the Sonderforschungsbereich 1060 *The Mathematics of Emergent Effects* and the Hausdorff Center for Mathematics.

This volume of LNCSE now comprises selected contributions of attendees of the workshop. The selected papers cover a wide range of topics from applied mathematics to physics and engineering and even industrial applications which clearly indicates the maturity meshfree methods have reached in recent years. Meshfree methods have a diverse and rich mathematical background and their flexibility renders them particularly interesting for challenging applications in which classical mesh-based approximation techniques struggle or even fail.

Bonn, Germany
Bonn, Germany
October 2016

Michael Griebel
Marc Alexander Schweitzer

Contents

A Two-Level Additive Schwarz Domain Decomposition Preconditioner for a Flat-Top Partition of Unity Method

Susanne C. Brenner, Christopher B. Davis, and Li-yeng Sung

Abstract We investigate a two-level additive Schwarz domain decomposition preconditioner for a flat-top partition of unity method. We establish condition number estimates for the biharmonic problem and present numerical results that confirm our analysis.

1 Introduction

Let Ω be a polygonal domain in $\mathbb{R}^2$ and $f \in L_2(\Omega)$. Consider the following model problem: Find $u \in H_0^2(\Omega)$ such that

$$a(u, v) = (f, v) \qquad \text{for all } v \text{ in } H_0^2(\Omega) \tag{1}$$

where

$$a(w, v) = \int_\Omega (D^2 w : D^2 v) dx = \sum_{i,j=1}^{2} \int_\Omega \frac{\partial^2 w}{\partial x_i \partial x_j} \frac{\partial^2 v}{\partial x_i \partial x_j} dx \tag{2}$$

and $(f, v) = \int_\Omega f v \, dx$.

S.C. Brenner (✉) • L.-Y. Sung
Department of Mathematics, Center for Computation & Technology, Louisiana State University, Baton Rouge, LA 70803, USA
e-mail: brenner@math.lsu.edu; sung@math.lsu.edu

C.B. Davis
Department of Mathematics, Tennessee Technological University, Cookeville, TN 38505, USA
e-mail: cbdavis@tntech.edu

M. Griebel, M.A. Schweitzer (eds.), *Meshfree Methods for Partial Differential Equations VIII*, Lecture Notes in Computational Science and Engineering 115,
DOI 10.1007/978-3-319-51954-8_1

Let V_h be a finite dimensional subspace of $H_0^2(\Omega)$. The Ritz-Galerkin method for (1) is to find $u_h \in V_h$ such that

$$a(u_h, v) = (f, v) \qquad \forall\, v \in V_h. \tag{3}$$

In this paper we will investigate a two-level additive Schwarz domain preconditioner [8, 30] for the discrete problem (3), where V_h is constructed by a flat-top partition of unity method.

The conditioning of partition of unity methods is an important topic that has recently received some attention. Stable generalized finite element methods whose condition numbers are comparable to standard finite element methods are discussed in [1, 22, 23, 29, 33]. Preconditioners for extended finite element methods have also been investigated. (See [4, 24, 31] and references therein for a non-exhaustive list.) The focus of the aforementioned work is on the ill-conditioning of the discrete problem due to the choice of the enrichment functions. As far as we know, there is only one paper [14] in the literature where an additive Schwarz preconditioner for partition of unity methods is treated, and the preconditioner considered there is a hierarchical multilevel preconditioner.

One of the important features of the partition of unity method is its ability to generate a smooth approximation space with ease, making it a good candidate for higher order problems. While there is a substantial literature on domain decomposition preconditioners for finite element methods for fourth order problems [6, 7, 9, 10, 18, 19, 25, 26, 32], to our knowledge domain decomposition preconditioners for the partition of unity method have not been studied. Our goal is to fill this gap.

The rest of the paper is organized as follows. We present the flat-top partition of unity method and the additive Schwarz preconditioner in Sects. 2 and 3. The condition number estimates are carried out in Sect. 4, followed by numerical results in Sect. 5. The paper ends with some concluding remarks in Sect. 6.

2 A Flat-Top Partition of Unity Method

In this section we describe the construction of V_h using a flat-top partition of unity.

2.1 *Partition of Unity*

First we recall the definition of a W^2_∞ partition of unity.

Definition 1 Let $\Lambda = \{\Omega_i\}_{i=1}^n$ be an open cover of $\bar{\Omega}$ satisfying a pointwise overlap condition

$$\exists M \in \mathbb{N} \quad \text{such that} \quad \text{card}\{i | x \in \Omega_i\} \le M \quad \forall\, x \in \Omega.$$

Let $\{\varphi_i\}_{i=1}^n$ be a family of functions in $W^2_\infty(\mathbb{R}^2)$ satisfying

$$\text{supp}\,\varphi_i \subset \bar{\Omega}_i \quad 1 \le i \le n,$$

$$\sum_{i=1}^{n} \varphi_i \equiv 1 \text{ on } \Omega,$$

$$|\varphi_i|_{W^m_\infty(\mathbb{R}^2)} \le \frac{C_m}{(\text{diam }\Omega_i)^m} \quad 0 \le m \le 2,\ 1 \le i \le n,$$

where C_m are constants. We will refer to $\{\varphi_i\}_{i=1}^n$ as a W^2_∞ partition of unity subordinate to the cover Λ and the covering sets $\Omega_i \in \Lambda$ as patches.

We will use a variant of the partition of unity in [11–13, 16, 28] and we refer the interested reader to these articles for a more thorough description of the construction. Below we briefly describe our approach for a rectangular domain. Other domains can be treated in a similar fashion.

We begin by choosing two small positive parameters γ_1 and γ_2, and construct the domain Ω_γ by enlarging Ω by a distance of γ_j in the $\pm x_j$ directions for $j = 1$ and 2. We then subdivide Ω_γ into congruent rectangles R_i for $1 \le i \le n$. The lengths of the sides of these rectangles are denoted by h_1 and h_2, which are proportional to γ_1 and γ_2 respectively. The mesh parameter h is the maximum of h_1 and h_2.

The patches Ω_i are formed by enlarging the rectangles R_i by a distance of γ_j in the $\pm x_j$ directions for $j = 1$ and 2. There is a rectangular region in the center of each Ω_i denoted by Ω_i^{flat}. The partition of unity function φ_i is a C^1 piecewise polynomial function such that $\varphi_i = 1$ on Ω_i^{flat} and smoothly decreases to 0 on $\partial\Omega_i$. The construction of the flat-top partition of unity is illustrated in Fig. 1.

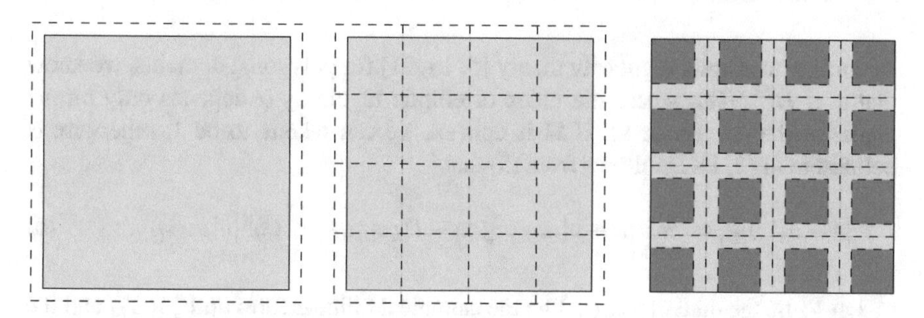

Fig. 1 The construction of the flat-top partition of unity: Ω is expand to Ω_γ (*left*); Ω_γ is subdivided into congruent rectangles (*middle*); Ω_i^{flat} are the dark shaded regions (*right*)

2.2 Generalized Finite Element Space

Let V_i be a subspace of the tensor product Lagrange finite element space $\mathbb{Q}_2$ defined on Ω_i whose members satisfy the homogeneous Dirichlet conditions on $\partial\Omega$. The interpolation nodes for V_i are placed inside the flat-top region Ω_i^{flat}. The generalized finite element space $V_h \subset V$ is given by

$$V_h = \sum_{i=1}^{n} \varphi_i V_i.$$

The interpolation operator $\Pi_h : C(\bar{\Omega}) \to V_h$ is defined by

$$\Pi_h v = \sum_{i=1}^{n} (\Pi_i v)\varphi_i, \tag{4}$$

where Π_i is the local nodal interpolation operator for V_i. The following interpolation error estimate can be established by combining standard interpolation error estimates for the $\mathbb{Q}_2$ finite element and the estimates for the partition of unity functions $\varphi_1, \ldots, \varphi_n$ (cf. [27, 28] for details):

$$\sum_{m=0}^{2} h^m |v - \Pi_h v|_{H^m(\Omega)} \le C h^s |v|_{H^s(\Omega)} \qquad \forall\, v \in H^s(\Omega) \quad \text{and} \quad 2 \le s \le 3. \tag{5}$$

2.3 Discretization Error Estimate and Conditioning

According to elliptic regularity theory [5, 15, 21] for polygonal domains, we know that $u \in H^{2+\alpha}(\Omega)$, where the index of elliptic regularity α depends only on the angles at the corners of Ω. If Ω is convex, we can take α to be 1, otherwise α belongs to $(1/2, 1)$. It follows from (5) that

$$|u - u_h|_{H^2(\Omega)} = \inf_{v \in V_h} |u - v|_{H^2(\Omega)} \le |u - \Pi_h u|_{H^2(\Omega)} \le C h^\alpha |u|_{H^{2+\alpha}(\Omega)}. \tag{6}$$

Let V_h' be the dual of V_h, $\langle \cdot, \cdot \rangle$ be the canonical bilinear form on $V_h' \times V_h$, and the linear operator $A_h : V_h \to V_h'$ be defined by

$$\langle A_h w, v \rangle = a(w, v) \quad \forall\, w, v \in V_h. \tag{7}$$

We can then rewrite (3) as

$$A_h x = f_h, \tag{8}$$

where $f_h \in V_h'$ is defined by

$$\langle f_h, v\rangle = \int_\Omega fv\,dx \qquad \forall\, v \in V_h.$$

It can be shown that the condition number of A_h satisfies

$$\kappa(A_h) \approx O(h^{-4}), \tag{9}$$

which is similar to the condition number for standard finite element methods.

3 A Two-Level Additive Schwarz Preconditioner

The two-level additive Schwarz preconditioner was introduced by Dryja and Widlund in [17]. It involves a coarse problem and local problems.

3.1 Coarse Problem

Let $V_0 \subset V$ be the generalized finite element space associated with a coarse mesh with mesh parameter H. We assume there are J coarse patches $\Omega_{j,H}$ $(1 \leq j \leq J)$ in the construction of V_0.

The coarse space V_0 is connected to V_h by the operator $I_0 : V_0 \longrightarrow V_h$, which is the restriction of the interpolation operator Π_h to V_0. The operator $A_0 : V_0 \longrightarrow V_0'$ is then given by

$$\langle A_0 w, v\rangle = a(I_0 w, I_0 v) \quad \forall w, v \in V_0. \tag{10}$$

3.2 Local Problems

The overlapping subdomains $\tilde{\Omega}_j$ of Ω are obtained by enlarging the coarse patch $\Omega_{j,H}$ $(1 \leq j \leq J)$ by the amount of $\delta_j(\geq 0)$ in the $\pm x_j$ directions for $j = 1$ and 2. This means that the overlap of the subdomains is given by $\delta = \max\{\delta_1 + \gamma_{1,H}, \delta_2 + \gamma_{2,H}\}$. By adjusting δ_j and $\gamma_{j,H}$, we can align $\partial\tilde{\Omega}_j$ with the boundaries of the patches for V_h and also control the overlap among the subdomains.

The local space $V_i \subset V_h$ is taken to be

$$V_j = \{v \in V_h : v = 0 \text{ on } \Omega\backslash\tilde{\Omega}_j\}$$

and it is connected to V_h by the natural injection $I_j : V_j \longrightarrow V_h$.

The operator $A_j : V_j \longrightarrow V_j'$ is given by

$$\langle A_j w, v\rangle = a(w,v) \quad \forall w, v \in V_j. \tag{11}$$

3.3 The Preconditioner

The two-level additive Schwarz preconditioner is defined by

$$B_{TL} = \sum_{j=0}^{J} I_j A_j^{-1} I_j^T,$$

where the transpose operator $I_j^T : V_h' \longrightarrow V_j'$ is given by

$$\langle I_j^T \mu, v\rangle = \langle \mu, I_j v\rangle \qquad \forall\, \mu \in V_h', v \in V_j.$$

Since $V_h = \sum_{j=0}^{J} I_j V_j$, the operator B_{TL} is symmetric positive definite and we have the following characterizations of the maximum and minimum eigenvalues of $B_{TL}A_h$ (cf. [8, Theorem 7.1.20]).

$$\lambda_{\max}(B_{TL}A_h) = \max_{\substack{v\in V_h\\ v\neq 0}} \frac{\langle A_h v, v\rangle}{\min_{\substack{v=\sum_{j=0}^{J} I_j v_j\\ v_j \in V_j}} \sum_{j=0}^{J} \langle A_j v_j, v_j\rangle} \tag{12}$$

$$\lambda_{\min}(B_{TL}A_h) = \min_{\substack{v\in V_h\\ v\neq 0}} \frac{\langle A_h v, v\rangle}{\min_{\substack{v=\sum_{j=0}^{J} I_j v_j\\ v_j \in V_j}} \sum_{j=0}^{J} \langle A_j v_j, v_j\rangle} \tag{13}$$

4 Condition Number Estimates

To avoid the proliferation of constants, we will use the notation $A \lesssim B$ (or $B \gtrsim A$) to represent the inequality $A \leq (\text{constant})B$, where the positive constant is independent of h, H, δ and J. The notation $A \approx B$ is equivalent to $A \lesssim B$ and $A \gtrsim B$.

4.1 Estimate for $\lambda_{\max}(B_{TL}A_h)$

The following lemma will lead to an upper bound for $\lambda_{\max}(B_{TL}A_h)$.

Lemma 1 *Let* $v_j \in V_j$ *for* $0 \le j \le J$ *and* $v = \sum_{j=0}^{J} I_j v_j$. *Then the following estimate holds*:

$$\langle A_h v, v\rangle \lesssim \sum_{j=0}^{J} \langle A_j v_j, v_j\rangle. \tag{14}$$

Proof Since I_j for $1 \le j \le J$ are natural injections, we derive from (2), (7) and (10)

$$\begin{aligned}
\langle A_h v, v\rangle &= \int_\Omega D^2\Big(\sum_{j=0}^{J} I_j v_j\Big) : D^2\Big(\sum_{k=0}^{J} I_k v_k\Big)\,dx \\
&\le 2\int_\Omega |D^2 I_0 v_0|^2\,dx + 2\int_\Omega D^2\Big(\sum_{j=1}^{J} I_j v_j\Big) : D^2\Big(\sum_{k=1}^{J} I_k v_k\Big)\,dx \\
&= 2\langle A_0 v_0, v_0\rangle + 2\sum_{j,k=1}^{J}\int_\Omega D^2 v_j : D^2 v_k\,dx.
\end{aligned} \tag{15}$$

Let the constant $c_{j,k}$ $(1 \le j, k \le J)$ be defined by

$$c_{j,k} = \begin{cases} 1 & \text{if } \tilde{\Omega}_j \cap \tilde{\Omega}_k \ne \emptyset, \\ 0 & \text{otherwise.} \end{cases}$$

Note that $c_{j,k} = c_{k,j}$.

Let N be the maximum number of subdomains that can have nonempty intersection with a subdomain. Then we have

$$\sum_{k=1}^{J} c_{j,k} \le N \quad \text{for} \quad 1 \le j \le J$$

and hence, in view of (11),

$$\begin{aligned}
\sum_{j,k=1}^{J}\int_\Omega (D^2 v_j : D^2 v_k)\,dx &= \sum_{j,k=1}^{J} c_{j,k}\int_\Omega (D^2 v_j : D^2 v_k)\,dx \\
&= \sum_{j,k=1}^{J} c_{j,k} |v_j|_{H^2(\Omega)} |v_k|_{H^2(\Omega)} \\
&\le \Big(\sum_{j,k=1}^{J} c_{j,k} |v_j|^2_{H^2(\Omega)}\Big)^{\frac{1}{2}} \Big(\sum_{j,k=1}^{J} c_{j,k} |v_k|^2_{H^2(\Omega)}\Big)^{\frac{1}{2}}
\end{aligned}$$

$$= \sum_{j=1}^{J} |v_j|^2_{H^2(\Omega)} \sum_{k=1}^{J} c_{j,k}$$

$$\leq N \sum_{j=1}^{J} |v_j|^2_{H^2(\Omega)} = N \sum_{j=1}^{J} \langle A_j v_j, v_j \rangle. \quad (16)$$

The estimate (14) follows by combining (15) and (16). □

Combining (12) and (14), we have an upper bound for the eigenvalues of $B_{TL}A_h$:

$$\lambda_{\max}(B_{TL}A_h) \lesssim 1. \quad (17)$$

4.2 Estimate for $\lambda_{\min}(B_{TL}A_h)$

The following lemma will lead to a lower bound for $\lambda_{\min}(B_{TL}A_h)$.

Lemma 2 *Given any $v \in V_h$, there exists a decomposition*

$$v = \sum_{j=0}^{J} I_j v_j \quad (18)$$

where $v_j \in V_j$ for $1 \leq j \leq J$ and

$$\sum_{j=0}^{J} \langle A_j v_j, v_j \rangle \lesssim \left[1 + \left(\frac{H}{\delta} \right)^3 \right] \langle A_h v, v \rangle. \quad (19)$$

Proof It follows from (5) (with $s = 2$) that

$$\sum_{k=0}^{1} h^k |v - \Pi_h v|_{H^k(\Omega)} + h^2 |\Pi_h v|_{H^2(\Omega)} \lesssim h^2 |v|_{H^2(\Omega)} \qquad \forall\, v \in H^2(\Omega). \quad (20)$$

Similarly, we have

$$\sum_{k=0}^{1} H^k |v - \Pi_H v|_{H^k(\Omega)} + H^2 |\Pi_H v|_{H^2(\Omega)} \lesssim H^2 |v|_{H^2(\Omega)} \qquad \forall\, v \in H^2(\Omega), \quad (21)$$

where Π_H is the analog of Π_h for V_H.

Let $v_0 = \Pi_H v \in V_0$, $w = v - I_0 v_0 = v - \Pi_h v_0$ and $v_j = \Pi_h(\theta_j w)$, where $\{\theta_j\}_{j=1}^J$ is a W_∞^2 partition of unity subordinate to the overlapping subdomains $\tilde{\Omega}_j$ such that

$$|\theta_j|_{W_\infty^k(\mathbb{R}^2)} \lesssim \delta^{-k} \qquad \text{for} \quad 0 \le k \le 2. \tag{22}$$

It is easy to check that $v_j \in V_j$ for $0 \le j \le J$ and (18) holds.

In view of (2), (7), (10), (20) and (21), we have

$$\begin{aligned} \langle A_0 v_0, v_0 \rangle = |I_0 v_0|_{H^2(\Omega)}^2 &= |\Pi_h v_0|_{H^2(\Omega)}^2 \\ &\lesssim |v_0|_{H^2(\Omega)}^2 \\ &= |\Pi_H v|_{H^2(\Omega)}^2 \lesssim |v|_{H^2(\Omega)}^2 = \langle A_h v, v \rangle. \end{aligned} \tag{23}$$

Next we consider

$$\langle A_j v_j, v_j \rangle = |v_j|_{H^2(\tilde{\Omega})}^2 = |\Pi_h(\theta_j w)|_{H^2(\Omega)}^2$$

for $1 \le j \le J$. In view of (20), we have

$$\begin{aligned} \langle A_j v_j, v_j \rangle &\lesssim |\theta_j w|_{H^2(\Omega)}^2 \\ &\lesssim \int_{\tilde{\Omega}_j} (w)^2 |D^2 \theta_j|^2 \, dx + \int_{\tilde{\Omega}_j} |D\theta_j|^2 |Dw|^2 \, dx \\ &\quad + \int_{\tilde{\Omega}_j} (\theta_j)^2 |D^2 w|^2 \, dx \end{aligned} \tag{24}$$

and it only remains to estimate the three terms on the right-hand side of (24).

Observe that (20) and (21) imply

$$\begin{aligned} \|w\|_{L_2(\Omega)} &= \|v - \Pi_h v_0\|_{L_2(\Omega)} \\ &\le \|v - \Pi_h v\|_{L_2(\Omega)} + \|v - \Pi_H v\|_{L_2(\Omega)} \\ &\qquad + \|(v - \Pi_H v) - \Pi_h(v - \Pi_H v)\|_{L_2(\Omega)} \\ &\lesssim h^2 |v|_{H^2(\Omega)} + H^2 |v|_{H^2(\Omega)} + h^2 |v - \Pi_H v|_{H^2(\Omega)} \\ &\lesssim H^2 |v|_{H^2(\Omega)}, \end{aligned} \tag{25}$$

and similarly

$$|w|_{H^1(\Omega)} \lesssim H |v|_{H^2(\Omega)}, \tag{26}$$

$$|w|_{H^2(\Omega)} \lesssim |v|_{H^2(\Omega)}. \tag{27}$$

It follows from (22) that

$$\int_{\tilde{\Omega}_j} |D\theta_j|^2 |Dw|^2\,dx \lesssim \delta^{-2}|w|^2_{H^1(\tilde{\Omega}_j)}, \tag{28}$$

$$\int_{\tilde{\Omega}_j} (\theta_j)^2 |D^2 w|^2\,dx \lesssim |w|^2_{H^2(\tilde{\Omega}_j)}. \tag{29}$$

Note that $D^2\theta_j$ vanishes except on a strip near $\partial\tilde{\Omega}_j$ with width $\approx \delta$. Therefore it follows from (22) and [6, Lemma 8.1] that

$$\int_{\tilde{\Omega}_j} (w)^2 |D^2\theta_j|^2\,dx \lesssim \left(\frac{1}{\delta^3 H}\right)\|w\|^2_{L_2(\tilde{\Omega}_j)} + \left(\frac{H}{\delta}\right)^3 |w|^2_{H^2(\tilde{\Omega}_j)}. \tag{30}$$

Combining (2), (7), (24), (25) and (28)–(30), we find

$$\begin{aligned}
\sum_{j=1}^{J}\langle A_j v_j, v_j\rangle &\lesssim \left(\frac{1}{\delta^3 H}\right)\sum_{j=1}^{J}\|w\|^2_{L_2(\tilde{\Omega}_j)} + \left(\frac{1}{\delta^2}\right)\sum_{j=1}^{J}|w|^2_{H^1(\tilde{\Omega}_j)} \\
&\quad + \left[1 + \left(\frac{H}{\delta}\right)^3\right]\sum_{j=1}^{J}|w|^2_{H^2(\tilde{\Omega}_j)} \\
&\lesssim \left(\frac{1}{\delta^3 H}\right)\|w\|^2_{L_2(\Omega)} + \left(\frac{1}{\delta^2}\right)|w|^2_{H^1(\Omega)} + \left[1 + \left(\frac{H}{\delta}\right)^3\right]|w|^2_{H^2(\Omega)} \\
&\lesssim \left[1 + \left(\frac{H}{\delta}\right)^3\right]|v|^2_{H^2(\Omega)} = \left[1 + \left(\frac{H}{\delta}\right)^3\right]\langle A_h v, v\rangle.
\end{aligned} \tag{31}$$

The estimate (19) follows from (23) and (31) □

Combining (13) and (19), we have a lower bound for the eigenvalues of $B_{TL}A_h$:

$$\lambda_{\min}(B_{TL}A_h) \gtrsim \left[1 + \left(\frac{H}{\delta}\right)^3\right]^{-1}. \tag{32}$$

4.3 Estimate for $\kappa(B_{TL}A_h)$

Putting (17) and (32) together, we have the following result on the condition number of the preconditioned system.

Theorem 1 *There exists a constant C independent of h, H, δ and J such that*

$$\kappa(B_{TL}A_h) = \frac{\lambda_{\max}(B_{TL}A_h)}{\lambda_{\min}(B_{TL}A_h)} \leq C\left[1 + \left(\frac{H}{\delta}\right)^3\right].$$

5 Numerical Results

We have applied the two-level Schwarz preconditioner to the model problem on the unit square and an L-shaped domain. The numerical results presented here were obtained using PETSc [2, 3] and Supermike II, one of the high performance supercomputers at the Louisiana State University.

Throughout these numerical examples we will use the following notation:

- κ, the estimated condition number of the preconditioned system
- its, the number of iterations until the relative residual falls below 10^{-6}
- t_{solve}, amount of (wall) time, in seconds, required to solve the preconditioned system
- H, the maximum width of the coarse mesh
- δ, the amount of overlap among the overlapping subdomains
- h, the maximum width of the fine mesh
- $\|e_h\|_{\text{energy}}$, the error in energy norm given by $|u_h - \pi_h u|_{H^2(\Omega)}$

We run two numerical experiments for each domain to observe the scalability. The first experiment, the case of small overlap, measures strong scalability. This is carried out by keeping the amount of overlap among the subdomains fixed and small, and then refining the coarse mesh. The second experiment, the case of generous overlap, measures weak scalability. This is carried out by keeping the quantity H/δ bounded, and then refining the fine mesh.

The local and coarse problems are solved by using a Cholesky factorization (on their own processors) and the global problem is solved using the preconditioned conjugate gradient method.

5.1 *Results for the Unit Square*

Let Ω be the unit square $(-0.5, 0.5)^2$. We take the exact solution to be

$$u(x) = \frac{35}{2}(x_1^2 - 0.25)^2(x_2^2 - 0.25)^2.$$

The generalized finite element space V_h is constructed through a flat-top partition unity with a background uniform mesh. An example of a fine mesh, a coarse mesh, and typical overlapping subdomains are shown in Fig. 2.

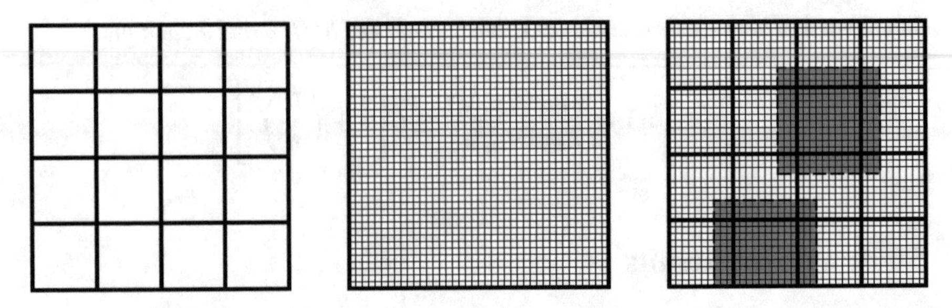

Fig. 2 A coarse mesh (*left*), a fine mesh (*middle*) and overlapping subdomains (*right*) for the unit square

Table 1 Small overlap results for the unit square

n_{sub}	H	κ	Rate	Its	t_{solve}
4	6.0000×10^{-1}	$9.1306\times10^{+6}$	–	1498	$3.8989\times10^{+3}$
16	2.7273×10^{-1}	$6.5953\times10^{+5}$	3.33	1409	$5.5002\times10^{+2}$
64	1.3043×10^{-1}	$7.4148\times10^{+4}$	2.96	400	$4.2519\times10^{+1}$
256	6.3830×10^{-2}	$1.0755\times10^{+4}$	2.70	165	$7.5584\times10^{+0}$
1024	3.1579×10^{-2}	$1.4248\times10^{+3}$	2.87	63	$4.2420\times10^{+0}$

5.1.1 Small Overlap

In the case of small overlap, the number of fine elements is fixed so that $h \approx 3.9113 \times 10^{-3}$. The amount of overlap among the subdomains is also fixed so that $\delta \approx 6.5189 \times 10^{-4}$.

The numerical results are presented in Table 1. We observe that

$$\kappa(B_{TL}A_h) \approx (H/\delta)^{\text{rate}}$$

where the rate is roughly 3, which agrees with Theorem 1. The scalability of the algorithm is also evidenced by the data in the last column.

5.1.2 Generous Overlap

In the case of generous overlap, the total number of subdomains is kept fixed ($n_{sub} = 64$) so that $H = 1.3043 \times 10^{-1}$. The fine mesh is then refined in such a way that $H/\delta \le 3$.

The numerical results are presented in Table 2. We observe that $\kappa(B_{TL}A_h)$ is uniformly bounded, which agrees with Theorem 1. We also observe $O(h)$ convergence in the energy error, which agrees with the estimate (6).

Table 2 Generous overlap results for the unit square

h	κ	Its	t_{solve}	$\|e_h\|_{\text{energy}}$	Rate
3.1579×10^{-2}	$3.6247\times10^{+1}$	23	4.6185×10^{-1}	5.6660×10^{-2}	–
1.5707×10^{-2}	$3.1383\times10^{+1}$	23	9.2236×10^{-1}	2.6070×10^{-2}	1.11
7.8329×10^{-3}	$2.7025\times10^{+1}$	21	$5.8779\times10^{+0}$	1.2454×10^{-2}	1.06
3.9113×10^{-3}	$2.5732\times10^{+1}$	21	$7.6505\times10^{+1}$	6.0795×10^{-3}	1.03
1.9544×10^{-3}	$2.4748\times10^{+1}$	21	$1.1662\times10^{+3}$	3.0027×10^{-3}	1.01

5.2 An L-Shaped Domain

Let Ω be the L-shaped domain $(-0.5, 0.5)^2 \setminus [0, 0.5]^2$. The exact solution u of the biharmonic problem is given by

$$u = (r^2 \cos^2(\theta) - 0.25)^2 (r^2 \sin^2(\theta) - 0.25)^2 r^{1+\alpha} g(\theta),$$

where the polar coordinate system (r, θ) is centered at the origin so that $\theta = 0$ corresponds to the positive y–axis and $\theta = 3\pi/2$ corresponds to the positive x–axis, and the function g (cf. [20, pp. 107–108]) is given by

$$\begin{aligned} g(\theta) = {} & [\cos((\alpha-1)\omega) - \cos((\alpha+1)\omega)] \\ & \times [(\alpha+1)\sin((\alpha-1)\theta) - (\alpha-1)\sin(\alpha+1)\theta)] \\ & - [\cos((\alpha-1)\theta) - \cos((\alpha+1)\theta)] \\ & \times [(\alpha+1)\sin((\alpha-1)\omega) - (\alpha-1)\sin(\alpha+1)\omega)] \,. \end{aligned}$$

Here $\omega = 3\pi/2$ is the angle of the reentrant corner and

$$\alpha \approx 0.544483736782464 \tag{33}$$

is the index of elliptic regularity.

The generalized finite element space V_h is constructed using a background mesh of quasi-uniform rectangles such that the reentrant corner is inside one of the rectangles. An example of a fine mesh, a coarse mesh, and typical overlapping subdomains are shown in Fig. 3.

5.2.1 Small Overlap

In the case of small overlap, the number of fine elements is fixed so that $h \approx 3.9164 \times 10^{-3}$. The amount of overlap between the subdomains is also fixed so that $\delta \approx 6.5104 \times 10^{-4}$.

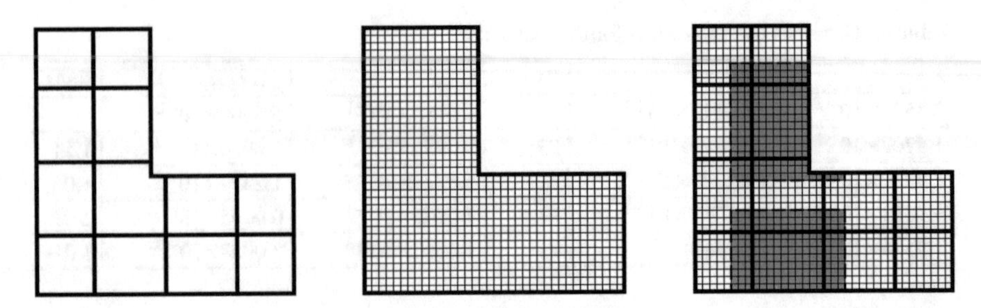

Fig. 3 A coarse mesh (*left*), a fine mesh (*middle*) and overlapping subdomains (*right*) for the *L*-shaped domain

Table 3 Small overlap results for the *L*-shaped domain

n_{sub}	H	κ	Rate	Its	t_{solve}
3	6.6667×10^{-1}	$8.8975\times10^{+6}$	–	1974	$4.5449\times10^{+3}$
12	2.9167×10^{-1}	$5.5286\times10^{+6}$	0.58	4816	$1.5500\times10^{+3}$
48	1.3542×10^{-1}	$1.0507\times10^{+6}$	2.16	1587	$1.3981\times10^{+2}$
192	6.5104×10^{-2}	$1.3655\times10^{+5}$	2.79	529	$1.8855\times10^{+1}$
768	3.1901×10^{-2}	$1.4414\times10^{+4}$	3.15	199	$9.8830\times10^{+0}$

Table 4 Generous overlap results for the *L*-shaped domain

h	κ	Its	t_{solve}	$\|e_h\|_{energy}$	Rate
3.1901×10^{-2}	$9.5077\times10^{+1}$	46	6.4991×10^{-1}	3.1239×10^{-3}	–
1.5788×10^{-2}	$1.1376\times10^{+2}$	47	$1.2841\times10^{+0}$	2.0686×10^{-3}	0.59
7.8532×10^{-3}	$1.2383\times10^{+2}$	41	$6.8940\times10^{+0}$	1.3966×10^{-3}	0.56
3.9164×10^{-3}	$1.2052\times10^{+2}$	37	$8.0453\times10^{+1}$	9.5070×10^{-4}	0.55
1.9557×10^{-3}	$1.2187\times10^{+2}$	37	$1.1779\times10^{+3}$	6.4955×10^{-4}	0.55

The numerical results are presented in Table 3. Again we observe that

$$\kappa(B_{TL}A_h) \approx (H/\delta)^3.$$

The scalability of the algorithm is also supported by the data in the last column.

5.2.2 Generous Overlap

In the case of generous overlap, the total number of subdomains is kept fixed ($n_{sub} = 48$) so that $H = 1.3542 \times 10^{-1}$. The fine mesh is then refined in such a way that $H/\delta \leq 3$.

The numerical results in Table 4 show that $\kappa(B_{TL}A_h)$ is uniformly bounded as predicted by Theorem 1, and that the energy error is $O(h^{0.55})$ as predicted by (6) and (33).

6 Concluding Remarks

We have extended the classical results for two-level additive Schwarz preconditioners to a flat-top partition of unity method for the biharmonic problem.

In the case of nonconvex domains, optimal convergence for the partition of unity method can be restored by including known corner singularities in the local approximation spaces. The preconditioner developed in this paper is also relevant for such methods.

The extension of the results in this paper to partition of unity methods for variational inequalities [11–13] and to partition of unity methods for sixth order problems are ongoing projects.

Acknowledgements The work of the first and third authors was supported in part by the National Science Foundation under Grant No. DMS-13-19172. Portions of this research were conducted with high performance computing resources provided by Louisiana State University (http://www.hpc.lsu.edu).

References

1. I. Babuška, U. Banerjee, Stable generalized finite element method (SGFEM). Comput. Methods Appl. Mech. Eng. **201–204**, 91–111 (2012)
2. S. Balay, W.D. Gropp, L.C. McInnes, B.F. Smith, Efficient management of parallelism in object-oriented numerical software libraries, in *Modern Software Tools in Scientific Computing* (Birkhauser, Boston, 1997), pp. 163–202
3. S. Balay, S. Abhyankar, M.F. Adams, J. Brown, P. Brune, K. Buschelman, V. Eijkhout, W.D. Gropp, D. Kaushik, M.G. Knepley, L.C. McInnes, K. Rupp, B.F. Smith, H. Zhang, PETSc users manual. Technical Report ANL-95/11 - Revision 3.5, Argonne National Laboratory (2014)
4. L. Berger-Vergiat, H. Waisman, B. Hiriyur, R. Tuminaro, D. Keyes, Inexact Schwarz-algebraic multigrid preconditioners for crack problems modeled by extended finite element methods. Int. J. Numer. Methods Eng. **90**, 311–328 (2012)
5. H. Blum, R. Rannacher, On the boundary value problem of the biharmonic operator on domains with angular corners. Math. Methods Appl. Sci. **2**, 556–581 (1980)
6. S.C. Brenner, A two-level additive Schwarz preconditioner for nonconforming plate elements. Numer. Math. **72**, 419–447 (1996)
7. S.C. Brenner, Two-level additive Schwarz preconditioners for nonconforming finite element methods. Math. Comput. **65**, 897–921 (1996)
8. S.C. Brenner, L.R. Scott, *The Mathematical Theory of Finite Element Methods*. Texts in Applied Mathematics, 3rd edn. (Springer, New York, 2008)
9. S.C. Brenner, K. Wang, Two-level additive Schwarz preconditioners for C^0 interior penalty methods. Numer. Math. **102**, 231–255 (2005)
10. S.C. Brenner, K. Wang, An iterative substructuring algorithm for a C^0 interior penalty method. Electron. Trans. Numer. Anal. **39**, 313–332 (2012)
11. S.C. Brenner, C.B. Davis, L.-Y. Sung, A partition of unity method for a class of fourth order elliptic variational inequalities. Comput. Methods Appl. Mech. Eng. **276**, 612–626 (2014)
12. S.C. Brenner, C.B. Davis, L.-Y. Sung, A partition of unity method for the displacement obstacle problem of clamped kirchhoff plates. J. Comput. Appl. Math. **265**, 3–16 (2014)

13. S.C. Brenner, C.B. Davis, L.-Y. Sung, A partition of unity method for the obstacle problem of simply supported Kirchhoff plates, in *Meshfree Methods for Partial Differential Equations VII*, ed. by M. Griebel, M.A. Schweitzer. Lecture Notes in Computational Science and Engineering, vol. 100 (Springer International Publishing, Berlin, 2015), pp. 23–41
14. W. Dahmen, S. Dekel, P. Petrushev, Multilevel preconditioning for partition of unity methods: some analytic concepts. Numer. Math. **107**, 503–532 (2007)
15. M. Dauge, *Elliptic Boundary Value Problems on Corner Domains*, Lecture Notes in Mathematics, vol. 1341 (Springer, Berlin/Heidelberg, 1988)
16. C.B. Davis, A partition of unity method with penalty for fourth order problems. J. Sci. Comput. **60**, 228–248 (2014)
17. M. Dryja, O.B. Widlund, An additive variant of the Schwarz alternating method in the case of many subregions. Technical Report 339, Department of Computer Science, Courant Institute (1987)
18. C. Farhat, P.-S. Chen, J. Mandel, F.-X. Roux, The two-level FETI method for static and dynamic plate problems Part I: an optimal iterative solver for biharmonic systems. Comput. Methods Appl. Mech. Eng. **155**, 129–151 (1998)
19. X. Feng, O.A. Karakashian, Two-level non-overlapping Schwarz preconditioners for a discontinuous Galerkin approximation of the biharmonic equation. J. Sci. Comput. **22/23**, 289–314 (2005)
20. P. Grisvard, *Singularities in Boundary Value Problems* (Masson, Paris, 1992)
21. P. Grisvard, *Elliptic Problems in Nonsmooth Domains* (Society for Industrial and Applied Mathematics, Providence, RI, 2011)
22. V. Gupta, C.A. Duarte, I. Babuška, U. Banerjee, A stable and optimally convergent generalized {FEM} (SGFEM) for linear elastic fracture mechanics. Comput. Methods Appl. Mech. Eng. **266**, 23–39 (2013)
23. V. Gupta, C.A. Duarte, I. Babuška, U. Banerjee, Stable {GFEM} (SGFEM): improved conditioning and accuracy of GFEM/XFEM for three-dimensional fracture mechanics. Comput. Methods Appl. Mech. Eng. **289**, 355–386 (2015)
24. C. Lang, D. Makhija, A. Doostan, K. Maute, A simple and efficient preconditioning scheme for heaviside enriched XFEM. Comput. Mech. **54**, 1357–1374 (2014)
25. P. LeTallec, J. Mandel, M. Vidrascu, Balancing domain decomposition for plates, in *Domain Decomposition Methods in Scientific and Engineering Computing*, ed. by D.E. Keyes, J. Xu. Contemporary Mathematics, vol. 180. (American Mathematical Society, Providence, RI, 1994), pp. 515–524
26. P. LeTallec, J. Mandel, M. Vidrascu, A Neumann-Neumann domain decomposition algorithm for solving plate and shell problems. SIAM J. Numer. Anal. **35**, 836–867 (1998)
27. J.M. Melenk, I. Babuška, The partition of unity finite element method: basic theory and applications. Comput. Methods Appl. Mech. Eng. **139**, 289–314 (1996)
28. H.-S. Oh, J.G. Kim, W.-T. Hong, The piecewise polynomial partition of unity functions for the generalized finite element methods. Comput. Methods Appl. Mech. Eng. **197**, 3702–3711 (2008)
29. M.A. Schweitzer, Stable enrichment and local preconditioning in the particle-partition of unity method. Numer. Math. **118**, 137–170 (2011)
30. A. Toselli, O.B. Widlund, *Domain Decomposition Methods - Algorithms and Theory* (Springer, New York, 2005)
31. H. Waisman, L. Berger-Vergiat, An adaptive domain decomposition preconditioner for crack propagation problems modeled by XFEM. Int. J. Multiscale Comput Eng. **11**, 633–654 (2013)
32. X. Zhang, Two-level Schwarz methods for the biharmonic problem discretized conforming C^1 elements. SIAM J. Numer. Anal. **33**, 555–570 (1996)
33. Q. Zhang, U. Banerjee, I. Babuška. Higher order stable generalized finite element method. Numer. Math. **128**, 1–29 (2014)

Extraction of Fragments and Waves After Impact Damage in Particle-Based Simulations

Patrick Diehl, Michael Bußler, Dirk Pflüger, Steffen Frey, Thomas Ertl, Filip Sadlo, and Marc Alexander Schweitzer

Abstract The analysis of simulation results and the verification against experimental data is essential to develop and interpret simulation models for impact damage. We present two visualization techniques to post-process particle-based simulation data, and we highlight new aspects for the quantitative comparison with experimental data. As the underlying simulation model we consider the particle method Peridynamics, a non-local generalization of continuum mechanics. The first analysis technique is an extended component labeling algorithm to extract the fragment size and the corresponding histograms. The distribution of the fragment size can be obtained by real-world experiments as demonstrated in Schram and Meyer (Simulating the formation and evolution of behind armor debris fields. ARL-RP 109, U.S. Army Research Laboratory, 2005), Vogler et al. (Int J Impact Eng 29:735–746, 2003). The second approach focuses on the visualization of the stress after an impact. Here, the particle-based data is re-sampled and rendered with standard volume rendering techniques to address the interference pattern of the

P. Diehl (✉)
Institute for Numerical Simulation, Wegelerstraße 6, 53115 Bonn, Germany
e-mail: diehl@ins.uni-bonn.de

M. Bußler • S. Frey • T. Ertl
University of Stuttgart Visualisation Research Centre, Allmandring 19, 70569 Stuttgart, Germany
e-mail: Michael.Bussler@visus.uni-stuttgart.de; Steffen.Frey@visus.uni-stuttgart.de; Thomas.Ertl@visus.uni-stuttgart.de

D. Pflüger
IPVS/SGS, Universitätsstraße 38, 70569 Stuttgart, Germany
e-mail: Dirk.Pflueger@ipvs.uni-stuttgart.de

F. Sadlo
Interdisciplinary Center for Scientific Computing, Im Neuenheimer Feld 368, 69120 Heidelberg, Germany
e-mail: filip.sadlo@iwr.uni-heidelberg.de

M.A. Schweitzer
Institut für Numerische Simulation, Rheinische Friedrich-Wilhelms-Universität Bonn, Wegelerstr. 6, 53115 Bonn, Germany

Fraunhofer Institute for Algorithms and Scientific Computing SCAI Sankt Augustin, Germany
e-mail: schweitzer@ins.uni-bonn.de

M. Griebel, M.A. Schweitzer (eds.), *Meshfree Methods for Partial Differential Equations VIII*, Lecture Notes in Computational Science and Engineering 115,
DOI 10.1007/978-3-319-51954-8_2

stress wave after reflection at the boundary. For the extraction and visual analysis, we used the widely-used Stanford bunny as a complex geometry. For a quantitative study with a simple geometry, the edge-on impact experiment (Schradin, Scripts German Acad Aeronaut Res 40:21–68, 1939; Strassburger, Int J Appl Ceram Technol 1:1:235–242, 2004; Kawai et al., Procedia Eng 103:287–293, 2015) can be applied. With these new visualization approaches, new insights for the quantitative comparison of fragmentation and wave propagation become intuitively accessible.

1 Introduction

Peridynamics is a generalization of traditional continuum mechanics. It is a particle-based approach and targeted towards the modeling of fractures and similar phenomena. The fundamental equations are integral equations. Damage is modeled by a force function that acts between two particles each.

Using peridynamic simulations, a vast range of materials with different properties, from polymethyl methacrylate to titanium alloy can be simulated [5]. The results significantly depend on the models and numerical schemes that are employed. Different models for bonds, time integration schemes and the summation of forces between particles are employed. Furthermore, several parameters such as the interaction horizon of particles can be tuned.

To study a certain material, the simulation results have to be validated and verified [3]. However, this is not a trivial task in itself. A common experimental benchmark is the Kalthoff-Winkler experiment [7], which was studied in [18, 23]. First, and where available or possible, simulation results can be verified against experiments [1, 2, 4]. Second, and much more frequent in the literature, metrics obtained in the simulation are compared against analytic results [11]—once again, only where possible—or against simulation results using different modeling approaches. It is obvious that characteristic numbers that can be measured by experiments cannot necessarily be obtained by analytic equations and vice versa.

Simulations, in contrast, have the advantage that metrics for both worlds can be obtained. Of course, this poses extra challenges. It is easy to obtain the size of fragments in scattering scenarios; characterizing their shapes requires significantly more effort. And the extraction of continuous measures such as the speed of stress propagation in a discrete and three-dimensional simulation world requires good interpolation techniques and careful analysis.

To provide a first, fast validation to simulation results, instructive visual feedback is more than helpful. A typical approach to quickly examine the effect of changes in the model or its parameters is to plot the particles and to color them depending on the local damage, for example. While this gives a quick impression on the damage behavior, it is restricted to the surface. It neither shows where a crack penetrates through the whole material, nor provides insight into what happens within the material. This basic visualization approach is already sufficient to study crack branching or the velocity at the crack tip, as demonstrated in the experiments in [9, 21]. For complex structures and other metrics, more advanced techniques are required.

In this work, we have developed and adapted methods to analyze impact damage in particle-based simulations [10, 12, 15]. Furthermore, we show their illustrative visualization and demonstrate its instructive power. This is a major step towards the future verification and validation of our models. We have focused on the analysis of fragmentation via fragment analysis and on impact damage via the propagation of stress waves.

1.1 Fragmentation

An important validation for particle-based methods is the identification of fragments after impacts, like a stone impacting the front window of a car. Here the size of the fragments is an important metric to estimate the medical harm to the occupants. Obtaining the relevant quantities, such as the size, mass and velocity of each fragment is still difficult in experiments. The experiment [26] provides a distribution for the number of fragments with respect to the fragment mass. The setup is a tungsten-alloy projectile perforating a steel armor plate with a velocity of $1020\,\mathrm{m\,s^{-1}}$. This experimental data is used as a benchmark for the smoothed particle hydrodynamics (SPH) model in [29] and for a Lagrangian approach in [17]. Another experiment covered in the literature is the fragmentation of a cylindrical steel tube using a gas gun impact with the velocity of $1920\,\mathrm{m\,s^{-1}}$ [28]. This experimental data is used as a benchmark for SPH [13]. All of these benchmarks show that fragmentation obtained by the simulations are qualitatively reasonable, but that the quantitative modeling of the material needs some improvement.

1.2 Impact Damage and Wave Propagation

For the safety of crashes with electric/hybrid cars, the impact damage in the ceramic core of the battery is essential. A common benchmark for the impact damage and wave propagation in ceramic material is the edge-on impact (EOI) experiment [8, 20, 27], which was developed in the 1980s for the visualization of impact damage and wave propagation. Different particle-based material models were verified against this experiment [2, 19]. Here, the reflection and interference of the impact shock wave is of high importance to understand where the damage in the ceramic core occurs.

In the remainder, we will first describe peridynamics in Sect. 2 and our visualization approaches in Sect. 3. In Sect. 4 we introduce experiments and explain their (visual) analysis. Finally, Sect. 5 concludes the work.

2 Peridynamics

As a reference particle-based method we consider peridynamics (PD), which is a non-local generalization of continuum mechanics, with a focus on discontinuous solutions as they arise in fracture mechanics. In this section, we present the essential ingredients that are important for the visualization techniques. The principle of this theory is that particles interact with other particles at a finite horizon δ by exchanging forces. This is very similar to SPH and MD approaches. The bond-based peridynamics equation of motion [22] for the acceleration at time t is given by the integral equation

$$\varrho(X)A(t,X) = \int\limits_{B_\delta(X)} f(t, x(t,X') - x(t,X), X' - X)dX' + b(t,X), \tag{1}$$

with the mass density $\varrho(X)$, f as the pair-wise force function which models the interaction of particles X and X' with respect to the initial reference configuration Ω_0, and with $b(t,X)$ denoting the external force. The internal forces between particles are exchanged within the finite interaction zone $B_\delta(X)$, see Fig. 1. As the constitutive law we use the Prototype Brittle Microelastic (PMB) material law [24] and for the simulations we use LAMMPS [16]. The bond-based PD, where a bond between two particles X and X' responds independently of all other bonds inside the interaction zone $B_\delta(X)$, implies that the Poisson ratio of the isotropic linear elastic solid is restricted to $\nu = 1/4$ for 3D and $1/3$ for 2D. Within the state-based PD, all other bonds connected to the endpoints of particle X influence the stretch and thus, any material, which is described by the classical continuum mechanic, can be

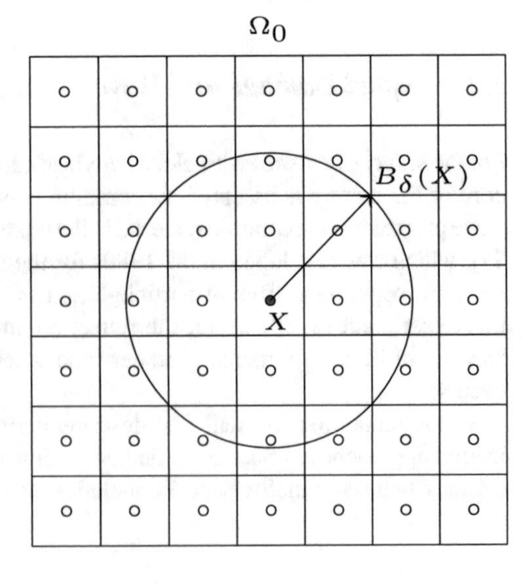

Fig. 1 The reference configuration Ω_0 at time $t = 0$ with the finite interaction zone of length δ for particle at position X. All particle inside the interaction zone $B_\delta(X)$ of particle X are connected with bonds to exchange forces

Table 1 Attributes per particle X, which where used in the post-processing pipeline for the extraction of the fragments and fracture surface

Attribute	Symbol	Unit
Acceleration	$a(t, X) \in \mathbb{R}^3$	ms^{-2}
Actual position	$x(t, X) \in \mathbb{R}$	m
Density	$\varrho(X) \in \mathbb{R}$	kg m^{-3}
Damage	$c(t, X) \in \mathbb{R}$	%
Displacement	$D(t, X) := \lvert X - x(t, X)\rvert \in \mathbb{R}^3$	m
Initial position	$X \in \mathbb{R}^3$	m
Force	$f(t, X) \in \mathbb{R}^3$	N
Velocity	$v(t, X) \in \mathbb{R}^3$	ms^{-1}
Volume	$V(X) \in \mathbb{R}$	m^3
Stress	$\sigma(t, X) \in \mathbb{R}^6$	Pa

modeled [25]. For the state-based PD, the equation of motion (1) is replaced by the following integral equation

$$\varrho(X)A(t, X) = \int_{B_\delta} \left\{\mathbf{T}[t, X]\langle X' - X\rangle - \mathbf{T}[t, X']\langle X - X'\rangle\right\} dX' + b(t, X) \qquad (2)$$

where $\mathbf{T}$ is the force vector state field. Cracks and fractures are modeled by the breaking of bonds between particles. Here a critical stretch s_c for bond breaking is predefined for all particles, and the bond between two particle breaks irreversibly if the stretch surpasses the predefined critical value. The ratio of the existing bonds inside the interaction zone $B_\delta(X)$ and the amount of bonds inside $B_\delta(x)$ at the reference configuration Ω_0 describes the damage $c(t, X)$ of a particle. For the simulation we used an extended version of Peridigm [14] and the elastic material model.

Table 1 shows for each particle at position X in the reference configuration Ω_0 the attributes available at each time step for the post-processing in the visualization pipeline. We extended Peridigm with a compute class for the adjacency matrix M_t, so that all bonds between particles are available as an additional information for the post-processing pipeline. Fore more details about peridynamics and material models we refer to [22, 24, 25].

3 Visualization Techniques

We present two visualization techniques for the analysis of data resulting from peridynamics simulations. The first technique discusses how the data describing fractures can be separated into fragments, revealing the shape and size of the fragments. The second technique describes the visualization of the stress tensor and is used to visualize the wave propagation after impact damage.

3.1 Clustering

For the visualization, it is important to identify each individual fragment. We therefore apply a connected components labeling (Algorithm 1). It iterates over all particles in a given time step t and identifies fragments by specifying a label for each particle, so that particles that belong to the same fragment share a common label. We use two criteria in the algorithm to identify connectivity between particles, the maximum damage value s and the maximum bond length r. Thus, the initial displacement between two particles X and X' is not larger than $\|X' - X\| \leq r$ for connected components. Note, that the maximum bond length and critical damage s are parameters to influence the fragmentation of the algorithm and the horizon δ and the critical stretch s_c are independent model parameters.

Algorithm 1: Component labeling

```
Input: particles P, bonds B, max damage value s, max bond length r, time t
Output: every particle labeled by piece id of connected component
for every particle n do
    label[n] ← 0 ;
    visited[n] ← false ;
n ← 0 ;
while n < |P| and c(t, X_n) > s do
    visited[n] ← true;
    n ← n + 1;
currentLabel = 1;
Stack S;
while n < |P| do
    label[n] ← currentLabel ;
    visited[n] ← true;
    S ← n ;
    while S ≠ Ø do
        n' ← S;
        C ← getConnectedParticles(n', B);
        for n'' ∈ C do
            if not visited[n''] and distance(n',n'') < r and c(t, X_n'') < s then
                label[n''] ← currentLabel;
                visited[n''] ← true;
                S ← n'';
    while n < |P| and (visited[n] or c(t, X_n) > s) do
        visited[n] ← true;
        n ← n + 1;
    currentLabel ← currentLabel + 1;
```

By specifying the maximum bond length, we can restrict the search of connected components to those particles that are in direct neighborhood to each other. The maximum damage value s is required, as the given set of bonds contains both active and broken bonds. Therefore, we could not rely for our analysis of connected components solely on the connections of particles through bonds, as this would require a set of bonds containing only the active part. Still, we found that filtering out particles by a user-specified maximum damage value gives good results for the identification of fragments. For our experiments, we used a maximum damage value of $s = 0.2$ and a maximum bond length of $r = 0.001$.

All particle labels are initialized with id 0 (Line 1). This id is also used to identify particles that do not belong to a fragment, as their label id is not altered by the algorithm. We also store a flag for each particle (Line 1), which specifies whether the particle was already visited to avoid infinite loops while traversing through the connections. In Lines 1–1, the algorithm iterates over all particles to find the first particle whose damage value is below the given threshold s. This particle belongs to the first fragment and is labeled with the current label id (Line 1). Next, the particle is pushed on a stack (Line 1), which is used to traverse all particles connected to the given particle. The traversal, Lines 1–1, is done by testing for each connected particle to see whether it has not been visited and meets the criteria for maximum distance r and damage threshold s (Line 1). If so, the particle is labeled with the current label id and also pushed on the stack. This step is repeated until the stack is empty and all connected particles have been visited. In Lines 1–1, the next component is identified by searching for a particle that has not been visited yet and that meets the damage criterion s. These steps are repeated until all particles have been visited and a label specified for each particle.

3.2 *Visualization of the Stress Tensor*

To highlight the waves after an impact damage, we visualize the stress tensor $\sigma(t, X)$, defined for each particle position X and time step t, by means of the spectral norm of the stress tensor. The scalar-valued spectral norm of a tensor A is given as the square root of the largest eigenvalue $\lambda_{\max}$ of A^*A,

$$\|A\|_2 = \sqrt{\lambda_{\max}(A^*A)}, \tag{3}$$

where A^* denotes the transpose of A.

The first step in the visualization pipeline is to calculate and store the spectral norm of the stress tensor for each particle. In the next step, the resulting scalar values given at the particle positions are resampled on a Cartesian grid using inverse distance weighting to obtain a continuous distribution of the spectral norm over the whole domain. For the resampling step, we use a grid resolution of $200 \times 200 \times 200$ and a maximum distance of 0.005. The resulting scalar field is visualized using standard volume rendering techniques.

4 Experiments and Their Visual Analysis

4.1 Fragments and Histograms

Figure 2 shows the geometry of a thin plate with a material density ϱ of $2200\,\mathrm{kg\,m^{-3}}$, a bulk modulus K of 14.9×10^9 Pa, and a shear modulus G of 8.9×10^9 Pa. The spherical projectile is modeled as steal with a material density ϱ of $7700\,\mathrm{kg\,m^{-3}}$, a bulk modulus K of 160×10^9 Pa, and a shear modulus G of 78.3×10^9 Pa, and it hits the target with a velocity of $200\,\mathrm{m\,s^{-1}}$. For the simulation with Peridigm, we use the elastic material model, the critical stretch damage model with a critical stretch $s_c = 0.0025$, and for the interaction of the sphere and the plate a contact model with a spring constant of 1×10^{12}.

Figure 3 shows a common visualization of particle-based simulations in the first column. Here, spheres are placed at the center of the actual position $x(t, X)$ of the particles over the time t. These spheres are colorized with the scalar damage value $c(t, X)$ of the particle. Blue indicates that there is no damage, and red indicates that all bonds inside the interaction zone $B_\delta(X)$ are broken. After the impact, the damage develops radially from the center of the plate and starts to bifurcate twice before hitting the boundary of the plate. At the final time step t =8.27×10^{-7}, there are plenty of "free" particles in the center of the plate that have no neighbors inside the interaction zone any more. Additionally, there are some "free" particles between these non-damaged particles. This somehow indicates that the crack resides and that the plate is scattered in different parts. With this particle-based visualization approach we can study how the damage develops through the material and observe different bifurcations of the cracks.

A plain visualization of the particles does not expose information about the shape of fragments or where exactly the crack path develops. Therefore Algorithm 1 is applied on the particle data to cluster the particles to fragments and label them. In the second column of Fig. 3, fragments that have been extracted this way are colored by their label. This exposes the shape of the fragments and visualizes the crack pattern in a clear way, see Fig. 3d.

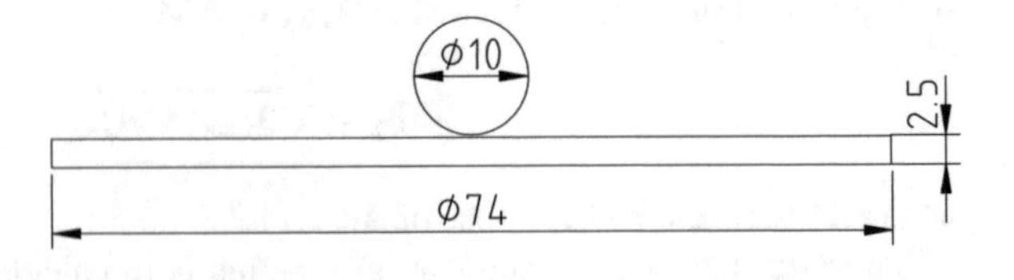

Fig. 2 Blueprint of the thin plate and the spherical projectile

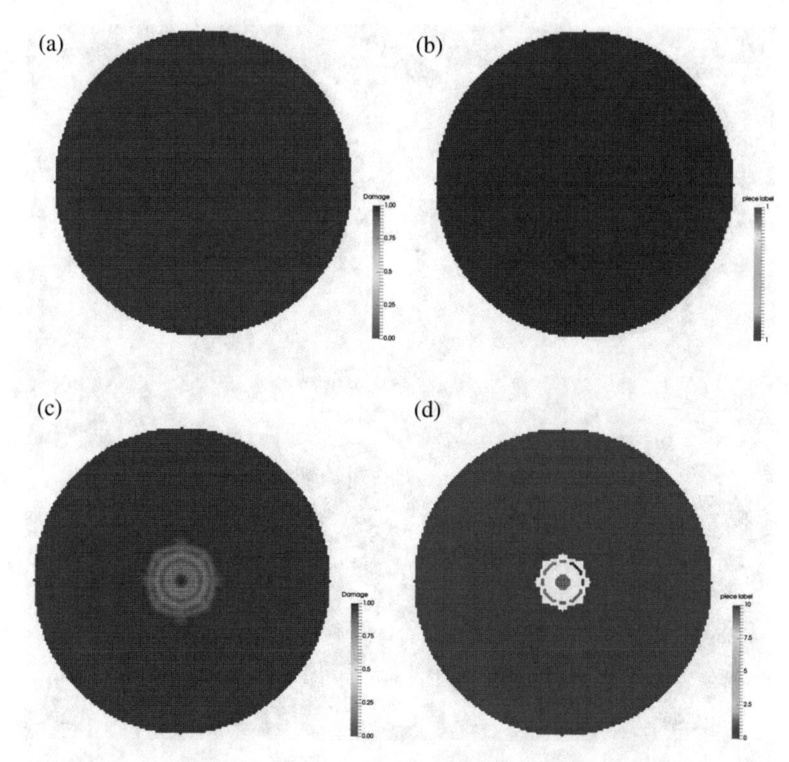

Fig. 3 Impact of a spherical projectile on a thin plate with a velocity of $200\,\mathrm{m\,s^{-1}}$. The nine figures at the *left-hand side* (*above* and *below*) show the visualization of the particles as spheres *colorized* with their damage $c(t, X)$. The *right-hand side* shows the particles *colorized* according to the extracted fragments. (**a**) $t = 0$. (**b**) $t = 0$ (**c**) $t = 1.27 \times 10^{-7}$ (**d**) $t = 1.27 \times 10^{-7}$. (**e**) $t = 1.91 \times 10^{-7}$ s. (**f**) $t = 1.91 \times 10^{-7}$ s. (**g**) $t = 2.55 \times 10^{-7}$ s. (**h**) $t = 2.55 \times 10^{-7}$ s. (**i**) $t = 3.18 \times 10^{-7}$ s. (**j**) $t = 3.18 \times 10^{-7}$ s. (**k**) $t = 3.82 \times 10^{-7}$ s. (**l**) $t = 3.82 \times 10^{-7}$ s. (**m**) $t = 4.45 \times 10^{-7}$ s. (**n**) $t = 4.45 \times 10^{-7}$ s. (**o**) $t = 8.27 \times 10^{-7}$ s. (**p**) $t = 8.27 \times 10^{-7}$ s

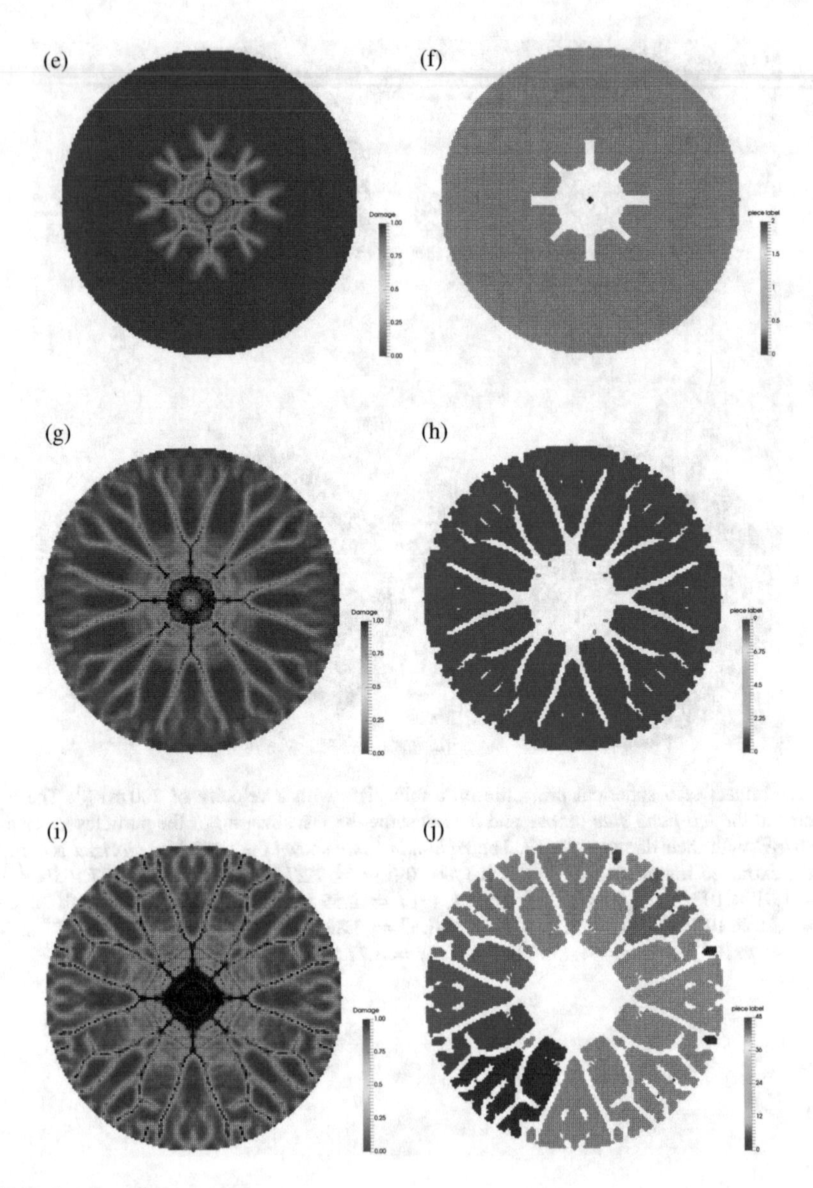

Fig. 3 (continued)

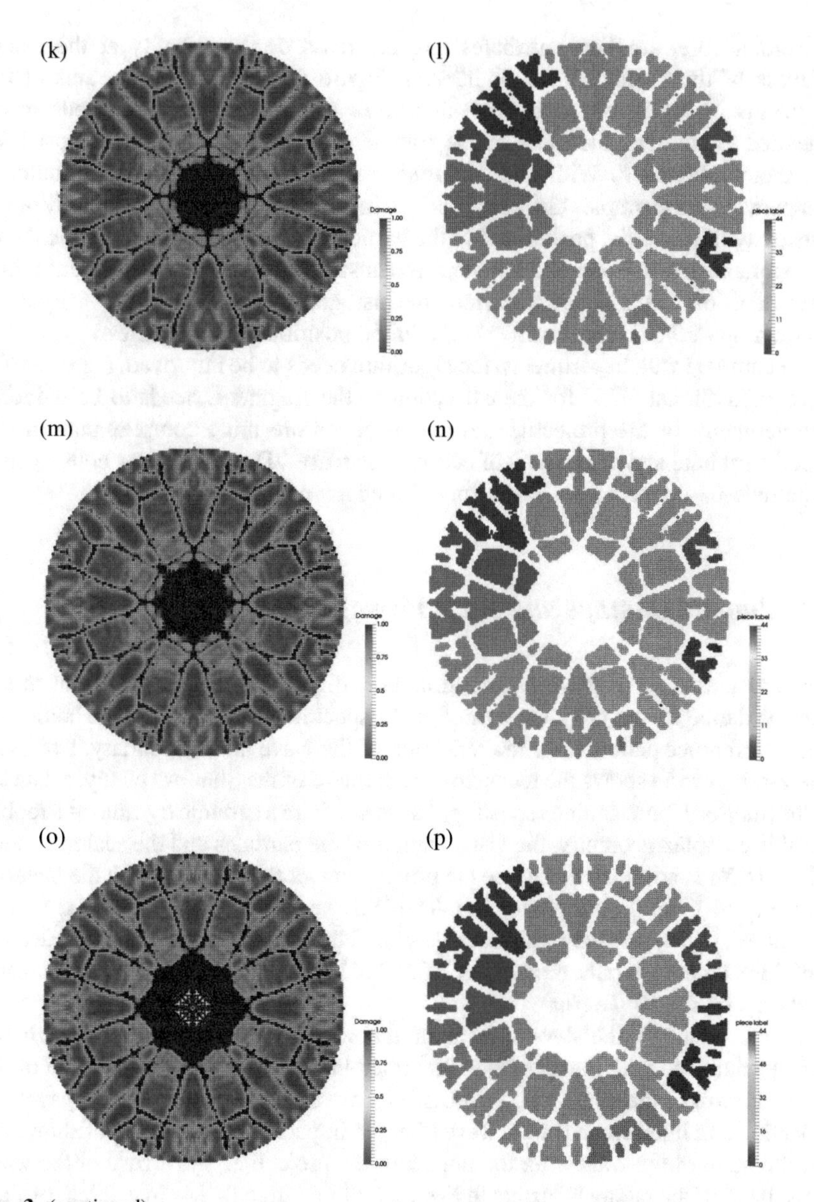

Fig. 3 (continued)

Furthermore, auxiliary attributes like the mass or the velocity at the center of mass of the fragments are of interest. Figure 4 shows the histograms of the fragments' sizes delivered by the algorithm. Here, the algorithm needs to be extended to estimate the mass of the fragment via the density ϱ of the particles and their volumes V. With this additional information, the histograms could be compared to the examples [26, 28] described in Sect. 1.1. Furthermore, a study of the sensitivity of the initial positions X of the particles with respect to the fragment size is important, because the crack pattern looks slightly different for different initial placements of the particles X [6]. Here, the histograms could be used to verify if the distribution of the size is sensitive to the initial position of the particles.

To compare with experiments, the algorithm needs to be improved. For example, the computational effort for the extraction of the fragments needs to be reduced. The geometry of the projectile and the specimen are more complex than in our experiment here and the amount of particles increases. Due to the slow convergence of particle-based methods, the run-time is then not negligible any more.

4.2 *Impact Damage and Wave Propagation*

The understanding of wave propagation after impact damage is important to see how the damage front propagates through the specimen and reflects at the boundary. The interference pattern after the reflection of the wave at the boundary is of great interest. Figure 5a shows the reconstructed surfaces of the Stanford bunny, a data set of the Stanford 3D scanning repository,[1] scanned from a ceramic figurine of a rabbit. For this complex geometry, the visualization of the particles and the scalar damage value $c(t, X)$ is not sufficient to see the propagation of the wave through the material (See, Fig. 5b). For the simulation in LAMMPS, we scattered the surface data set of the bunny with 1,787,245 particles and defined the material with a material density ϱ of 3369 kg m^{-3}, a bulk modulus K of 210×10^9 Pa and a critical stress intensity factor K_{Ic} of $2 \times 10^6\, Pa\sqrt{m}$.

Figure 6 shows the development of the wave after the impact through the bunny visualized with the technique described in Sect. 3.2. The visualization of the spectral norm of the stress $\sigma(t, X)$ provides a more global view of the propagation, reflection and interference of the wave after the impact damage. Figure 6f shows the development of the wave after the impact of the projectile. The arrival of the wave at the back of the bunny is visible in Fig. 6h (colored in red). The interference of the reflected wave is shown in Fig. 6l, and in Fig. 6u artifacts at the occiput and the lugs of the bunny are clearly evident.

[1]http://graphics.stanford.edu/data/3Dscanrep/.

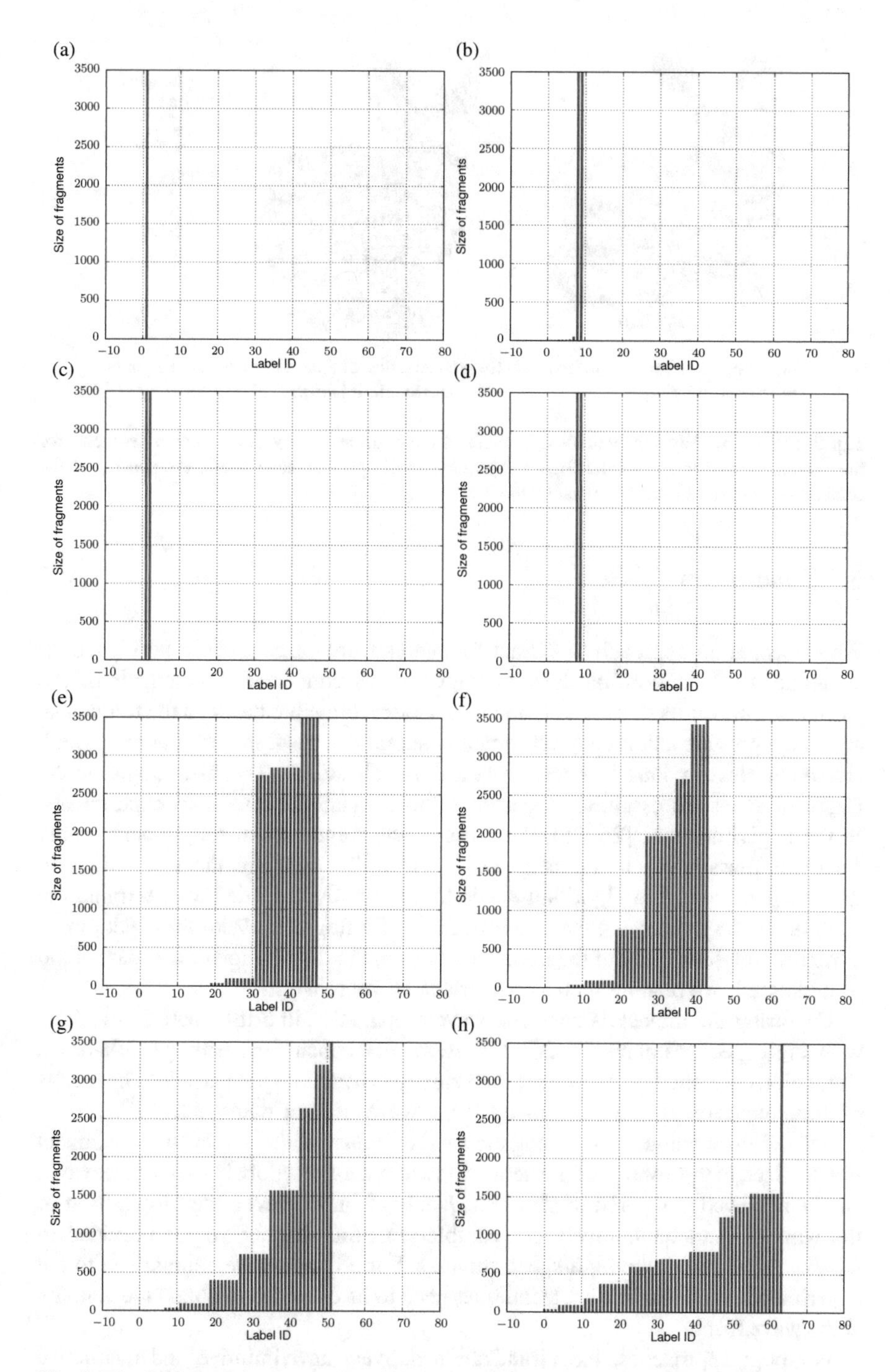

Fig. 4 Distribution of the fragments with respect to the amount of particles per fragment. **(a)** $t = 6.36 \times 10^{-8}$ s. **(b)** $t = 1.27 \times 10^{-7}$ s. **(c)** $t = 1.91 \times 10^{-7}$ s. **(d)** $t = 2.55 \times 10^{-7}$ s. **(e)** $t = 3.18 \times 10^{-7}$ s. **(f)** $t = 3.82 \times 10^{-7}$ s. **(g)** $t = 4.45 \times 10^{-7}$ s. **(h)** $t = 8.27 \times 10^{-7}$ s

(a) The reconstructed surfaces of the Stanford bunny. (b) Visualization of the impact damage per particle of the Stanford bunny.

Fig. 5 Figure (**a**) shows the extracted surfaces of the Stanford bunny as a complex geometry for the simulation and analysis, and Figure (**b**) shows the impact damage where the spheres at the actual positions $x(t, X)$ are *colorized* by the damage $c(t, X)$

5 Conclusion

We presented an approach to extract fragments from particle-based peridynamics simulations. The algorithm delivers fragments as connected components of the particles. Our results show details about the cracks' branches between the fragments, and they provide additional information about the resulting fragments, which are much less obvious in straightforward visualizations. The histograms of the fragment sizes are essential to compare the simulations results to experiments. In the experiments in [26, 28], the mass of the fragments or the velocity at the center of mass of each fragment are provided. To compare the simulations to these experiments, the algorithm needs to be slightly extended to determine the mass of the fragments and the center of mass of a fragment. With these additional attributes, the sensitivity of the initial positions to the distribution of the mass could be addressed. We planing such a comparison as future work.

Capturing the impact damage and wave propagation in brittle materials is done with high-speed cameras [8, 20, 27]. Here, the benchmark with particle-based simulations is a challenge, because the velocity is available per particle. The results of these experiments are the velocity at the wave front or damage front. With our visualization technique , the propagation and inference of the waves after the impact is visualized in the volume of the geometry. Thus, a more "global" view of the waves can be achieved compared to standard visualization. However, the propagation of the waves is not qualitatively comparable with the shadow graphs provided in the experiment with the high-speed cameras. For a quantitative comparison to the experiments, the visualization techniques need to be extended to obtain the velocity at the wave front.

For both approaches, the visualization delivers new intuitive and instructive aspects for analyzing the simulation results qualitatively with new insights in

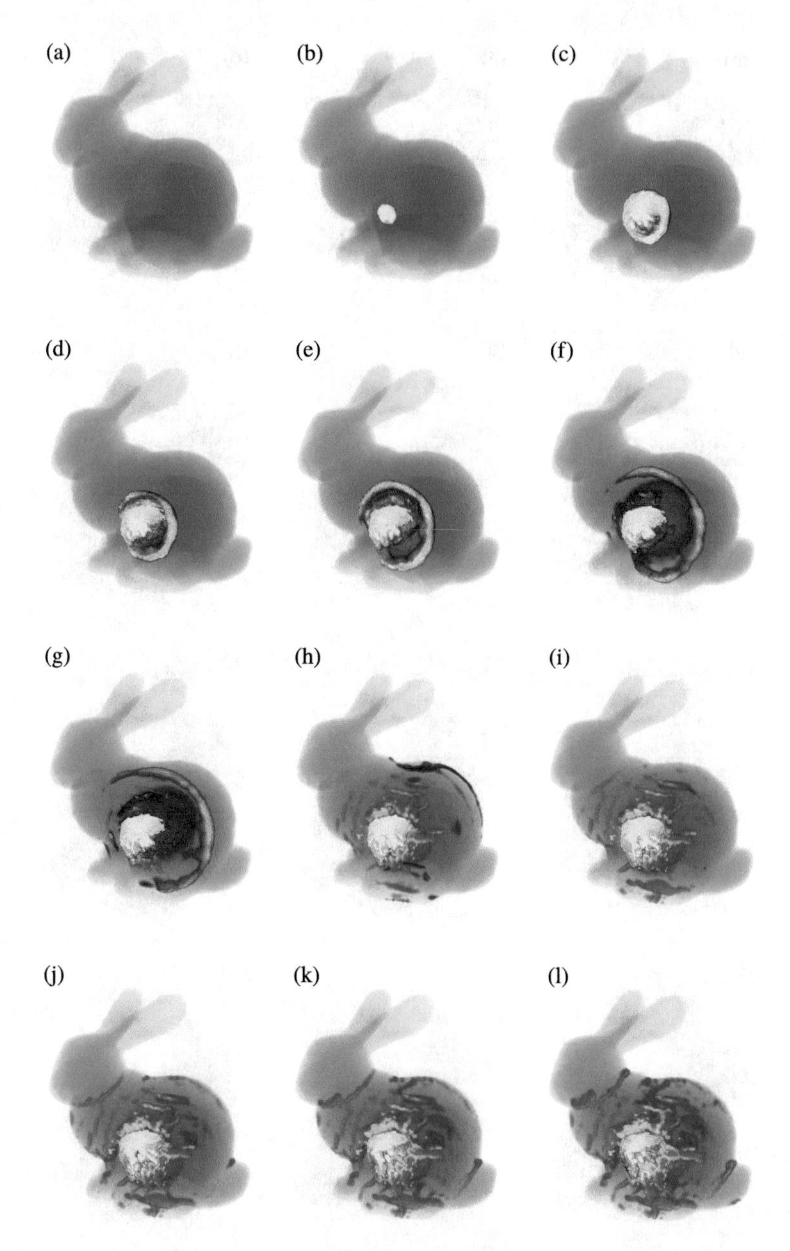

Fig. 6 Visualization of the stress in the Stanford bunny (1,787,245 particles) with a spherical projectile with an impact velocity of $100\,\mathrm{m\,s^{-1}}$ and a time step size of 10^{-8}. (**a**) $t = 0$ s. (**b**) $t = 10^{-8}$ s. (**c**) $t = 5 \times 10^{-8}$ s. (**d**) $t = 7 \times 10^{-8}$ s. (**e**) $t = 9 \times 10^{-8}$ s. (**f**) $t = 12 \times 10^{-8}$ s. (**g**) $t = 14 \times 10^{-8}$ s (**h**) $t = 20 \times 10^{-8}$ s. (**i**) $t = 25 \times 10^{-8}$ s. (**j**) $t = 30 \times 10^{-8}$ s. (**k**) $t = 35 \times 10^{-8}$ s. (**l**) $t = 40 \times 10^{-8}$ s. (**m**) $t = 45 \times 10^{-8}$ s. (**n**) $t = 50 \times 10^{-8}$ s. (**o**) $t = 55 \times 10^{-8}$ s. (**p**) $t = 60 \times 10^{-8}$ s. (**q**) $t = 65 \times 10^{-8}$ s. (**r**) $t = 70 \times 10^{-8}$ s (**s**) $t = 75 \times 10^{-8}$ s (**t**) $t = 80 \times 10^{-8}$ s (**u**) $t = 85 \times 10^{-8}$ s (**v**) $t = 90 \times 10^{-8}$ s (**w**) $t = 95 \times 10^{-8}$ s (**x**) $t = 100 \times 10^{-8}$ s

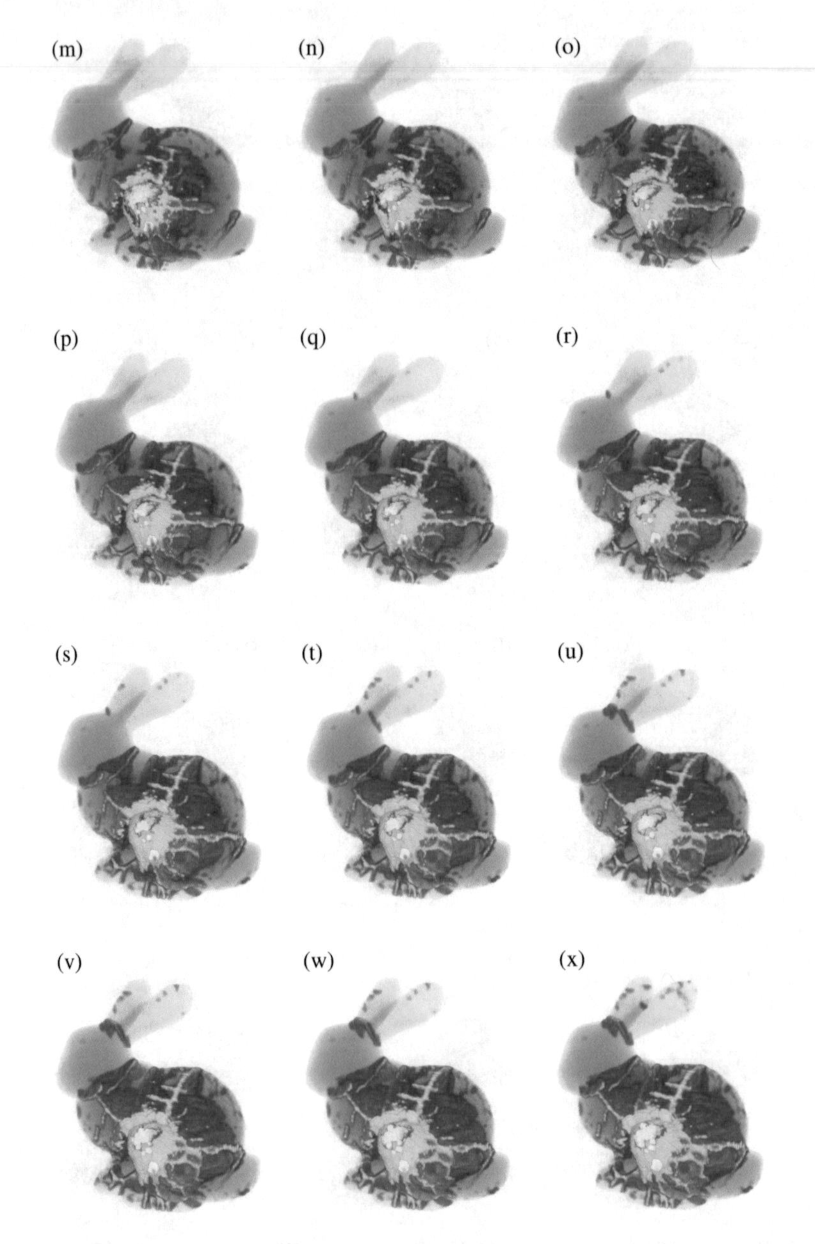

Fig. 6 (continued)

the fragmentation and the propagation of the wave after the impact damage. Both approaches have to be slightly extended for a quantitative comparison of experiments in the future.

References

1. F. Bobaru, G. Zhang, Why do cracks branch? a peridynamic investigation of dynamic brittle fracture. Int. J. Fract. **196**, 1–40 (2016)
2. P. Diehl, M.A. Schweitzer, Simulation of wave propagation and impact damage in brittle materials using peridynamics, in *Recent Trends in Computational Engineering – CE2014*, ed. by M. Mehl, M. Bischoff, M. Schäfer. Lecture Notes in Computational Science and Engineering (Springer, Berlin, 2015)
3. J. Fineberg, M. Marder, Instability in dynamic fracture. Phys. Rep. **313**(1–2), 1–108 (1999)
4. J.T. Foster, Dynamic crack initiation toughness: experiments and peridynamic modeling. Ph.D. thesis, Purdue University (2009)
5. F. Franzelin, P. Diehl, D. Pflüger, Non-intrusive uncertainty quantification with sparse grids for multivariate peridynamic simulations, in *Meshfree Methods for Partial Differential Equations VII*, ed. by M. Griebel, M.A. Schweitzer. Lecture Notes in Computational Science and Engineering, vol. 100 (Springer International Publishing, Berlin, 2015), pp. 115–143 (English)
6. S.F. Henke, S. Shanbhag, Mesh sensitivity in peridynamic simulations. Comput. Phys. Commun. **185**(1), 181–193 (2014)
7. J.F. Kalthoff, S. Winkler, Failure mode transition at high rates of shear loading, in *Impact Loading and Dynamic Behavior of Materials*, vol. 1, ed. by C.Y. Chiem, H.D. Kunze, L.W. Meyer (DGM Informationsgesellschaft Verlag, Oberursel, 1988), pp. 185–195
8. N. Kawai, S. Zama, W. Takemoto, K. Moriguchi, K. Arai, S. Hasegawa, E. Sato, Stress wave and damage propagation in transparent materials subjected to hypervelocity impact. Procedia Eng. **103**, 287–293 (2015). Proceedings of the 2015 Hypervelocity Impact Symposium (HVIS 2015)
9. A. Kobayashi, N. Ohtani, T. Sato, Phenomenological aspects of viscoelastic crack propagation. J. Appl. Polym. Sci. **18**(6), 1625–1638 (1974)
10. J.A. Levine, A.W. Bargteil, C. Corsi, J. Tessendorf, R. Geist, A peridynamic perspective on spring-mass fracture, in *Proceedings of the ACM SIGGRAPH/Eurographics Symposium on Computer Animation*, Aire-la-Ville, SCA ’14 (Eurographics Association, Copenhagen, 2014), pp. 47–55
11. E. Madenci, E. Oterkus, Benchmark problems, in *Peridynamic Theory and its Applications*, (Springer, Berlin, 2013), pp. 151–166
12. J.F. O’Brien, A.W. Bargteil, J.K. Hodgins, Graphical modeling and animation of ductile fracture, in *Proceedings of ACM SIGGRAPH 2002* (ACM Press, New York, Aug 2002), pp. 291–294
13. J.M. Owen, SPH and material failure, in *Proceedings from the 5LC 2005* (2005)
14. M.L. Parks, D.J. Littlewood, J.A. Mitchell, S.A. Silling, Peridigm users’ guide. Technical Report SAND2012-7800, Sandia National Laboratories (2012)
15. M. Pauly, R. Keiser, B. Adams, P. Dutré, M. Gross, L.J. Guibas, Meshless animation of fracturing solids. ACM Trans. Graph. **24**(3), 957–964 (2005)
16. S. Plimpton, Fast parallel algorithms for short-range molecular dynamics. J. Comput. Phys. **117**(1), 1–19 (1995)
17. G.R. Johnson, C.A. Gerlach, R.A. Stryk, T.J. Holmquist, N.L. Rowe, A quantitative assessment of computational results for behind armor Debris, in 23rd *International Symposium On Ballistics*, April 2007

18. S. Raymond, V. Lemiale, R. Ibrahim, R. Lau, A meshfree study of the Kalthoff–Winkler experiment in 3d at room and low temperatures under dynamic loading using viscoplastic modelling. Eng. Anal. Bound. Elem. **42**, 20–25 (2014). Advances on Meshfree and other Mesh reduction methods.
19. W. Riedel, S. Hiermaier, K. Thoma, Transient stress and failure analysis of impact experiments with ceramics. Mater. Sci. Eng. B **173**, 139–147 (2010), Elsevier
20. H. Schradin, Physikalische Vorgänge bei hohen Belastungen und Belastungsgeschwindigikeiten (Physical processes at high loadings and loading rates). Scripts German Acad. Aeronaut. Res. **40**, 21–68 (1939)
21. E. Sharon, J. Fineberg, Universal features of the microbranching instability in dynamic fracture. Philos. Mag. B **78**(2), 243–251 (1998)
22. S.A. Silling, Reformulation of elasticity theory for discontinuities and long-range forces. J. Mech. Phys. Solids **48**(1), 175–209 (2000)
23. S.A. Silling, Dynamic fracture modeling with a meshfree peridynamic code, in *Fluid and Solid Mechanics*, ed. by K.J. Bathe, vol. 1. Massachusetts Institute of Technology (Elsevier, Amsterdam, 2003)
24. S.A. Silling, E. Askari, A meshfree method based on the peridynamic model of solid mechanics. Comput. Struct. **83**, 1526–1535 (2005)
25. S.A. Silling, M. Epton, O. Weckner, J. Xu, E. Askari, Peridynamic states and constitutive modeling. J. Elast. **88**(2), 151–184 (2007) [English]
26. S.J. Schram, H.W. Meyer, Simulating the formation and evolution of behind armor debris fields. ARL-RP 109, U.S. Army Research Laboratory (2005)
27. E. Strassburger, Visualization of impact damage in ceramics using the edge-on impact technique. Int. J. Appl. Ceram. Technol. **1**, 1:235–242 (2004)
28. T.J. Vogler, T.F. Thornhill, W.D. Reinhart, L.C. Chhabildas, D.E. Grady, L.T. Wilson, O.A. Hurricane, A. Sunwoo, Fragmentation of materials in expanding tube experiments. Int. J. Impact Eng. **29**(1–10), 735–746 (2003). Hypervelocity Impact
29. X. Zhang, G. Jia, H. Huang, Fragment identification and statistics method of hypervelocity impact SPH simulation . Chin. J. Aeronaut. **24**(1), 18–24 (2011)

A Meshfree Semi-implicit Smoothed Particle Hydrodynamics Method for Free Surface Flow

Adeleke O. Bankole, Michael Dumbser, Armin Iske, and Thomas Rung

Abstract This work concerns the development of a meshfree semi-implicit numerical scheme based on the Smoothed Particle Hydrodynamics (SPH) method, here applied to free surface hydrodynamic problems governed by the shallow water equations. In *explicit* numerical methods, a severe limitation on the time step is often due to stability restrictions imposed by the CFL condition. In contrast to this, we propose a *semi-implicit* SPH scheme, which leads to an unconditionally stable method. To this end, the discrete momentum equation is substituted into the discrete continuity equation to obtain a linear system of equations for only one scalar unknown, the free surface elevation. The resulting system is not only sparse but moreover symmetric positive definite. We solve this linear system by a matrix-free conjugate gradient method. Once the new free surface location is known, the velocity can directly be computed at the next time step and, moreover, the particle positions can subsequently be updated. The resulting meshfree semi-implicit SPH method is validated by using a standard model problem for the shallow water equations.

A.O. Bankole • A. Iske (✉)
Department of Mathematics, University of Hamburg, Bundesstrasse 55, 20146 Hamburg, Germany
e-mail: adeleke.bankole@uni-hamburg.de; armin.iske@uni-hamburg.de; iske@math.uni-hamburg.de

M. Dumbser
Department of Civil, Environmental and Mechanical Engineering, University of Trento, Via Mesiano 77, I-38123 Trento, Italy
e-mail: michael.dumbser@unitn.it

T. Rung
Institute of Fluid Dynamics and Ship Theory, Hamburg University of Technology, Am Schwarzenberg-Campus 4, 21073 Hamburg, Germany
e-mail: thomas.rung@tuhh.de

M. Griebel, M.A. Schweitzer (eds.), *Meshfree Methods for Partial Differential Equations VIII*, Lecture Notes in Computational Science and Engineering 115,
DOI 10.1007/978-3-319-51954-8_3

1 Introduction

In this work, we propose a meshfree semi-implicit SPH scheme for two-dimensional inviscid hydrostatic free surface flows. These flows are governed by the *shallow water equations* which can be derived either vertically or laterally averaged from the three dimensional incompressible Navier-Stokes equations with the assumption of a hydrostatic pressure distribution (see [5, 6]).

Several methods have been developed for both structured and unstructured meshes using finite difference, finite volume and finite element schemes [5–8, 19]. Explicit schemes are often limited by a severe time step restriction, due to the Courant-Friedrichs-Lewy (CFL) condition. In contrast, semi-implicit methods lead to stable discretizations allowing large time steps at reasonable computational costs. In staggered grid methods for finite differences and finite volumes, discrete variables are often defined at different (staggered) locations. The pressure term, which is the free surface elevation, is defined in the cell center, while the velocity components are defined at the cell interfaces. In the momentum equation, both the pressure term, due to the gradients in the free surface elevations, and the velocity term, in the mass conservation, are discretized implicitly, whereas the nonlinear convective terms are discretized explicitly. In mesh-based schemes, the semi-Lagrangian method discretizes these terms explicitly (see [3, 12, 13]).

In this work a new semi-implicit *Smoothed Particle Hydrodynamics* (SPH) scheme for the numerical solution of the shallow water equations in two space dimensions is proposed, where the flow variables are the particle free surface elevation, the particle total water depth, and the particle velocity. The discrete momentum equations are substituted into the discretized mass conservation equation to give a discrete equation for the free surface leading to a system in only one single scalar quantity, the free surface elevation location. Solving for one scalar quantity in a single equation distinguishes our method, in terms of efficiency, from other methods. The system is solved for each time step as a linear algebraic system. The components of the momentum equation at the new time level can directly be computed from the new free surface, which we conveniently solve by a matrix-free version of the conjugate gradient (CG) algorithm [11, 17]. Consequently, the particle velocities are computed at the new time step and the particle positions are then updated. In this semi-implicit SPH method, the stability is independent of the wave celerity. Therefore, large time steps can be permitted to enhance the numerical efficiency [5].

The rest of this paper is organized as follows. The problem formulation, including the two-dimensional shallow water equations and the utilized models for the particle approximations, is given in Sect. 2. Our meshfree semi-implicit SPH scheme is constructed in Sect. 3. Numerical results, to validate the proposed semi-implicit SPH scheme, are presented in Sect. 4. Concluding remarks are given in Sect. 5.

2 Problem Formulation and Models

This section briefly introduces the utilized models and particle approximations. Vectors are defined by reference to Cartesian coordinates. Latin subscripts are used to identify particle locations, where subscript i refers to the focal particle and subscript j denotes the neighbor of particle i.

2.1 The Kernel Function

We use a mollifying function W, a positive decreasing radially symmetric function with compact support, of the generic form

$$W(r,h) = \frac{1}{h^d} W\left(\frac{\|r\|}{h}\right) \quad \text{for} \quad r \in [0,\infty) \quad \text{and} \quad h > 0.$$

In our numerical examples, we work with the B-spline kernel of degree 3 [15], given as

$$W(r,h) = W_{ij} = K \times \begin{cases} 1 - \frac{3}{2}\left(\frac{r}{h}\right)^2 + \frac{3}{4}\left(\frac{r}{h}\right)^3 & \text{for} \quad 0 \le \frac{r}{h} \le 1 \\ \frac{1}{4}\left(2 - \frac{r}{h}\right)^3 & \text{for} \quad 1 \le \frac{r}{h} \le 2 \\ 0 & \text{for} \quad \frac{r}{h} > 2 \end{cases}$$

where the normalisation coefficient K takes the value $2/3$ (for dimension $d = 1$), $10/(7\pi)$ (for $d = 2$), or $1/\pi$ (for $d = 3$). For the mollifier $W \in W^{3,\infty}(\mathbb{R}^d)$, $h > 0$ is referred to as the *smoothing length*, being related to the particle spacing Δ_P by $h = 2\Delta_P$. The smoothing length h can vary locally according to

$$h_{ij} = \frac{1}{2}[h_i + h_j] \qquad \text{where } h_i = \sigma \sqrt[d]{\frac{m_j}{\rho_j}}. \tag{1}$$

In this study, we use the smoothing length in (1). Moreover, σ is in $[1.5, 2.0]$, which ensures approximately a constant number of particle neighbors of between 40–50 in the compact support of each kernel. A popular approach for the kernel's normalisation is by Shepard interpolation [18], where

$$W'_{ij} = \frac{W_{ij}}{\sum_{j=1}^{N} \frac{m_j}{\rho_j} W_{ij}}.$$

Normalisation is of particular importance for particles close to free surfaces, since this will reduce numerical instabilities and other undesired effects near the boundary.

The gradient of the kernel function is corrected by using the formulation proposed by Belytschko et al. [1]. For the sake of notational convenience, we will from now refer to the kernel function W'_{ij} as W_{ij} and to its gradient $\nabla W'_{ij}$ as ∇W_{ij}.

2.2 Governing Equations

The governing equations considered in this work are nonlinear hyperbolic conservation laws of the form

$$L_b(\boldsymbol{\Phi}) + \nabla \cdot (\boldsymbol{F}(\boldsymbol{\Phi}, \boldsymbol{x}, t)) = 0 \qquad \text{for } t \in \mathbb{R}^+, \boldsymbol{\Phi} \in \mathbb{R} \tag{2}$$

together with the initial condition

$$\boldsymbol{\Phi}(\boldsymbol{x}, 0) = \boldsymbol{\Phi}_0(\boldsymbol{x}) \qquad \text{for } \boldsymbol{x} \in \Omega \subset \mathbb{R}^d, \boldsymbol{\Phi}_0 \in \mathbb{R}$$

where L_b is the transport operator given by

$$L_b(\boldsymbol{\Phi}) = \frac{\partial \boldsymbol{\Phi}}{\partial t} + \nabla \cdot (b\boldsymbol{\Phi})$$

and

$$\boldsymbol{x} = (x^1, \ldots, x^d), \quad \boldsymbol{F} = (F^1, \ldots, F^d), \quad \boldsymbol{b} = (b^1, \ldots, b^d),$$

where $\boldsymbol{b}$ is a regular vector field in $\mathbb{R}^d$, $\boldsymbol{F}$ is a flux vector in $\mathbb{R}^d$, and $\boldsymbol{x}$ is the position.

Figure 1 gives a sketch of the flow domain, i.e., the free surface elevation and the bottom bathymetry. In this configuration, the vertical variation is much smaller

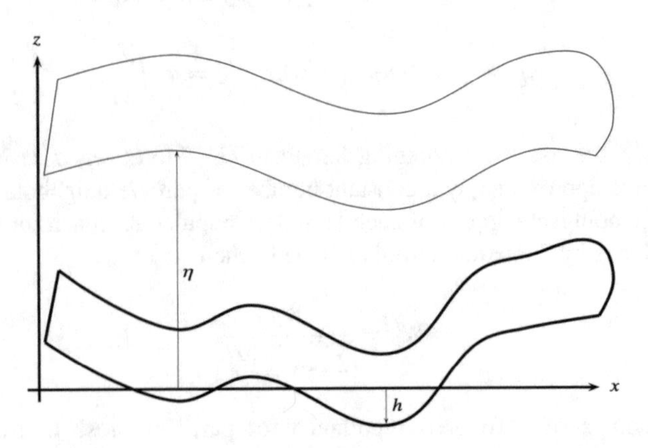

Fig. 1 Sketch of the flow domain: the free surface (*light*) and the bottom bathymetry (*thick*)

than the horizontal variation, as typical for rivers flowing over long distances of e.g. hundreds or thousands of kilometers. We consider the frictionless, inviscid two dimensional shallow water equations in Lagrangian derivatives, given as

$$\frac{D\eta}{Dt} + \nabla \cdot (H\boldsymbol{v}) = 0 \tag{3}$$

$$\frac{D\boldsymbol{v}}{Dt} + g\nabla\eta = 0 \tag{4}$$

$$\frac{D\boldsymbol{r}}{Dt} = \boldsymbol{v} \tag{5}$$

where $\eta = \eta(x, y, t)$ is the free surface location,

$$H(x, y, t) = h(x, y) + \eta(x, y, t)$$

is the total water depth with bottom bathymetry $h(x, y)$, and where $\boldsymbol{v} = v(x, y, t)$ is the particle velocity, $\boldsymbol{r} = r(x, y, t)$ the particle position, and g the gravity acceleration.

2.3 Hydrostatic Approximation

In geophysical flows, the vertical acceleration is often small when compared to the gravitational acceleration and to the pressure gradient in the vertical direction. This is the case in our flow model shown in Fig. 1. If we consider, for instance, tidal flows in the ocean, the velocity in the horizontal direction is of the order of 1 m/s, whereas the velocity in the vertical direction is only of the order of one meter per tidal cycle. Therefore, the advective and viscous terms in the vertical momentum equation of the Navier-Stokes equation are neglected, in which case the pressure equation becomes

$$\frac{dp}{dz} = -g, \tag{6}$$

with normalised pressure, i.e., the pressure is divided by a constant density. The solution of (6) is given by the hydrostatic pressure

$$p(x, y, z, t) = p_0(x, y, t) + g[\eta(x, y, t) - z],$$

where $p_0(x, y, t)$ is the atmospheric pressure at the free surface, taken as constant.

3 Construction of a Meshfree Semi-implicit SPH Scheme

There are several numerical methods for solving Eqs. (3)–(5), including finite differences, finite volumes or finite elements, explicit or implicit methods, conservative or non-conservative schemes, mesh-based or meshfree methods. The meshfree SPH scheme of this work relies on the semi-implicit finite difference method of Casulli [4].

Explicit numerical methods are often, for the sake of numerical stability, limited by the CFL condition. The resulting stability restrictions are usually leading to very small time steps, in contrast to implicit methods. In fact, fully implicit discretisations lead to unconditionally stable methods. On the down side, they typically require solving a large number of coupled nonlinear equations. Moreover, for the sake accuracy, the time step size in implicit methods cannot be chosen arbitrarily large. Semi-implicit methods, e.g. that of Casulli [4], aim to reduce the shortcomings of explicit and fully implicit methods. Following along the lines of [4], we achieve to balance accuracy and stability, at reasonable time step sizes, by a semi-implicit SPH scheme for the two-dimensional shallow water equations, as supported by our numerical results.

3.1 The Smoothed Particle Hydrodynamics Method

Let us briefly recall the basic features of the smoothed particle hydrodynamics (SPH) method. The SPH method is regarded as a powerful tool in computational fluid dynamics. Due to the basic concept of SPH, numerical simulations for fluid flow are obtained by discretisations of the flow equations with using finite sets of particles. Moreover, the target flow quantity, say $A(t,\boldsymbol{x})$, e.g., the velocity field or water height, is smoothed by a suitable kernel function $W(\boldsymbol{x},\boldsymbol{x}',h)$, by smoothing parameter $h>0$, w.r.t. the measure that is associated with the mass density $\rho(t,\boldsymbol{x})$ of the flow, i.e.,

$$A(t,\boldsymbol{x}) = \int_{\Omega} \frac{A(t,\boldsymbol{x}')}{\rho(t,\boldsymbol{x}')} W(\boldsymbol{x}-\boldsymbol{x}',h)\rho(t,\boldsymbol{x}')d\boldsymbol{x}' \qquad \text{for } h>0.$$

Due to the Lagrangian description of SPH, the smoothed quantities are approximated by a set of Lagrangian particles, each carrying an individual mass m_i, density ρ_i and field property A_i. Accordingly, for a given point $\boldsymbol{x}$ in space, the field property A_i, defined at the particles, located at $\boldsymbol{x}_j$, can be interpolated from neighboring points:

$$A(t,\boldsymbol{x}) \approx \sum_{j=1}^{N} m_j \frac{A_j(t)}{\rho_j(t)} W(\boldsymbol{x}-\boldsymbol{x}_j,h),$$

i.e., the field property A at point $\boldsymbol{x}$ is approximated by the sum of contributions from particles at $\boldsymbol{x}_j$ surrounding $\boldsymbol{x}$, being weighted by the distance from each particle. The smoothing kernel $W(\boldsymbol{x}-\boldsymbol{x}', h)$ is required to satisfy the following properties.

- **Unit mass:**

$$\int_{\Omega} W(\boldsymbol{x}-\boldsymbol{x}', h)d\boldsymbol{x}' = 1 \qquad \text{for all } \boldsymbol{x} \text{ and } h > 0.$$

- **Compact support:**

$$W(\boldsymbol{x}-\boldsymbol{x}', h) = 0 \qquad \text{for } |\boldsymbol{x}-\boldsymbol{x}'| > \alpha h,$$

  where the scaling factor $\alpha > 0$ determines the shape (i.e., flatness) of W.
- **Positivity:**

$$W(\boldsymbol{x}-\boldsymbol{x}', h) \geq 0 \qquad \text{for all } \boldsymbol{x}, \boldsymbol{x}' \text{ and } h > 0.$$

- **Decay:** $W(\boldsymbol{x}-\boldsymbol{x}', h)$ should, for any $h > 0$, be monotonically decreasing.
- **Localisation:**

$$\lim_{h \searrow 0} W(\boldsymbol{x}-\boldsymbol{x}', h) = \delta(\boldsymbol{x}-\boldsymbol{x}') \qquad \text{for all } \boldsymbol{x}, \boldsymbol{x}',$$

  where δ denotes the usual Dirac point evaluation functional.
- **Symmetry:** $W(\boldsymbol{x}-\boldsymbol{x}', h)$ should, for any $h > 0$, be an even function.
- **Smoothness:** W should be sufficiently smooth (yet to be specified).

3.2 *Classical SPH Formulation*

The standard SPH formulation discretizes the computational domain $\Omega(t)$ by a finite set of N particles, with positions $\boldsymbol{r}_i$. According to Gingold and Monaghan [10], the SPH discretization of the shallow water equations (3)–(5) are given as

$$\frac{\eta_i^{n+1} - \eta_i^n}{\Delta t} + \sum_{j=1}^{N} \frac{m_j}{\rho_j} H_{ij}^n \boldsymbol{v}_j^n \nabla W_{ij} = \boldsymbol{0} \tag{7}$$

$$\frac{\boldsymbol{v}_i^{n+1} - \boldsymbol{v}_i^n}{\Delta t} + g \sum_{j=1}^{N} \frac{m_j}{\rho_j} \eta_j^n \nabla W_{ij} = \boldsymbol{0} \tag{8}$$

$$\frac{\boldsymbol{r}_i^{n+1} - \boldsymbol{r}_i^n}{\Delta t} = \boldsymbol{v}_i^n \tag{9}$$

where the particles are advected by (9), with Δt being the time step size, m_j the particle mass, ρ_j the particle density, and ∇W_{ij} is the gradient of kernel W_{ij} w.r.t. x_i. In the scheme [10, 15] of Gingold and Monaghan, $\nabla \cdot (H\boldsymbol{v})$ and $\nabla \eta$ are explicitly computed. We remark that Eqs. (7)–(9) follow from a substitution of the flow variable with corresponding derivatives, using integration by parts, and the divergence theorem.

3.3 SPH Formulation of Vila and Ben Moussa

In the construction of our proposed semi-implicit SPH scheme, we use the concept of Vila and Ben Moussa [2, 21], whose basic idea is to replace the centered approximation

$$(F(v_i, x_i, t) + F(v_j, x_j, t)) \cdot n_{ij}$$

of (2) by a numerical flux $G(n_{ij}, v_i, v_j)$, from a conservative finite difference scheme, satisfying

$$G(n(x), v, v) = F(v, x, t) \cdot n(x)$$
$$G(n, v, u) = -G(-n, u, v).$$

With using this formalism, the SPH discretization of Eqs. (7)–(8) becomes

$$\frac{\eta_i^{n+1} - \eta_i^n}{\Delta t} + \sum_{j=1}^{N} \frac{m_j}{\rho_j} 2H_{ij}^n \boldsymbol{v}_{ij}^n \nabla W_{ij} = \mathbf{0},$$

$$\frac{\boldsymbol{v}_i^{n+1} - \boldsymbol{v}_i^n}{\Delta t} + g \sum_{j=1}^{N} \frac{m_j}{\rho_j} 2\eta_{ij}^n \nabla W_{ij} = \mathbf{0}.$$

In this way, we define for a pair of particles, i and j, the free surface elevation η_i, η_j and the velocity $\boldsymbol{v}_i$, $\boldsymbol{v}_j$, respectively (see Fig. 2). In our approach, we, moreover,

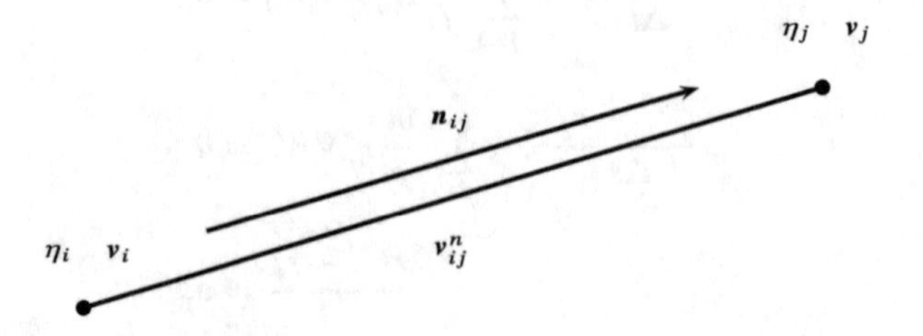

Fig. 2 Staggered velocity defined at the midpoint of two pair of interacting particles i and j

use a staggered velocity $\boldsymbol{v}_{ij}$ between two interacting particles i and j as

$$\boldsymbol{v}_{ij} = \frac{1}{2}(\boldsymbol{v}_i + \boldsymbol{v}_j) \cdot \boldsymbol{n}_{ij}$$

in the normal direction $\boldsymbol{n}_{ij}^{d=1,2}$ at the midpoint of the two interacting particles, where

$$n_{ij}^1 = \frac{x_j - x_i}{\|x_j - x_i\|} \qquad \text{and} \qquad n_{ij}^2 = \frac{y_j - y_i}{\|y_j - y_i\|}$$

for the two components of vector $\boldsymbol{n}_{ij}$. Moreover,

$$\delta_{ij}^1 = \|x_j - x_i\| \qquad \text{and} \qquad \delta_{ij}^2 = \|y_j - y_i\|$$

gives the distance between particles i and j. Since the velocities at the particles' midpoint are known, we can use kernel summation for velocity updates.

3.4 Semi-implicit SPH Scheme

For the derivation of the semi-implicit SPH scheme, let us regard the governing equations (3)–(5). Writing Eqs. (3)–(5) in a non-conservative quasi-linear form by expanding derivatives in the continuity equation and momentum equations (with assuming smooth solutions), this yields

$$u_t + uu_x + vu_y + g\eta_x = 0 \tag{10}$$

$$v_t + uv_x + vv_y + g\eta_y = 0 \tag{11}$$

$$\eta_t + u\eta_x + v\eta_y + H(u_x + v_y) = -uh_x - vh_y. \tag{12}$$

Rewriting (10)–(12) in matrix form, we obtain

$$\mathbf{Q}_t + \mathbf{A}\mathbf{Q}_x + \mathbf{B}\mathbf{Q}_y = \mathbf{C}, \tag{13}$$

where

$$\mathbf{A} = \begin{pmatrix} u & 0 & \boxed{g} \\ 0 & u & 0 \\ \boxed{H} & 0 & u \end{pmatrix} \qquad \mathbf{B} = \begin{pmatrix} v & 0 & 0 \\ 0 & v & \boxed{g} \\ 0 & \boxed{H} & v \end{pmatrix}$$

$$\mathbf{Q} = \begin{pmatrix} u \\ v \\ \eta \end{pmatrix} \qquad \mathbf{C} = \begin{pmatrix} 0 \\ 0 \\ -uh_x - vh_y \end{pmatrix}.$$

Equation (13) is a strictly hyperbolic system with real and distinct eigenvalues. The characteristic equation, given by

$$det(q\mathbf{I} + r\mathbf{A} + s\mathbf{B}) = 0\,, \tag{14}$$

can be simplified as

$$(q + ru + sv)\left[(q + ru + sv)^2 - gH(r^2 + s^2)\right] = 0\,, \tag{15}$$

where the solution (r, s, q) of Eq. (15) are the directions normal to a characteristic cone at the cone's vertex. We split Eq. (15), whereby we obtain

$$q + ru + sv = 0$$

and

$$(q + ru + sv)^2 - gH(r^2 + s^2) = 0\,, \tag{16}$$

with the characteristic curves $u = dx/dt$ and $v = dy/dt$. If the characteristic cone has a vertex at $(\overline{x}, \overline{y}, \overline{t})$, then this cone consist of the line passing through vertex $(\overline{x}, \overline{y}, \overline{t})$ and parallel to the vector $(u, v, 1)$, satisfying

$$((x - \overline{x}) - u(t - \overline{t}))^2 + ((y - \overline{y}) - v(t - \overline{t}))^2 - gH(t - \overline{t})^2 = 0. \tag{17}$$

In particular, the gradient of the left hand side of (17) satisfies (16) on the cone surface. After solving (14), the solution yields

$$\lambda_1 = \boldsymbol{v} - \sqrt{gH}, \qquad \lambda_2 = \boldsymbol{v}, \qquad \lambda_3 = \boldsymbol{v} + \sqrt{gH}.$$

When the particle velocity $\boldsymbol{v}$ is far smaller than the particle celerity $\sqrt{gH}$, i.e., $|\boldsymbol{v}| \ll \sqrt{gH}$, the particle flow is said to be strictly subcritical and thus the characteristic speeds λ_1 and λ_3 have opposite directions. The maximum wave speed is given as

$$\lambda_{\max} = \max(\sqrt{gH_i}, \sqrt{gH_j}).$$

In this case, $\sqrt{gH}$ represents the dominant term which originates from the off diagonal terms g and H in the matrix $\mathbf{A}$ and $\mathbf{B}$.

We now have tracked back where the term $\sqrt{gH}$ originates from in the governing equations. We remark that the first part of the characteristic cone in (15) depends only on the particle velocity u and v. Equation (16), defining the second part of the characteristic cone, depends only on the celerity $\sqrt{gH}$. As we can see, gH in (15) comes from the off-diagonal terms g and H in the matrices **A** and **B**. The terms g and H represent the coefficients of the derivative of the free surface elevation η_x in (10), the coefficient of the derivative η_y in (11) for the momentum equations, and the coefficient of velocity u_x and v_y in the volume conservation Eq. (12). We want to avoid the stability to depend on the celerity $\sqrt{gH}$, therefore we discretize the derivatives η_x, η_y and u_x, v_y implicitly.

Further along the lines of the above analysis, we now develop a semi-implicit SPH scheme for the two-dimensional shallow water equations. To this end, the derivatives of the free surface elevation η_x and η_y in the momentum equation and the derivative of the velocity in the continuity equation are discretized *implicitly*. The remaining terms, such as the nonlinear advective terms in the momentum equation, are discretized *explicitly*, so that the resulting equation system is linear.

Let us consider the continuity equation in the original conservative form, given as

$$\eta_t^n + \nabla \cdot (H^n \boldsymbol{v}^{n+1}) = 0.$$

The velocity **v** is discretized implicitly, whereas the total water depth H is discretized explicitly. In our following notation, for implicit and explicit discretization, we use $n + 1$ and n for the superscript, respectively, i.e.,

$$\begin{aligned} \boldsymbol{v}_t^n + g \cdot \nabla \eta^{n+1} &= 0 \\ \eta_t^n + \nabla \cdot (H^n \boldsymbol{v}^{n+1}) &= 0. \end{aligned}$$

We discretize the particle velocities and free surface elevation in time by the Θ method, for the sake of time accuracy and computational efficiency, i.e., $n + 1 = n + \Theta$, and so

$$\boldsymbol{v}_t^n + g \cdot \nabla \eta^{n+\Theta} = 0 \tag{18}$$

$$\eta_t^n + \nabla \cdot (H^n \boldsymbol{v}^{n+\Theta}) = 0 \tag{19}$$

where the Θ-method notation reads as

$$\begin{aligned} \eta^{n+\Theta} &= \Theta \eta^{n+1} + (1 - \Theta)\eta^n \\ \boldsymbol{v}^{n+\Theta} &= \Theta \boldsymbol{v}^{n+1} + (1 - \Theta)\boldsymbol{v}^n. \end{aligned}$$

The *implicitness factor* Θ should be in $[1/2, 1]$, according to Casulli and Cattani [5]. The general semi-implicit SPH discretization of (18)–(19) then takes the form

$$\frac{\boldsymbol{v}_{ij}^{n+1} - \boldsymbol{F}\boldsymbol{v}_{ij}^{n}}{\Delta t} + \frac{g}{\boldsymbol{\delta}_{ij}}\Theta(\eta_j^{n+1} - \eta_i^{n+1}) + \frac{g}{\boldsymbol{\delta}_{ij}}(1-\Theta)(\eta_j^{n} - \eta_i^{n}) = 0 \tag{20}$$

$$\begin{aligned} \frac{\eta_i^{n+1} - \eta_i^{n}}{\Delta t} &+ \Theta \sum_{j=1}^{N} \frac{m_j}{\rho_j}(2H_{ij}^{n}\boldsymbol{v}_{ij}^{n+1})\nabla \boldsymbol{W}_{ij} \cdot \boldsymbol{n}_{ij} \\ &+ (1-\Theta)\sum_{j=1}^{N} \frac{m_j}{\rho_j}(2H_{ij}^{n}\boldsymbol{v}_{ij}^{n})\nabla \boldsymbol{W}_{ij} \cdot \boldsymbol{n}_{ij} = 0 \end{aligned} \tag{21}$$

where

$$H_{ij}^{n} = \max(0, h_{ij}^{n} + \eta_i^{n}, h_{ij}^{n} + \eta_j^{n}).$$

In a Lagrangian formulation, the explicit operator $\boldsymbol{F}\boldsymbol{v}_{ij}^{n}$ in (20) has the form

$$\mathbf{F}\mathbf{v}_{ij}^{n} = \frac{1}{2}(\mathbf{v}_i + \mathbf{v}_j),$$

where $\boldsymbol{v}_i$ and $\boldsymbol{v}_j$ denote the velocity of particles i and j at time t^n. The velocity at time t^{n+1} is obtained by summation,

$$\mathbf{v}_i^{n+1} = \mathbf{v}_i^{n} + \sum_{j=1}^{N} \frac{m_j}{\rho_j}(\mathbf{v}_{ij}^{n+1} - \mathbf{v}_i^{n})W_{ij}. \tag{22}$$

Note that in (20) we have *not* used the gradient of the kernel function for the discretization of the gradient of η. We rather used a finite difference discretization for the pressure gradient. This increases the accuracy, since $\boldsymbol{F}$ in (20) corresponds to an explicit spatial discretization of the advective terms. Since SPH is a Lagrangian scheme, the nonlinear convective term is discretized by the Lagrangian (material) derivative contained in the particle motion in (9). Equation (22) is used to interpolate the particle velocities from the particle location to the staggered velocity location.

3.5 The Free Surface Equation

Let the particle volume ω_i in (21) be given as $\omega_i = m_i/\rho_i$. Irrespective of the form imposed on $\boldsymbol{F}$, Eqs. (20)–(21) constitute a linear system of equations with unknowns $\boldsymbol{v}_i^{n+1}$ and η_i^{n+1} over the entire particle configuration. We solve this system at each time step for the particle variables from the prescribed initial and boundary conditions. To this end, the discrete momentum equation is substituted into the discrete continuity equation. This reduces the model to a smaller model, where η_i^{n+1} is the only unknown.

Multiplying (21) by ω_i and inserting (20) into (21), we obtain

$$\omega_i \eta_i^{n+1} - g\Theta^2 \frac{\Delta t^2}{\delta_{ij}} \sum_{j=1}^{N} 2\omega_i \omega_j \left[H_{ij}^n (\eta_j^{n+1} - \eta_i^{n+1}) \nabla \boldsymbol{W}_{ij} \cdot \boldsymbol{n}_{ij} \right] = \boldsymbol{b}_i^n, \tag{23}$$

where the right hand side $\boldsymbol{b}_i^n$ represents the known values at time level t^n given as

$$\begin{aligned} \boldsymbol{b}_i^n &= \omega_i \eta_i^n - \Delta t \sum_{j=1}^{N} 2\omega_i \omega_j H_{ij}^n \boldsymbol{F}\boldsymbol{v}_{ij}^{n+\Theta} \nabla \boldsymbol{W}_{ij} \cdot \boldsymbol{n}_{ij} \\ &+ g\Theta(1-\Theta) \frac{\Delta t^2}{\delta_{ij}} \sum_{j=1}^{N} 2\omega_i \omega_j \left[H_{ij}^n (\eta_j^n - \eta_i^n) \nabla \boldsymbol{W}_{ij} \cdot \boldsymbol{n}_{ij} \right], \end{aligned} \tag{24}$$

with $\boldsymbol{F}\boldsymbol{v}_{ij}^{n+\Theta} = \Theta \boldsymbol{F}\boldsymbol{v}_{ij}^n + (1-\Theta)\boldsymbol{v}_{ij}^n$. Since H_{ij}^n, ω_i, ω_j are non-negative numbers, Eqs. (23)–(24) constitute a linear system of N equations for η_i^{n+1} unknowns.

The resulting system is symmetric and positive definite. Therefore, the system has a unique solution, which can be computed efficiently by an iterative method. We obtain the new free surface location by (23), and (20) yields the particle velocity $\boldsymbol{v}_i^{n+1}$.

3.6 Neighboring Search Technique

The geometric search for neighboring particles j around a focal particle i at some specific position $\boldsymbol{x}_i$ can be done efficiently. To this end, we create a background Cartesian grid (see Fig. 3). This background grid contains the fluid with a mesh size of $2L$, and the grid is kept fixed throughout the simulation. The grid comprises macrocells which consist of particles (see [16] for computational details), quite similar to the book-keeping cells used in [14].

To compute the free surface elevation η and the fluid velocity $\boldsymbol{v}$, only particles inside the same macro cell or in the surrounding macro cells contribute. Ferarri et al. [9] explain the neighboring search in detail: The idea is to build a list of particles in a given macro cell and, vice versa, to keep a list of indices, one for each particle, pointing to macro cells containing that particle. We store the coordinates of each particle to reduce the time required for the neighbor search. In our neighbor search,

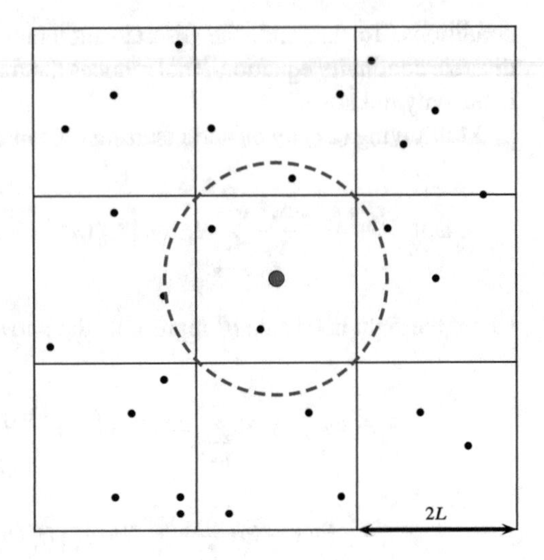

Fig. 3 Fictitious Cartesian grid: neighboring search is done within the nine cells in a two-dimensional space. The smoothing length is constant and the support domain for the particles is $2L$

a particle can only interact with particles in its macro cell or in neighboring macro cells. For the two-dimensional case of the present study we only need to loop over the bounding box of nine macro cells (see Fig. 3).

4 Numerical Results

Now we evaluate the performance of the proposed semi-implicit SPH scheme. This is done by employing a standard test problem for the 2d shallow water equations. In this model problem, we assume a smooth solution, i.e., a collapsing Gaussian bump.

4.1 A Collapsing Gaussian Bump

We consider a smooth free surface wave propagation, by the initial value problem

$$\eta(x, y, 0) = 1 + 0.1e^{-\frac{1}{2}\left(\frac{r^2}{\sigma^2}\right)},$$

$$u(x, y, 0) = v(x, y, 0) = h(x, y) = 0,$$

in the domain $\Omega = [-1, 1] \times [-1, 1]$ with a prescribed flat bottom bathymetry, i.e., $h(x, y) = 0$, where $\sigma = 0.1$ and $r^2 = x^2 + y^2$. The computational domain Ω is discretized with 124,980 particles. The final simulation time is $t = 0.15$, and the time step is chosen to be $\Delta t = 0.0015$. We have used the implicitness factor $\Theta =$

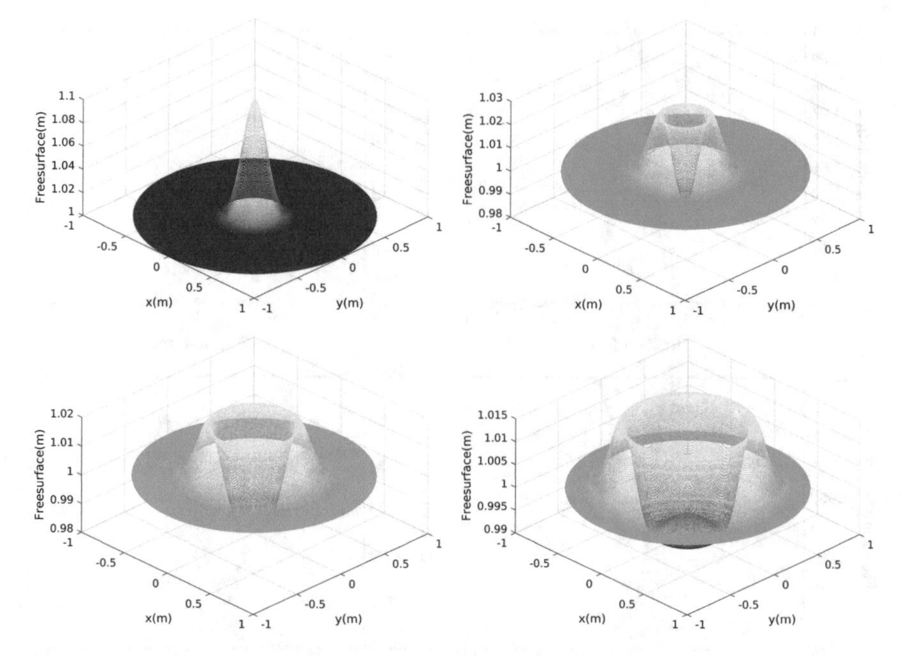

Fig. 4 3d surface plot of the free-surface: SISPH solution at times t =0.0 s, 0.05 s, 0.10 s, 0.15 s with 124,980 particles

0.65. The smoothing length is taken as $h_i = \alpha(\omega_i)^{1/d}$, where $\alpha = [1.5, 2]$ and $d = 2$. The obtained numerical solution is shown in Fig. 5. The profiles in Fig. 4 show the three dimensional surface plots of the free surface elevation at times t =0.0 s, 0.05 s, 0.10 s, 0.15 s. Due to the radial symmetry of the problem, we obtain a reference solution by solving the one-dimensional shallow water equations with a geometric source term in radial direction: a method based on the high order classical shock capturing total variation diminishing (TVD) finite volume scheme is employed for computing the reference solution using 5000 points and the Osher-type flux for the Riemann solver, see [20] for details. The comparison between our numerical results obtained with semi-implicit SPH scheme and the reference solution is shown. A good agreement between the two solutions is observed in Fig. 5. We attribute the (rather small) differences in the plots to the fact that the SPH method has a larger effective stencil, which may increase the numerical viscosity. The cross section of the free surface elevation and the velocity in the x-direction is shown in Fig. 5. We have used a higher resolution of particle numbers of 195,496, the cross section of the free surface elevation and the velocity at final time $t = 0.15$ s can be seen in Fig. 6. We observe similar results compared to particle numbers 124,980.

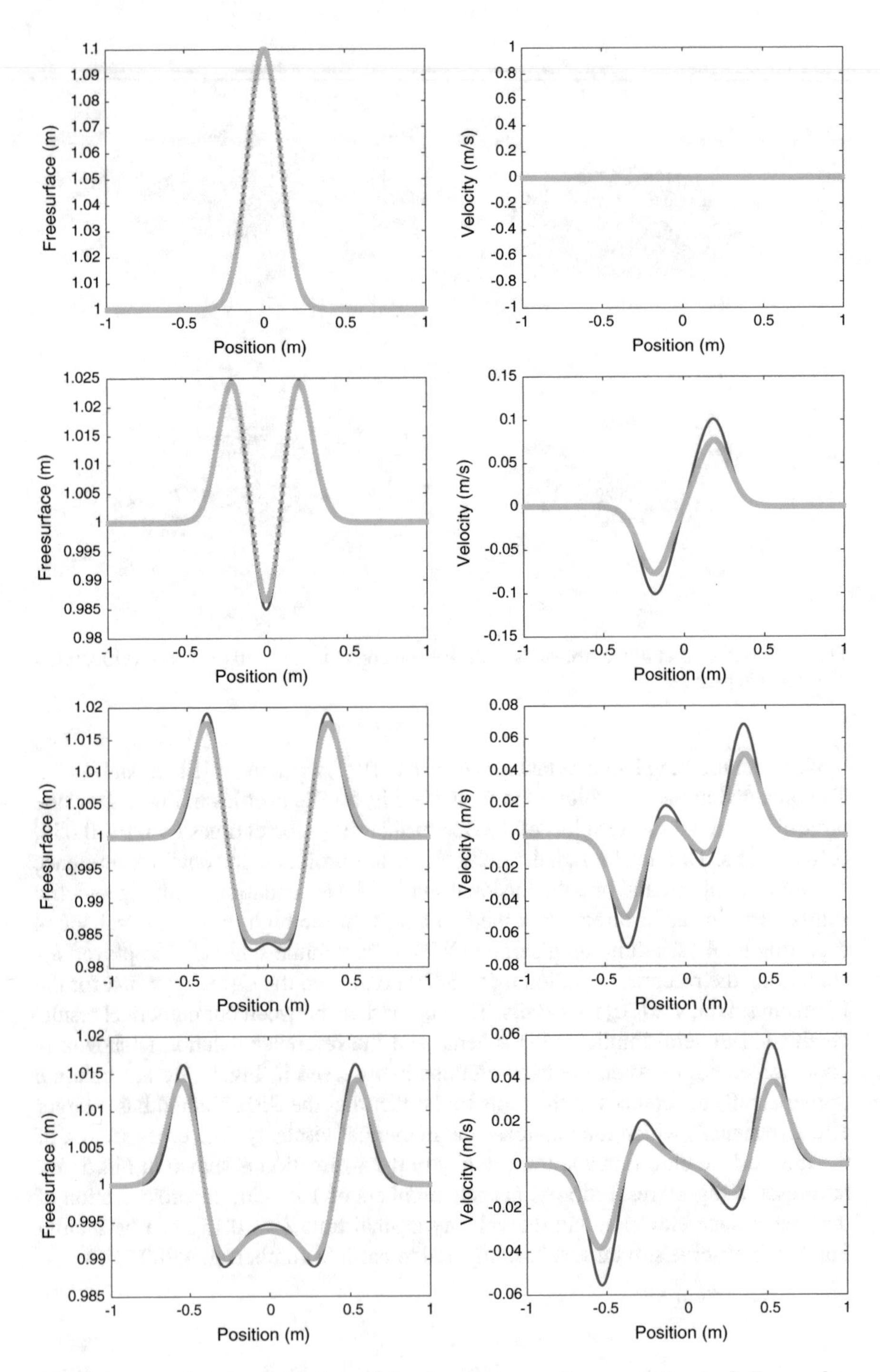

Fig. 5 Cross section of semi-implicit solution (*green*) versus reference solution (*red*): Free-surface (*left*), velocity (*right*) in the x-direction at times $t = 0.0\,\mathrm{s}$, $0.05\,\mathrm{s}$, $0.10\,\mathrm{s}$, $0.15\,\mathrm{s}$

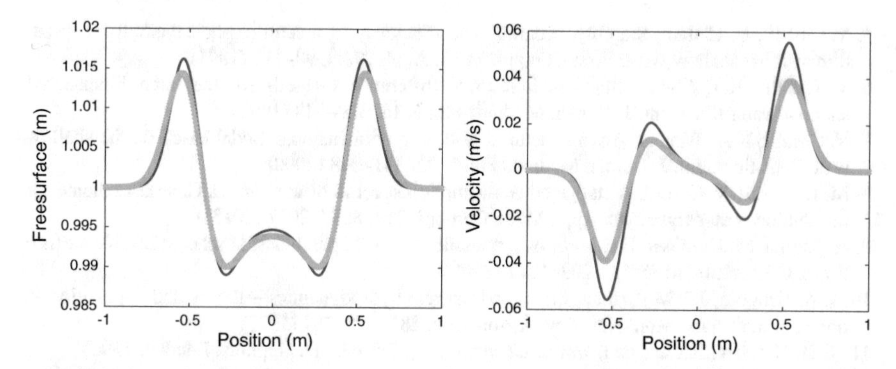

Fig. 6 Cross section of semi-implicit solution (*green*) versus reference solution (*red*): Free-surface (*left*), velocity (*right*) in the x-direction at times $t = 0.15$ s with a higher resolution of 195,496 particles

5 Conclusion

We have proposed a meshfree semi-implicit smoothed particle hydrodynamics (SPH) method for the shallow water equations in two space dimensions. In our scheme, the momentum equation is discretized by a finite difference approximation for the gradient of the free surface and the SPH approximation for the mass conservation equation. By the substitution of the discrete momentum equations into the discrete mass conservation equations, this leads to a sparse linear system for the free surface elevation. We solve this system efficiently by a matrix-free version of the conjugate gradient (CG) algorithm.

The key features of the proposed semi-implicit SPH method are briefly as follows: The method is mass conservative; efficient; time steps are not restricted by a stability condition (coupled to the surface wave speed), thus large time steps are permitted.

Ongoing research is devoted to nonlinear wetting and drying problems, application to shock problems, and extension of the scheme to the fully three-dimensional case.

References

1. T. Belytschko, Y. Krongauz, J. Dolbow, C. Gerlach, On the completeness of meshfree particle methods. Int. J. Numer. Meth. Eng. **43**(5), 785–819 (1998)
2. B. Ben Moussa, J.P. Villa, Convergence of SPH method for scalar nonlinear conservation laws. SIAM J. Numer. Anal. **37**(3), 863–887 (2000)
3. L. Bonaventura, A. Iske, E. Miglio, Kernel-based vector field reconstruction in computational fluid dynamic models. Int. J. Numer. Methods Fluids **66**(6), 714–729 (2011)
4. V. Casulli, Semi-implicit finite difference methods for the two-dimensional shallow water equations. J. Comput. Phys. **86**, 56–74 (1990)

5. V. Casulli, E. Cattani, Stability, accuracy and efficiency of a semi-implicit method for three-dimensional shallow water flow. Comput. Math. Appl. **27**(4), 99–112 (1994)
6. V. Casulli, R.T. Cheng, Semi-implicit finite difference methods for the three-dimensional shallow water flows. Int. J. Numer. Methods Fluids **15**, 629–648 (1992)
7. V. Casulli, R.A. Walters, An unstructured grid, three-dimensional model based on the shallow water equations. Int. J. Numer. Methods Fluids **32**, 331–348 (2000)
8. M. Dumbser, V. Casulli, A staggered semi-implicit spectral discontinuous Galerkin scheme for the shallow water equations. Appl. Math. Comput. **219**, 8057–8077 (2013)
9. A. Ferarri, M. Dumbser, E.F. Toro, A. Armanini, A new 3d parallel SPH scheme for free surface flows. Comput. Fluid **38**(6), 1203–1217 (2009)
10. R.A. Gingold, J.J. Monaghan, Smoothed particle hydrodynamics – theory and application to non-spherical stars. Mon. Not. Roy. Astron. Soc. **181**, 375–389 (1977)
11. G.H. Golub, C.F. van Loan, *Matrix Computations*, 3rd edn. (J. Hopkins, London, 1996)
12. A. Iske, M. Käser, Conservative semi-Lagrangian advection on adaptive unstructured meshes. Numer. Methods Partial Differ. Equ. **20**(3), 388–411 (2004)
13. M. Lentine, J.T. Gretarsson, R. Fedkiw, An unconditionally stable fully conservative semi-Lagrangian method. J. Comput. Phys. **230**(8), 2857–2879 (2011)
14. J.J. Monaghan, Simulating free surface flows with SPH. J. Comput. Phys. **110**, 399–406 (1994)
15. J.J. Monaghan, Smoothed particle hydrodynamics. Rep. Progr. Phys. **68**, 1703–1759 (2005)
16. J.J. Monaghan, J.C. Lattanzio, A refined particle method for astrophysical problems. Astron. Astrophys. **149**, 135–143 (1985)
17. Y. Saad, *Iterative Methods for Sparse Linear Systems*, 2nd edn. (SIAM, Philadelphia, 2003)
18. D. Shepard, A two dimensional interpolation function for irregular spaced data. *Proceedings of the 23rd A.C.M. National Conference* (1968), pp. 517–524
19. M. Tavelli, M. Dumbser, A high order semi-implicit discontinuous Galerkin method for the two dimensional shallow water equations on staggered unstructured meshes. Appl. Math. Comput. **234**, 623–644 (2014)
20. E.F. Toro, *Shock-Capturing Methods for Free-Surface Shallow Flows*, 2nd edn. (Wiley, Chichester, 2001)
21. J.P. Villa, On particle weighted methods and smooth particle hydrodynamics. Math. Models Methods Appl. Sci. **9**(2),161–209 (1999)

A Meshfree Method for the Fractional Advection-Diffusion Equation

Yanping Lian, Gregory J. Wagner, and Wing Kam Liu

Abstract "Non-local" phenomena are common to problems involving strong heterogeneity, fracticality, or statistical correlations. A variety of temporal and/or spatial fractional partial differential equations have been used in the last two decades to describe different problems such as turbulent flow, contaminant transport in ground water, solute transport in porous media, and viscoelasticity in polymer materials.

The study presented herein is focused on the numerical solution of spatial fractional advection-diffusion equations (FADEs) via the reproducing kernel particle method (RKPM), providing a framework for the numerical discretization of spacial FADEs. However, our investigation found that an alternative formula of the Caputo fractional derivative should be used when adopting Gauss quadrature to integrate equations with fractional derivatives. Several one-dimensional examples were devised to demonstrate the effectiveness and accuracy of the RKPM and the alternative formula.

1 Introduction

A diversity of deterministic and stochastic partial differential equations [3, 12, 13] have been developed with fractional-order derivative operators built-in to describe "non-local" phenomena due to strong heterogeneity, fracticality, or statistical correlations. In particular, the spatial and/or temporal fractional advection-diffusion equations have been shown to to be useful in the description of anomalous diffusion phenomena [1, 8, 11].

Y. Lian • G.J. Wagner • W.K. Liu (✉)
Department of Mechanical Engineering, Northwestern University, Evanston, IL, USA
e-mail: lianyp@northwestern.edu; gregory.wagner@northwestern.edu; w-liu@northwestern.edu

M. Griebel, M.A. Schweitzer (eds.), *Meshfree Methods for Partial Differential Equations VIII*, Lecture Notes in Computational Science and Engineering 115,
DOI 10.1007/978-3-319-51954-8_4

The one dimensional spatial fractional advection-diffusion equation is defined as

$$\frac{\partial \phi(x,t)}{\partial t} + u\frac{\partial \phi(x,t)}{\partial x} - \nu\frac{\partial}{\partial x}\phi^{(\alpha)}(x,t) = f(x,t) \tag{1}$$

where ϕ represents the solute concentration, $f(x,t)$ is a source term, u stands for the advective velocity, ν is a coefficient of the fractional diffusion term, and $\alpha \in (0,1]$ stands for the fractional derivative order associated with a left-sided Caputo fractional derivative. The left-sided Caputo fractional derivative is defined for a general function $\psi(x)$ as

$$\psi^{(\alpha)}(x) = \frac{1}{\Gamma(1-\alpha)}\int_0^x (x-y)^{-\alpha}\psi'(y)dy \tag{2}$$

where $\Gamma(1-\alpha) = \int_0^\infty z^{-\alpha}e^{-z}dz$ is Gamma function, and prime notation stands for first order derivative with respect of y.

Equation (2) can be discretized by different numerical methods. Here, we focus on the ability of the reproducing kernel particle method (RKPM) [9, 10] to approximate the above equation by performing a parametric study. In RKPM the approximation of the general function $\psi(x)$ is made by a set of scattered particles used to discretize the material domain Ω as

$$\psi^R(x) = \int_\Omega \Psi_\rho(x; x-\bar{x})\psi(\bar{x})d\bar{x} \tag{3}$$

the superscript R stands for reconstruction or approximation, and subscript ρ is called the dilation parameter and used to determine the size of support domain of a point. $\Psi_\rho(x; x-\bar{x})$ is a kernel function defined as

$$\Psi_\rho(x; x-\bar{x}) = C_\rho(x; x-\bar{x})\omega_\rho(x-\bar{x}) \tag{4}$$

where $C_\rho(x; x-\bar{x})$ is the correction function, and $\omega_\rho(x-\bar{x})$ is the weight function, both of which are defined as follows.

$$C_\rho(x; x-\bar{x}) = \boldsymbol{P}^T(x-\bar{x})\boldsymbol{b}(x,\rho) \tag{5}$$

where $P^T(x) = \left[1\ x\ x^2\ \cdots\ x^k\right]$ is the kth-order polynomial basis with unknown coefficients collected by the vector function $\boldsymbol{b}(x,\rho) = [b_0\ b_1\ b_2 \cdots b_k]$. An example for the weight function is the quartic spline function as follows.

$$\omega_\rho(x-\bar{x}) = \frac{5}{4\rho}\begin{cases} 1-6r^2+8r^3-3r^4 & 0 \le r < 1 \\ 0 & r \ge 1 \end{cases}$$

with $r = |x-\bar{x}|/\rho$.

After some algebra and using the Einstein summation convention when repeated indices are present, Eq. (3) can be rewritten as:

$$\psi^R(x) = N_I(x)\psi(x_I) \tag{6}$$

with

$$N_I(x) = \boldsymbol{P}^T(0)\boldsymbol{M}(x)^{-1}\boldsymbol{P}(x-x_I)\omega_\rho(x-x_I)\Delta V_I \tag{7}$$

and $\boldsymbol{M}^{-1}$ being the inverse matrix of $\boldsymbol{M}$ defined by

$$\boldsymbol{M}(x) = \sum_{J=1}^{L} \omega_\rho(x-x_J)\boldsymbol{P}(x-x_J)\boldsymbol{P}^T(x-x_J)\Delta V_J \tag{8}$$

with L the total number of particles that have contribution to the particle I, and ΔV_J a measure of the sub-domain represented by particle J.

2 RKPM Approximation for the Fractional Derivative

Approximating the fractional derivative of the function ψ at a discretization point x_I through Eq. (6) requires the computation of the fractional derivative of the shape function $N_I(x)$. For simplicity and demonstration purposes, let us assume a uniformly spaced grid where n particles are evenly spaced by Δx. The position of particle I is then $x_I = I\Delta x$ for $I = 0, \cdots, n-1$. By choosing a first-order polynomial basis, $k = 1$, and $\rho = 2\Delta x$ the global internal particle I has a nonlinear shape function given by

$$N_I(x) = \frac{5}{8}\begin{cases} 1 - 6r^2 + 8r^3 - 3r^4 & 0 \le r < 1 \\ 0 & r \ge 1 \end{cases} \tag{9}$$

From Eq. (2) the fractional derivative for the above shape function becomes:

$$N_I^{(\alpha)}(x) = \frac{1}{\Gamma(1-\alpha)}\int_0^x (x-y)^{-\alpha}N_I'(y)dy \tag{10}$$

Substituting Eq. (9) into Eq. (10) yields

$$N_I^{(\alpha)}(x) = \lambda(\alpha)\begin{cases} 0 & x \in [x_0, x_{I-2}] \\ H(x-x_{I-2}; x-x_I) & x \in [x_{I-2}, x_I] \\ H_m(x) + G(x-x_I; x-x_I) - G(0; x-x_I) & x \in [x_I, x_{I+2}] \\ H_m(x) + G_m(x) & x \in [x_{I+2}, x_N] \end{cases} \tag{11}$$

with $\lambda(\alpha) = \frac{15}{2\rho^2\Gamma(2-\alpha)}$, $H_m(x) = H(x - x_{I-2}; x - x_I) - H(x - x_I; x - x_I)$, $G_m(x) = G(x - x_{I-2}; x - x_I) - G(x - x_I; x - x_I)$, and

$$\begin{aligned} H(\eta;\xi) = {} & \xi\eta^{1-\alpha} + \tfrac{1-\alpha}{2-\alpha}\eta^{2-\alpha} \\ & - \tfrac{1}{\rho^2}\left[\xi^2\eta^{1-\alpha} + \tfrac{2\xi(1-\alpha)}{2-\alpha}\eta^{2-\alpha} + \tfrac{1-\alpha}{3-\alpha}\eta^{3-\alpha}\right] \\ & + \tfrac{1}{\rho^2}\left[\xi^3\eta^{1-\alpha} + \tfrac{3\xi^2(1-\alpha)}{2-\alpha}\eta^{2-\alpha} + \tfrac{3\xi(1-\alpha)}{3-\alpha}\eta^{3-\alpha} + \tfrac{1-\alpha}{4-\alpha}\eta^{4-\alpha}\right] \end{aligned} \tag{12}$$

and

$$\begin{aligned} G(\eta;\xi) = {} & \xi\eta^{1-\alpha} + \tfrac{1-\alpha}{2-\alpha}\eta^{2-\alpha} \\ & + \tfrac{1}{\rho^2}\left[\xi^2\eta^{1-\alpha} + \tfrac{2\xi(1-\alpha)}{2-\alpha}\eta^{2-\alpha} + \tfrac{1-\alpha}{3-\alpha}\eta^{3-\alpha}\right] \\ & + \tfrac{1}{\rho^2}\left[\xi^3\eta^{1-\alpha} + \tfrac{3\xi^2(1-\alpha)}{2-\alpha}\eta^{2-\alpha} + \tfrac{3\xi(1-\alpha)}{3-\alpha}\eta^{3-\alpha} + \tfrac{1-\alpha}{4-\alpha}\eta^{4-\alpha}\right] \end{aligned} \tag{13}$$

Figure 1 depicts the $N_I^{(\alpha)}(x)$ for selected values of $\alpha \in (0, 1]$ within a material domain $x = [0, 15]$ with a particle spacing $\Delta x = 1$.

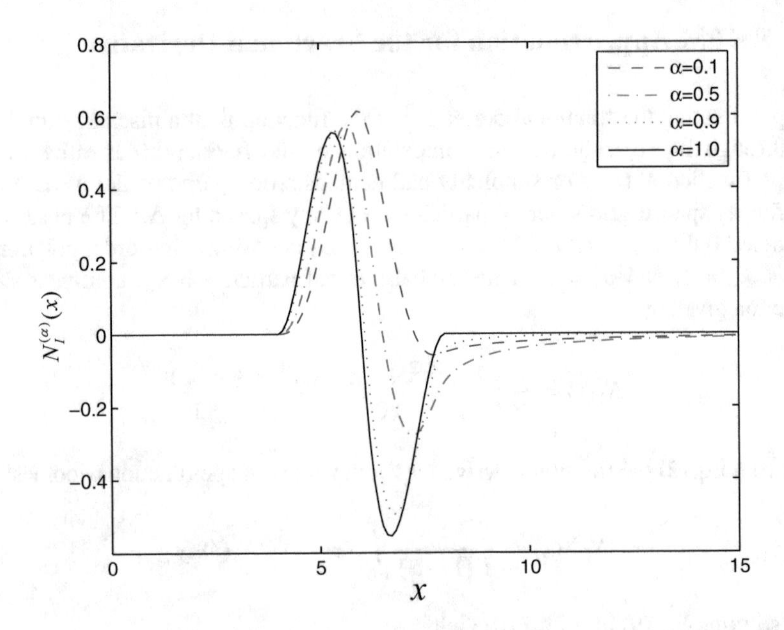

Fig. 1 Non-locality of $N_I^{(\alpha)}(x)$ with $\alpha = 0.1, 0.5$, and 0.9, where the support domain is extended to the *right side* of the domain, and can shrink back to compact domain for $\alpha = 1$, namely the support domain is equal to that of the shape function

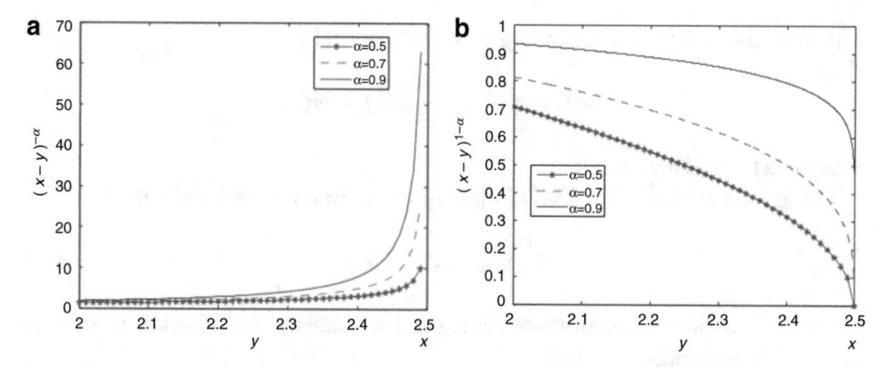

Fig. 2 The curves of kernel function with different value of α: (**a**) $(x-y)^{-\alpha}$, (**b**) $(x-y)^{1-\alpha}$

2.1 Alternative Approximation for the Fractional Derivative Using RKPM

The approximation of Eq. (2) requires numerical integration, such that a random distribution of particles and the effect of boundary conditions can be taken into account. However the presence of the kernel function $(x-y)^{-\alpha}$ inside the integral of Eq. (2) can lead to numerical issues when using Gauss quadrature. This can be explained by the fact that this kernel function becomes steeper as α approaches 1, see Fig. 2a. In practice, the problem could be alleviated by including a large quantity of quadrature points to guarantee accuracy, but there is an alternative solution that reduces the computational burden.

If we use integration by parts on Eq. (2) so that the order of the kernel function is increased by 1, then we get

$$\psi^{(\alpha)}(x) = \frac{1}{\Gamma(2-\alpha)}\left[\psi'(0)(x)^{\alpha} - \int_a^x (x-y)^{1-\alpha}\psi''(y)dy\right] \tag{14}$$

This was plotted in Fig. 2b, where it is clear that the kernel function $(x-y)^{1-\alpha}$ becomes relatively smoother than the original formulation given by Eq. (2). Equation (14) is denoted henceforth as the alternative formula of Caputo fractional derivative and is used in the remaining of this article.

3 Spatial Fractional Advection-Diffusion Equation via RKPM

As shown in the Introduction, the steady state spatial FADE can be written within a material domain $\Omega = [0, 1]$ as

$$\frac{d\phi(x)}{dx} - \nu\frac{d}{dx}\phi^{(\alpha)}(x) = f(x) \tag{15}$$

with essential or Dirichlet boundary conditions given by

$$\phi(0) = \bar{\phi}(0), \qquad \phi(1) = \bar{\phi}(1) \tag{16}$$

where $\bar{\phi}(x)$ is a given function.

Using a Petrov-Galerkin (P-G) formulation, the trial function takes the form

$$\phi^h(x,t) = N_I(x)\phi_I(t) \tag{17}$$

where $\phi_I(t)$ indicates the value of function ϕ at node x_I, and $N_I(x)$ is defined by Eq. (7). The test function is then,

$$w^* = w + \gamma w' \tag{18}$$

Here w' leads to the difference between test and trial function. The constant $\gamma = \beta\frac{\rho}{2}$ acts like a viscosity coefficient, and β is a dimensionless stabilization parameter that can affect the central difference scheme for the advection term such that it becomes closer to a forward difference scheme and achieving a stabilized solution. If γ is set as 0, then the P-G formulation is reduced to a Galerkin formulation.

Following the standard steps, a weak form of the FADE (15) yields

$$\int_\Omega uw^*\phi'(x))dx + \int_\Omega vw'\phi^{(\alpha)}(x))dx - \int_\Omega \gamma vw'\phi^{(1+\alpha)}(x))dx = \int_\Omega w^* f(x)dx \tag{19}$$

where $\phi^{(1+\alpha)}(x) = \frac{d}{dx}\phi^{(\alpha)}(x)$. Substituting the above trial function into Eq. (19), yields the Petrov-Galerkin RKPM discrete equation as

$$\left[u\left(\gamma \boldsymbol{K}^*(1) - \boldsymbol{K}^*(0)\right) + v\boldsymbol{K}^*(\alpha) - \gamma v\boldsymbol{K}^*(1+\alpha)\right]\boldsymbol{\Phi} = \boldsymbol{F} \tag{20}$$

where $\boldsymbol{\Phi} = \left[\phi_0 \cdots \phi_J \cdots\right]^T$ is the nodal value vector, $F_I = \int_\Omega \left[N_I + \gamma N_I'\right] f(x)dx$, and

$$K^*_{IJ}(\beta) = \int_\Omega N_I' N_J^{(\beta)} dx \tag{21}$$

where β takes the value of 0,1,α, and $1+\alpha$. It should be noted that the alternative formula of Caputo fractional derivative is used for $\boldsymbol{K}^*(\alpha)$, which reads:

$$\boldsymbol{K}^*(\alpha) = \boldsymbol{K}^b + \boldsymbol{K}^m(\alpha) \tag{22}$$

with

$$K^b_{IJ} = \frac{1}{\Gamma(2-\alpha)}\int_\Omega N_I'(x)N_J'(0)x^{1-\alpha}dx \tag{23}$$

and

$$K_{IJ}^{m} = \frac{1}{\Gamma(2-\alpha)} \int_{\Omega} N_I'(x) \left(\int_{\Omega}^{x} (x-y)^{1-\alpha} N_J''(y) dy \right) dx \tag{24}$$

Due to the non-local fractional differential operator, the induced stiffness matrix is no-longer compact but dense.

Before solving illustrative examples, the stabilization parameter β needs to be chosen appropriately, due to its crucial role in the Petrov-Galerkin formulation. For the integer derivative advection-diffusion equation [2, 4–7], the stabilization parameter is usually calculated by

$$\beta = \coth(\mathrm{Pe}) - 1/\mathrm{Pe} \tag{25}$$

where Pe is the element Peclet number, and defined by $\mathrm{Pe} = \frac{2u\Delta x}{\nu}$ which is used in the finite element method. However, by this definition, the element Peclet number is no longer a dimensionless number as $\alpha < 1$, because the dimension of ν takes the form of $L^{1+\alpha}/T$ [1], where L and T is the length and time dimensions, respectively. Therefore, Eq. (25) is not suitable to calculate the stabilization parameter for the case of $\alpha < 1$. A theoretical study about the correct definition of element Peclet number for FADE will be presented in an upcoming contribution. In the work presented herein, Eq. (25) is applied directly as an approximation of the stabilization parameter for the cases with strong advection propensity, meanwhile the element Peclet number is approximately determined by $\mathrm{Pe} = \frac{2u\rho}{\nu}$ to taking into account the effect of dilation parameter. For cases with lower advection propensity, Galerkin formulation is used.

4 Illustrative Examples

Several examples including steady state and time-dependent problems are studied. For the time dependent example, forward time integration is implemented with lumped mass matrix approximation. The Lagrange multiplier method is used to impose the Dirichlet boundary conditions.

4.1 Steady State Examples

Two cases will be considered: homogeneous and non-homogeneous boundary conditions. Starting with the homogeneous boundary conditions, and for a material domain of $x \in [0, 2]$, the steady state fractional advection-diffusion equation is given as

$$u\frac{d}{dx}\phi(x) - \nu\frac{d}{dx}\phi^{(\alpha)}(x) = f(x) \tag{26}$$

with

$$\phi(0) = 0, \qquad \phi(2) = 0 \tag{27}$$

and

$$f(x) = -\nu\left(\frac{24x^{3-\alpha}}{\Gamma(4-\alpha)} - \frac{24x^{2-\alpha}}{\Gamma(3-\alpha)} + \frac{8x^{1-\alpha}}{\Gamma(2-\alpha)}\right) + u\left(4x^3 - 12x^2 + 8x\right) \tag{28}$$

The analytic solution reads

$$\phi(z) = x^2\,(2-x)^2 \tag{29}$$

Two different values of α=0.7 and 0.2 are considered herein, while the other parameters are chosen to be as $u = 0.2$, $\nu = 0.4$, $\Delta x = 0.1$, and $\rho = 3\Delta x$. For this homogeneous boundary condition problem, Galerkin formulation is used. Numerical results from RKPM with the alternative and original Caputo fractional formula denoted by RKPM1 and RKPM2, respectively, are shown in Figs. 3 and 4. Both figures demonstrate that the numerical results with the alternative Caputo

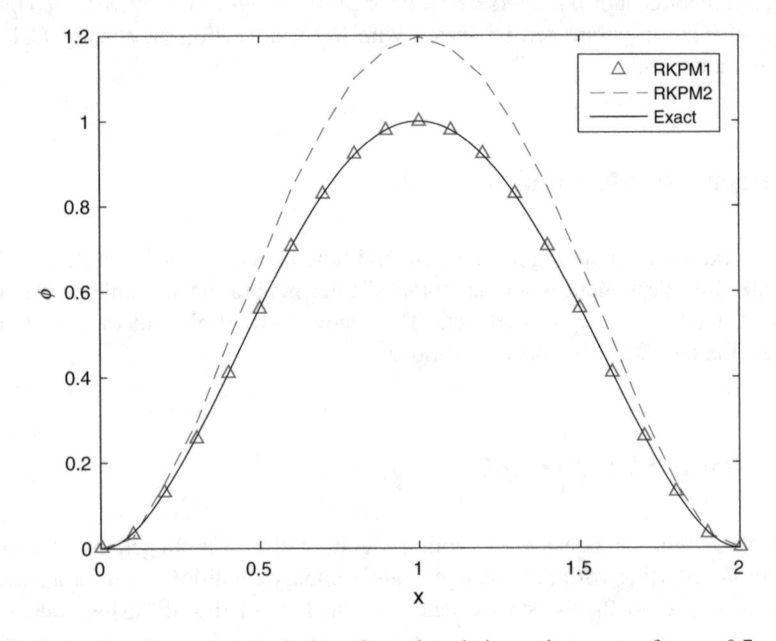

Fig. 3 Comparison between numerical results and analytic result to case of $\alpha = 0.7$, where RKPM1 and RKPM2 stand for the result from RKPM with alternative and original Caputo fractional formulas, respectively

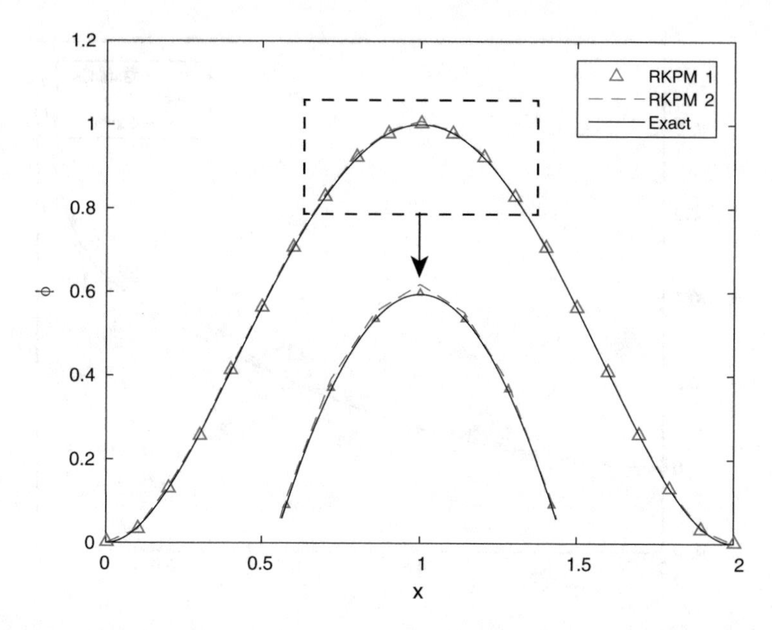

Fig. 4 Comparison between numerical results and analytic result to case of $\alpha = 0.2$, where RKPM1 and RKPM2 stand for the result from RKPM with alternative and original Caputo fractional formulas, respectively

fractional formula are in good agreement with the analytic solution, while the numerical results with the original formula deviate from the analytic solution when α increasing and using Gauss quadrature.

Now considering the problem with non-homogeneous boundary conditions and for a material domain of $x \in [0, 1]$, the Dirichlet boundary conditions are set as

$$\phi(0) = 0, \qquad \phi(1) = 1 \tag{30}$$

with $f(x) = x$.

The analytic solution reads

$$\phi(x) = \phi(0) + c\frac{\nu}{u}\left[E_{\alpha,1}(\frac{u}{\nu}x^{\alpha}) - 1\right] - \frac{1}{u}f(x) * \left[E_{\alpha,1}(\frac{u}{\nu}x^{\alpha}) - 1\right] \tag{31}$$

where c is determined by the right Dirichlet B.C, and $E_{\alpha,1}(z)$ is a two parameter Mittag-Leffler function defined by

$$E_{\alpha,1}(z) = \sum_{k=0}^{\infty} \frac{z^k}{\Gamma(\alpha k + 1)} \tag{32}$$

For illustrative purposes we chose two sets of parameters: (1) $\alpha = 0.9$, $\nu = 0.0439$, and $u = 1$; and (2) $\alpha = 0.1$, $\nu = 1.4515$ and $u = 1$. For both cases, two

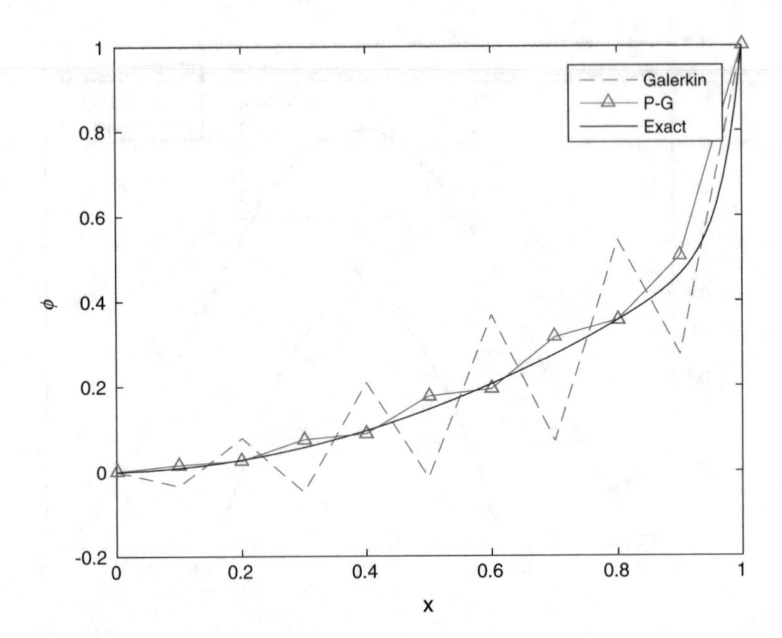

Fig. 5 Comparison between numerical results and analytic result for case of $\alpha = 0.9$ with $\Delta x = 0.1$, where Galerkin and P-G stand for the numerical result from Galerkin RKPM and Petrov-Galerkin RKPM, respectively

discretization spacings, $\Delta x = 0.1$ and $\Delta x = 0.01$, were considered with $\rho = 3\Delta x$. Both Petrov-Galerkin and Galerkin RKPM are used.

In the first case, $\Delta x = 0.1$ induces a strong advection propensity. Therefore, as shown in Fig. 5, spurious oscillations are present in numerical result by the Galerkin RKPM, while the numerical result utilizing Petrov-Galerkin RKPM is in better agreement with analytic solution, although showing some oscillatory behavior. The finer particle spacing of $\Delta x = 0.01$ reduces the advection propensity and then the numerical results become in good agreement with the analytic solution as shown in Fig. 6.

However, when considering the second set of parameters where $\alpha = 0.1$ the oscillations present in the solutions cannot be eliminated, regardless of using the stabilization formulation given by the Petrov-Galerkin RKPM or even by the particle spacing refinement, as shown in Figs. 7 and 8. The fractional order α dictates the difference in the observed behavior. Note that the stabilization parameter used here is only taking into account the advection term. Hence, decreasing the fractional order controls how strong the advection effect is in the fractional diffusion term. This explains why the stabilization parameter formula from the integer advection-diffusion equation does not work for this case, although improving the numerical solution from the Galerkin RKPM when considering high element Peclet number. A suitable stabilization parameter formula which can take into account both advection and fractional diffusion terms will be presented in an upcoming contribution.

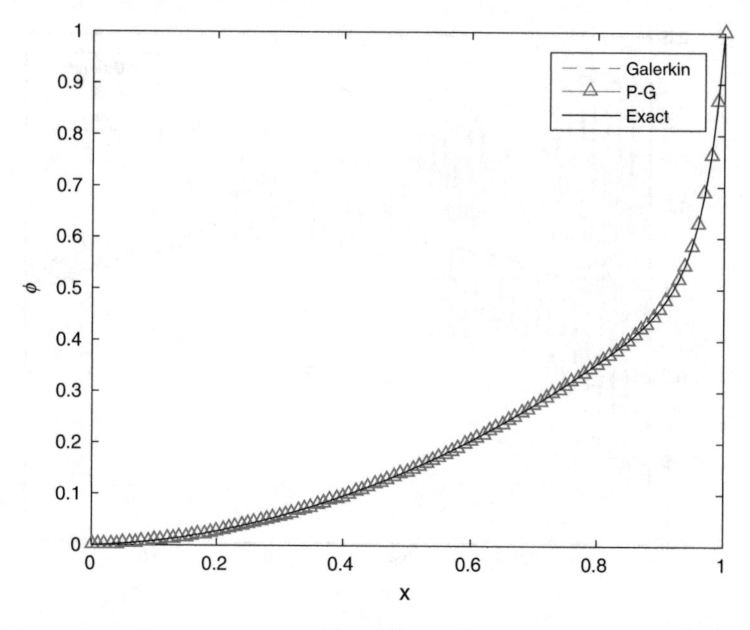

Fig. 6 Comparison between numerical results and analytic result for case of $\alpha = 0.9$ with $\Delta x = 0.01$, where Galerkin and P-G stand for the result from Galerkin RKPM and Petrov-Galerkin RKPM, respectively

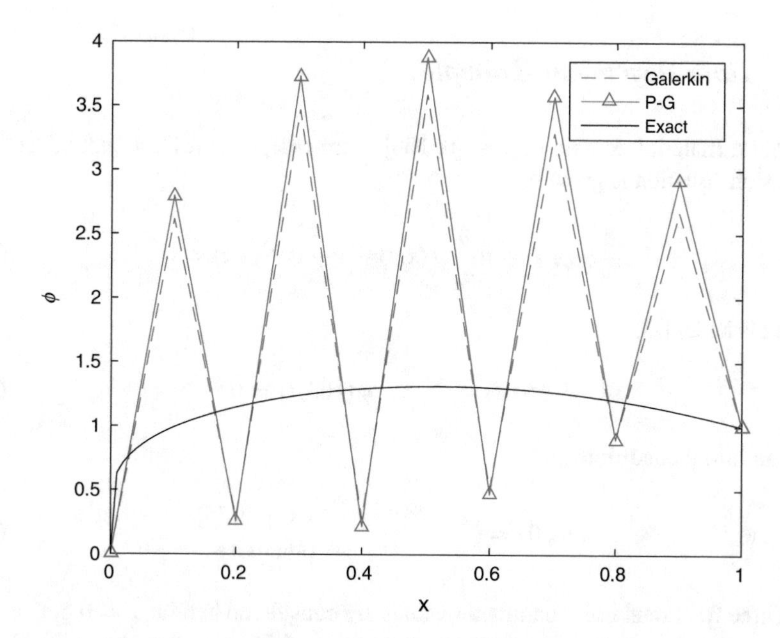

Fig. 7 Comparison between numerical results and analytic result for case of $\alpha = 0.1$ with $\Delta x = 0.1$, where Galerkin and P-G stand for the result from Galerkin RKPM and Petrov-Galerkin RKPM, respectively

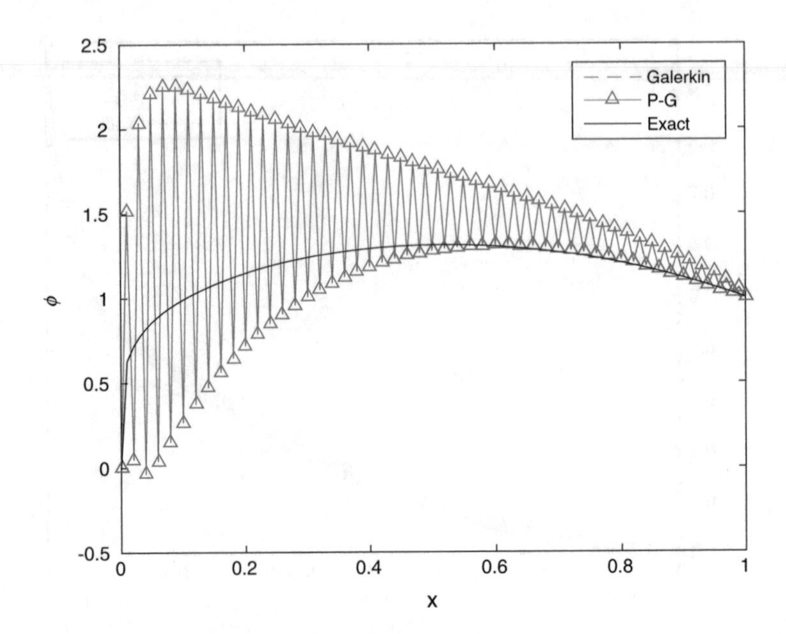

Fig. 8 Comparison between numerical results and analytic result for case of $\alpha = 0.1$ with $\Delta x = 0.01$, where Galerkin and P-G stand for the result from Galerkin RKPM and Petrov-Galerkin RKPM, respectively

4.2 *Time-Dependent Examples*

Within a material domain of $x \in [0, 100]$, a time-dependent fractional advection-diffusion equation is given by

$$\frac{\partial}{\partial t}\phi(x,t) + u\frac{\partial}{\partial x}\phi(x,t) - \nu\frac{\partial}{\partial x}\phi^{(\alpha)}(x,t) = 0 \tag{33}$$

with Dirichlet B.C.

$$\phi(0,t) = 0, \qquad \phi(100,t) = 0 \tag{34}$$

and an initial condition

$$\phi(x,0) = \{ \begin{matrix} e^{-0.05(x-32.5)^2} & x \in [30, 35] \\ 0 & \text{otherwise} \end{matrix} \tag{35}$$

Three fractional order parameter values are considered herein: $\alpha = 0.5$, $\alpha = 0.9$ and $\alpha = 1$. The remaining parameters are $u = 0.25, \nu = 0.1$, $\Delta x = 0.25$ and $\rho = 3\Delta x$ for all three values of α. We note that the parameters α and Δx are chosen so that oscillations will not be present in the numerical solutions for all

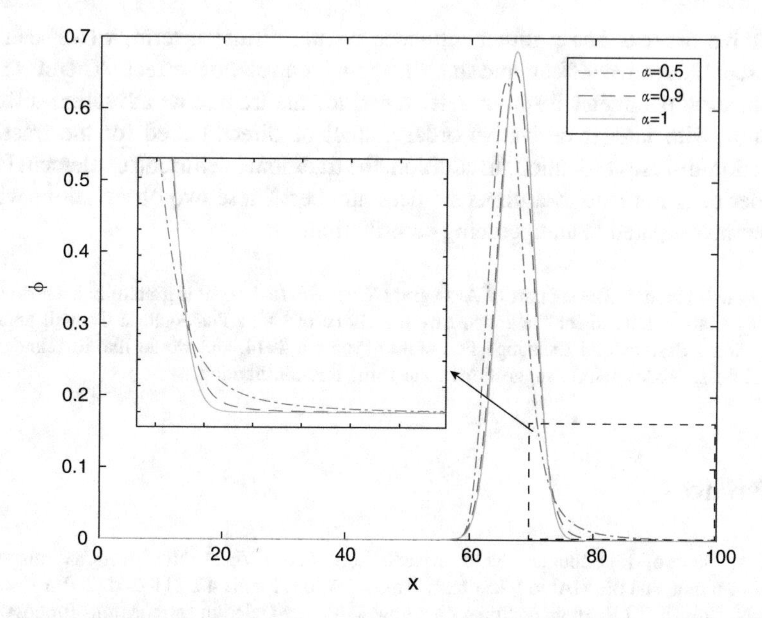

Fig. 9 Numerical results for time-dependent advection-diffusion equation at time of $t = 3$ with $\Delta x = 0.25$

three cases when using Galerkin RKPM. Numerical solutions of three cases at time $t = 3$ are shown in Fig. 9. Two characteristics of the FADE solution can be observed when considering the different fractional orders. First, the peak value is inversely proportional to the fractional order, while the peak position moves slower for decreasing values of the fractional order. This demonstrates that the advection effect arises from the fractional diffusion term and in the direction opposite to that of the given advection term. Second, the spread of the domain of the plume described by the FADE is larger than that of the traditional advection-diffusion equation as shown in the close-up portion of Fig. 9. This demonstrates that a so-called anomalous diffusion in terms of spatial correlation can be described by a non-locality of the fractional derivative operator, which also leads to the lower peak value.

5 Conclusion

The approximation of the spatial fractional advection-diffusion equation that arises in many physical and engineering applications is studied using the reproducing kernel particle method. An alternative form of the Caputo fractional derivative is recommended when using Gauss quadrature to integrate expressions with fractional derivative. The numerical results demonstrate the effectiveness and accuracy of the proposed alternative formula for RKPM. We also concluded that the fractional derivative order α has a direct influence on the diffusion term, which can also represent advective effect and has a spatial correlation effect. Therefore, the stabilization parameter formula determined for the traditional advection-diffusion equation with integer derivative order cannot be directly used for the fractional advection-diffusion equation. In addition, the traditional definition of element Peclet number does not lead to a dimensionless number. These two observations will be further investigated in an upcoming contribution.

Acknowledgements The support of ARO grant W911NF-15-1-0569 is gratefully acknowledged. Yanping Lian is grateful for the support by the Office of China Postdoctoral Council under the International Postdoctoral Exchange Fellowship Program 2014, and would like to acknowledge Miguel Bessa for his helpful suggestions concerning this contribution.

References

1. D.A. Benson, R. Schumer, M.M. Meerschaert, S.W. Wheatcraft, Fractional dispersion, Lev motion, and the MADE tracer tests. Transp. Porous Media **42**, 211–240 (2001)
2. A.N. Brooks, T.J.R. Hughes, Streamline upwind/Petrov-Galerkin formulations for convection dominated flows with particular emphasis on the incompressible Navier-Stokes equations. Comput. Methods Appl. Mech. Eng. **32**(1), 199–259 (1982)
3. D. Craiem, R.L. Magin, Fractional order models of viscoelasticity as an alternative in the analysis of red blood cell (RBC) membrane mechanics. Phys. Biol. **7**(1), 013001 (2010)
4. T.J.R. Hughes, A. Brooks, A theoretical framework for Petrov-Galerkin methods with discontinuous weighting functions: application to the streamline-upwind procedure. Finite Elem. Fluids **4**, 47–65 (1982)
5. T.J.R. Hughes, M. Mallet, M. Akira, A new finite element formulation for computational fluid dynamics: II. Beyond SUPG. Comput. Methods Appl. Mech. Eng. **54**(3), 341–355 (1986)
6. T.J.R. Hughes, L.P. Franca, M. Mallet, A new finite element formulation for computational fluid dynamics: VI. Convergence analysis of the generalized SUPG formulation for linear time-dependent multidimensional advective-diffusive systems. Comput. Methods Appl. Mech. Eng. **63**(1), 97–112 (1987)
7. T.J.R. Hughes, G.R. Feijóo, L. Mazzei, J.B. Quincy, The variational multiscale method - a paradigm for computational mechanics. Comput. Methods Appl. Mech. Eng. **166**(1), 3–24 (1998)
8. Y. Li, S. Tang, B.C. Abberton, M. Kroger, C. Burkhart, B. Jiang, G.J. Papakonstantopoulos, M. Poldneff, W.K. Liu, A predictive multiscale computational framework for viscoelastic properties of linear polymers. Polymer **53**, 5935–5952 (2012)
9. W.K. Liu, S. Jun, S.F. Li, J. Adee, T. Belytschko, Reproducing kernel particle methods for structural dynamics. Int. J. Numer. Methods Eng. **38**, 1655–1679 (1995)
10. W.K. Liu, S. Jun, Y.F. Zhang, Reproducing kernel particle methods, Int. J. Numer. Methods Fluids **20**, 1081–1106 (1995)
11. M.M. Meerschaert, H.P. Scheffler, C. Tadjeran, Finite difference methods for two-dimensional fractional dispersion equation. J. Comput. Phys. **211**, 249–261 (2006)
12. Y.A. Rossikhin, M.V. Shitikova, Application of fractional calculus for dynamic problems of solid mechanics: novel trends and recent results. Appl. Mech. Rev. **63**(1), 010801 (2009)
13. B.J. West, Colloquium: fractional calculus view of complexity: a tutorial. Rev. Mod. Phys. **86**, 1169–1184 (2014)

Meshless Multi-Point Flux Approximation

Alexander A. Lukyanov and Cornelis Vuik

Abstract The reservoir simulation of the complex reservoirs with anisotropic permeability, which includes faults and non-orthogonal grids, with a fully discontinuous permeability tensor in the discretization is a major challenge. Several methods have already been developed and implemented within industry standard reservoir simulators for non-orthogonal grids (e.g., Multi-Point Flux Approximation (MPFA) "O" method). However, it has been noticed that some of the numerical methods for elliptic/parabolic equations may violate the maximum principle (i.e., lead to spurious oscillations), especially when the anisotropy is particularly strong. It has been found that the oscillations are closely related to the poor approximation of the pressure gradient in the flux computation. Therefore, proposed methods must correctly approximate underlying operators, satisfy a discrete maximum principle and have coercivity properties. Furthermore, the method must be robust and efficient. This paper presents the meshless multi-point flux approximation of second order elliptic operators containing a tensor coefficient. The method is based on a pressure gradient approximation commonly used in meshless methods (or Smoothed Particle Hydrodynamics method—SPH method). The proposed discretization schemes can be written as a sum of sparse positive semidefinite matrix and perturbation matrix. We show that convergence rates are retained as for finite difference methods $O(h^{\alpha})$, $1 \leq \alpha < 2$, where h denotes the maximum particle spacing. The results are presented, discussed and future studies are outlined.

A.A. Lukyanov (✉)
Schlumberger-Doll Research, One Hampshire Street, Cambridge, MA 02139, USA
e-mail: alukyanov@slb.com

C. Vuik
Faculty of Electrical Engineering, Mathematics and Computer Science (EEMCS), Delft Institute of Applied Mathematics, Delft University of Technology, 2628 CD Delft, The Netherlands
e-mail: c.vuik@tudelft.nl

M. Griebel, M.A. Schweitzer (eds.), *Meshfree Methods for Partial Differential Equations VIII*, Lecture Notes in Computational Science and Engineering 115,
DOI 10.1007/978-3-319-51954-8_5

1 Introduction

The multi-point flux approximation, MPFA, is a discretization method developed by the oil industry to be the next generation method in reservoir simulations and it can be applied to different types of mesh, for example using quadrilateral meshes as in Aavatsmark et al. [2, 3], Aavatsmark [1], Edwards and Rogers [12], Klausen and Russell [17] or unstructured grids as in Edwards [11] to approximate the following operator:

$$\mathbf{L}(p(\mathbf{r})) = -\nabla\left(\mathbf{M}(\mathbf{r}, p(\mathbf{r}))\nabla p(\mathbf{r})\right) - g(\mathbf{r}), \ \forall \mathbf{r} \in \Omega \subset \mathbb{R}^n \tag{1}$$

where $p(\mathbf{r})$ is the pressure, $\mathbf{M}(\mathbf{r}, p(\mathbf{r})) = \left(m_{\alpha\beta}\right)$ is the mobility tensor, $g(\mathbf{r})$ is the known sink / source term, $n = 1, 2, 3$ is the spatial dimension. Consider the operator in the expression (1) with a piecewise constant mobility $\mathbf{M}(\mathbf{r}, p(\mathbf{r})) \in L_2(\Omega)$.

Several methods have already been developed and implemented within an industry standard reservoir simulator for non-orthogonal grids. The methods are known as the O-method, U-method and the L-method for quadrilateral meshes in two and three dimensions (see [1, 12, 18]. The MPFA methods are not restricted to quadrilateral meshes and have been investigated in Edwards [11]. It has been noticed that some of the numerical methods for elliptic/parabolic equations may violate the maximum principle (i.e. lead to spurious oscillations). Therefore, proposed methods must satisfy a discrete maximum principle to avoid any spurious oscillations. The discrete maximum principle for MPFA methods was discussed, e.g., in Edwards and Rogers [12], Mlacnik and Durlofsky [28], Lee et al. [19].

However, non-physical oscillations can appear in the developed multi-point flux approximations when the anisotropy is particularly strong. It has been found that the oscillations are closely related to the poor approximation of the pressure gradient in the flux computation. In this paper, the meshless multi-point flux approximation for the general fluid flow in porous media is proposed. The discretization scheme is based both on the generalized Laplace approximation and on a gradient approximation commonly used in the Smoothed Particle Hydrodynamics (SPH) community for thermal, viscous, and pressure projection problems and can be extended to include higher-order terms in the appropriate Taylor series. The proposed discretization scheme is combined with mixed corrections, which ensure linear completeness. The mixed correction utilizes Shepard Functions in combination with a correction to derivative approximations. Incompleteness of the kernel support combined with the lack of consistency of the kernel interpolation in conventional meshless methods results in fuzzy boundaries. In corrected meshless methods, the domain boundaries and field variables at the boundaries are approximated with the improved accuracy comparing to the conventional SPH method. The resulting schemes improve the particle deficiency (kernel support incompleteness) problem. Although, the analysis of the different discretization schemes in this paper is restricted to 2D (i.e., $n = 2$), the results can be applied in any space.

2 Fluid Flow Modelling Using SPH

To calculate the second derivatives of $\mathbf{F}$, several methods were proposed (Chen et al. [7]; Bonet and Kulasegaram [4]; Colin et al. [9]). However, second-order derivatives can often be avoided entirely if the PDE is written in a weak form. It is important to note that approximations using second-order derivatives of the kernel are often noisy and sensitive to the particle distributions, particularly for spline kernels of lower orders.

Brookshaw [6] proposed an approximation of the Laplacian for an inhomogeneous scalar field $m(\mathbf{r})$ that only includes first order derivatives:

$$\begin{aligned}&\langle \nabla \left(m\left(\mathbf{r}_I\right) \nabla \mathbf{F}\left(\mathbf{r}_I\right)\right)\rangle = \\ &= \sum_{\Omega_{\mathbf{r}_I,h}} V_{\mathbf{r}_J} \left[\mathbf{F}\left(\mathbf{r}_J\right) - \mathbf{F}\left(\mathbf{r}_I\right)\right] \frac{\left(\mathbf{r}_J - \mathbf{r}_I\right)\cdot\left(m_J + m_I\right)\nabla W\left(\mathbf{r}_J - \mathbf{r}_I, h\right)}{\left\|\mathbf{r}_J - \mathbf{r}_I\right\|^2}\end{aligned} \tag{2}$$

where $V_{\mathbf{r}_J}$ is the volume of a particle J, $\|\bullet\|$ is the Euclidean norm throughout this paper, $\mathbf{F}(\mathbf{r})$ is the unknown scalar or vector field (e.g., pressure p) $\forall \mathbf{r} \in \Omega \subset \mathbb{R}^n$, $m_I = m(\mathbf{r}_I)$, $\mathbf{r}_I \in \Omega \subset \mathbb{R}^n$ and $m_J = m(\mathbf{r}_J)$, $\mathbf{r}_J \in \Omega \subset \mathbb{R}^n$ are the field coefficients, $W(\mathbf{r}_J - \mathbf{r}_I, h)$ is the Kernel.

This Laplacian approximation was used by Brookshaw [6], Cleary and Monaghan [8], Jubelgas et al. [16] for thermal conduction, Morris et al. [29] for modelling viscous diffusion, Cummins and Rudman [10] for a vortex spin-down and Rayleigh-Taylor instability, Shao and Lo [33] for simulating Newtonian and non-Newtonian flows with a free surface, Moulinec et al. [20] for comparisons of weakly compressible and truly incompressible algorithms, Hu and Adams [15] for macroscopic and mesoscopic flows, Zhang et al. [34] for simulations of the solid-fluid mixture flow. There are several numerical SPH schemes commonly used in numerical simulations for a scalar inhomogeneous field $m(\mathbf{r})$. High order accuracy approximations can also be derived by using SPH discretization based on higher order Taylor series expansions [13, 14, 22, 31]. However, it is usually required that the discrete numerical schemes can reproduce linear fields [5, 23, 27, 30] or polynomials up to a given order [21].

The correction terms to Brookshaw's formulation, which improve the accuracy of the Laplacian operator near boundaries, were proposed by Schwaiger [31]:

$$\begin{aligned}&\langle \nabla \left(m\left(\mathbf{r}_I\right) \nabla \mathbf{F}\left(\mathbf{r}_I\right)\right)\rangle = \\ &\frac{\Gamma^{-1}_{\beta\beta}}{n}\left\{\sum_{\Omega_{\mathbf{r}_I,h}} V_{\mathbf{r}_J} \left[\mathbf{F}\left(\mathbf{r}_J\right) - \mathbf{F}\left(\mathbf{r}_I\right)\right] \frac{\left(\mathbf{r}_J - \mathbf{r}_I\right)\cdot\left(m_J + m_I\right)\nabla W\left(\mathbf{r}_J - \mathbf{r}_I, h\right)}{\left\|\mathbf{r}_J - \mathbf{r}_I\right\|^2}\right\} - \\ &-\frac{\Gamma^{-1}_{\beta\beta}}{n}\left\{\left[\langle \nabla_\alpha (m\left(\mathbf{r}_I\right)\mathbf{F}\left(\mathbf{r}_I\right))\rangle - \mathbf{F}\left(\mathbf{r}_I\right)\langle \nabla_\alpha m\left(\mathbf{r}_I\right)\rangle + m\left(\mathbf{r}_I\right)\langle \nabla_\alpha \mathbf{F}\left(\mathbf{r}_I\right)\rangle\right]\mathbf{N}^\alpha\right\}\end{aligned} \tag{3}$$

$$\mathbf{N}^{\alpha}(\mathbf{r}_I) = \left[\sum_{\Omega_{\mathbf{r}_I,h}} V_{\mathbf{r}_J} \nabla_{\alpha} W(\mathbf{r}_J - \mathbf{r}_I, h) \right] \tag{4}$$

$$\langle \nabla_{\alpha} \mathbf{F}(\mathbf{r}_I) \rangle = \sum_{\Omega_{\mathbf{r}_I,h}} V_{\mathbf{r}_J} \left[\mathbf{F}(\mathbf{r}_J) - \mathbf{F}(\mathbf{r}_I) \right] \nabla_{\alpha}^{*} W(\mathbf{r}_J - \mathbf{r}_I, h) \tag{5}$$

$$\nabla_{\alpha}^{*} W = \mathbf{A}_{\alpha\beta}^{-1} \nabla_{\beta} W, \quad \mathbf{A}_{\alpha\beta} = \left[\sum_{\Omega_{\mathbf{r}_I,h}} V_{\mathbf{r}_J} \left[\mathbf{r}_J^{\alpha} - \mathbf{r}_I^{\alpha} \right] \nabla_{\beta} W(\mathbf{r}_J - \mathbf{r}_I, h) \right] \tag{6}$$

where $n = 1, 2, 3$ is the spatial dimension, the gradient approximation $\langle \nabla_{\alpha} \mathbf{F}(\mathbf{r}_I) \rangle$ is computed using (5) and the tensor $\Gamma_{\alpha\beta}$ is defined by

$$\Gamma_{\alpha\beta}(\mathbf{r}_I) = \sum_{\Omega_{\mathbf{r}_I,h}} V_{\mathbf{r}_J} \frac{\left(\mathbf{r}_J^{\gamma} - \mathbf{r}_I^{\gamma}\right) \nabla_{\gamma} W(\mathbf{r}_J - \mathbf{r}_I, h)}{\|\mathbf{r}_J - \mathbf{r}_I\|^2} \left(\mathbf{r}_J^{\alpha} - \mathbf{r}_I^{\alpha}\right) \left(\mathbf{r}_J^{\beta} - \mathbf{r}_I^{\beta}\right) \tag{7}$$

Throughout this paper, the summation by repeated Greek indices is assumed. For multi-dimensional problems, the correction tensor $\Gamma_{\alpha\beta}(\mathbf{r}_I)$ is a matrix. If the particle $\mathbf{r}_I$ has entire stencil support (i.e., the domain support for all kernels $W(\mathbf{r}_J - \mathbf{r}_I, h)$ is entire and symmetric) then $\Gamma_{\alpha\beta}(\mathbf{r}_I) \approx \delta_{\alpha\beta}$, $\delta_{\alpha\beta}$ is the Kronecker symbol. Unfortunately, $\Gamma_{\alpha\beta}(\mathbf{r}_I)$ deviates from $\delta_{\alpha\beta}$ for the provided algorithm and, hence, it is important to minimize this deviation from $\delta_{\alpha\beta}$ in the new methods.

Remark 1 It is important to note that correction tensors $\Gamma_{\alpha\beta}$ and $\mathbf{A}_{\alpha\beta}$ are the same tensors. Indeed, using the following identity:

$$\begin{aligned} \left[\mathbf{r}_J^{\alpha} - \mathbf{r}_I^{\alpha} \right] \frac{\left(\mathbf{r}_J^{\gamma} - \mathbf{r}_I^{\gamma}\right) \nabla_{\gamma} W(\mathbf{r}_J - \mathbf{r}_I, h)}{\|\mathbf{r}_J - \mathbf{r}_I\|^2} &= \\ = \pm \frac{1}{h} \frac{dW}{dz} \frac{\left[\mathbf{r}_J^{\alpha} - \mathbf{r}_I^{\alpha} \right]}{\|\mathbf{r}_J - \mathbf{r}_I\|} &= \nabla_{\alpha} W(\mathbf{r}_J - \mathbf{r}_I, h), \quad \forall \alpha \end{aligned} \tag{8}$$

where $z = \|\mathbf{r}_J - \mathbf{r}_I\| / h$, $\forall \mathbf{r}_J, \mathbf{r}_I \in \Omega \subset \mathbb{R}^n$, the following relations can be established:

$$\begin{aligned} \Gamma_{\alpha\beta}(\mathbf{r}_I) &= \sum_{\Omega_{\mathbf{r}_I,h}} V_{\mathbf{r}_J} \frac{\left(\mathbf{r}_J^{\gamma} - \mathbf{r}_I^{\gamma}\right) \nabla_{\gamma} W(\mathbf{r}_J - \mathbf{r}_I, h)}{\|\mathbf{r}_J - \mathbf{r}_I\|^2} \left(\mathbf{r}_J^{\alpha} - \mathbf{r}_I^{\alpha}\right) \left(\mathbf{r}_J^{\beta} - \mathbf{r}_I^{\beta}\right) = \\ &= \sum_{\Omega_{\mathbf{r}_I,h}} V_{\mathbf{r}_J} \left[\mathbf{r}_J^{\alpha} - \mathbf{r}_I^{\alpha} \right] \nabla_{\beta} W(\mathbf{r}_J - \mathbf{r}_I, h) = \mathbf{A}_{\alpha\beta}(\mathbf{r}_I) \end{aligned} \tag{9}$$

To calculate coefficients in the scheme (3)–(7) is a trivial task. However, in general, it should be performed at each Newton-Raphson iteration in the non-linear case (i.e., $m = m(F)$). It also requires additional efforts to invert the correction matrix $\mathbf{A}_{\alpha\beta}$ (inversion of $n \times n$ matrices per each particle, where $n = 1, 2, 3$ is the spatial dimension) and storage cost of $\nabla_\alpha W(\mathbf{r}_J - \mathbf{r}_I, h)$, $\nabla_\alpha^* W(\mathbf{r}_J - \mathbf{r}_I, h)$, and corresponding $\Gamma_{\alpha\alpha}^{-1} = \mathbf{A}_{\alpha\alpha}^{-1}$ per each particle.

2.1 *Kernel Property*

A central point of the SPH formalism is the concept of the interpolating function (or kernel) through which the continuum properties of the medium are recovered from a discrete sample of N points with prescribed mass m_I (for conventional Lagrangian methods) or volume V_I (for fully Eulerian methods). In the Lagrangian description, these points move according to the specified governing laws, whereas these points are fixed in space for the Eulerian description. A good interpolating kernel must satisfy a few basic requirements: it must tend to the delta function in the continuum limit and has to be a continuous function with definite first derivatives at least. From a more practical point of view it is also advisable to deal with symmetric finite range kernels, the latter is to avoid N^2 calculations. In this paper, the cubic spline is used:

$$W(z,h) = \frac{\Xi}{h^n} \begin{cases} 1 - \frac{3}{2}z^2 + \frac{3}{4}z^3, \; 0 \le z \le 1 \\ \frac{1}{4}(2-z)^3, \; 1 \le z \le 2 \\ 0, \; z > 2 \end{cases} \tag{10}$$

where $z = \left\| \mathbf{r}' - \mathbf{r} \right\| / h$, $\forall \mathbf{r}, \mathbf{r}' \in \Omega \subset \mathbb{R}^n$ and $\Xi = \frac{3}{2}, \frac{10}{7\pi}, \frac{1}{\pi}$ in 1D (i.e., $n = 1$), 2D (i.e., $n = 2$) and 3D (i.e., $n = 3$), respectively.

3 Meshless Transmissibilities

The well-known two-point flux approximation (TPFA) is a numerical scheme used in most commercial reservoir simulators for the pressure Eq. (1): $\mathbf{L}(p) = 0$. The net flow rate of a fluid (single phase and component fluid) from a cell I into neighbouring cells is obtained by summing fluxes over the neighbouring cells J:

$$\mathbf{q} = \sum_J \widetilde{T}_{JI} \left[p(\mathbf{r}_J) - p(\mathbf{r}_I) \right], \; T_{JI} \ge 0 \tag{11}$$

where $\widetilde{T}_{JI}$ is the transmissibility between cells J and I, $\mathbf{q}$ is the total flux through the boundary of the control volume located at the point $\mathbf{r}_I$. The transmissibility $\widetilde{T}_{JI}$

defined at an interior face f between cells J and I is calculated as

$$\widetilde{T}_{JI} = \frac{1}{\left[\frac{\left\|\mathbf{r}_{f,J}\right\|^2}{\mathbf{S}_f \mathbf{M} \mathbf{r}_{f,J}} + \frac{\left\|\mathbf{r}_{f,I}\right\|^2}{\mathbf{S}_f \mathbf{M} \mathbf{r}_{f,I}}\right]} \tag{12}$$

where $\mathbf{r}_{f,J}$ and $\mathbf{r}_{f,I}$ are the vectors from centres of cells J and I to the face f respectively, $\mathbf{S}_f$ is the area vector of the face f. In the case of $\mathbf{M}$-orthogonal mesh, when $\mathbf{M}\mathbf{S}_f$ and $[\mathbf{r}_J - \mathbf{r}_I]$ are collinear, the expression (11) reduces to the form of the central finite difference scheme and approximates the flux with $O\left(h^2\right)$ order of accuracy for any mobility tensor field $\mathbf{M}$. The expression (12) ensures that the flux into the adjoining region is continuous [8]. The TPFA scheme (11) is unconditionally monotone scheme. It is clear that the expression (3) cannot be written in the form (11) due to terms $\langle \nabla_\alpha \left(m\left(\mathbf{r}_I\right) \mathbf{F}\left(\mathbf{r}_I\right)\right)\rangle \mathbf{N}^\alpha$ and $\mathbf{F}\left(\mathbf{r}_I\right) \langle \nabla_\alpha m\left(\mathbf{r}_I\right)\rangle \mathbf{N}^\alpha$. Hence, it is only possible in this case to introduce a definition of a partial meshless transmissibility between particles $\mathbf{r}_J$ and $\mathbf{r}_I$ as follows:

$$T\left(\mathbf{r}_J, \mathbf{r}_I\right) = T_{JI} = \frac{\Gamma_{\beta\beta}^{-1}}{n} \times \left\{ \sum_{\Omega_{\mathbf{r}_I,h}} V_{\mathbf{r}_J} \frac{\left(\mathbf{r}_J - \mathbf{r}_I\right) \cdot \left(m_J + m_I\right) \cdot \nabla W\left(\mathbf{r}_J - \mathbf{r}_I, h\right)}{\left\|\mathbf{r}_J - \mathbf{r}_I\right\|^2} - V_{\mathbf{r}_J} m_I \nabla W\left(\mathbf{r}_J - \mathbf{r}_I, h\right) \mathbf{N}^\alpha \right\} \tag{13}$$

It is important to note that transmissibilities T_{JI} and $\widetilde{T}_{JI}$ have different physical units. Furthermore, it raises the question wherever the proposed scheme (3) is monotone. Hence, let Ω be a bounded domain in $\mathbb{R}^n$ (a compact) with a piecewise boundary $\partial\Omega = \bar{\Gamma}_D \cup \bar{\Gamma}_N$, $\bar{\Gamma}_D \cap \bar{\Gamma}_N = \emptyset$, where measure $\mu\left(\Gamma_D\right) \neq 0$, Γ_D is the part of the boundary corresponding to the Dirichlet boundary condition, Γ_N is the part of the boundary corresponding to the Neumann boundary condition. In the following sections this question will be analysed in details for some modified schemes by stating that the solution of the equation for $\mathbf{M}\left(\mathbf{r}, p\left(\mathbf{r}\right)\right) = m\left(\mathbf{r}, p\left(\mathbf{r}\right)\right) \cdot \mathbf{I}$, $m\left(\mathbf{r}, p\left(\mathbf{r}\right)\right) \geq 0$:

$$-\nabla\left(\mathbf{M}\left(\mathbf{r}, p\left(\mathbf{r}\right)\right) \nabla p\left(\mathbf{r}\right)\right) = g\left(\mathbf{r}\right), \ \forall \mathbf{r} \in \Omega \subset \mathbb{R}^n \tag{14}$$

is non-negative subject to its existence and that the solution p^k for each kth—Picard iteration is a non-negative vector and the linear system is solved exactly, $\mathbf{I}$ is the unit tensor. Modifications were introduced due to the following theorem.

Theorem 1 *The discretization scheme ((3)–(7)) is at least $O\left(h^\omega\right)$, $1 \leq \omega < 2$ order of accuracy in average for any scalar mobility field $m\left(\mathbf{r}, p\left(\mathbf{r}\right)\right) \in C^2\left(\Omega\right) \geq 0$ everywhere within the numerical domain Ω sufficiently far away from the boundary $\partial\Omega$.*

Proof Using Taylor series expansions about a point $\mathbf{r}_I$ and the relation (5), the following relations can be written:

$$\begin{aligned}\mathbf{F}(\mathbf{r}_J) = \mathbf{F}(\mathbf{r}_I) + \mathbf{F}_{,\alpha}(\mathbf{r}_I)\left[\mathbf{r}_J^\alpha - \mathbf{r}_I^\alpha\right] + \\ + \frac{1}{2}\mathbf{F}_{,\alpha\gamma}(\mathbf{r}_I)\left[\mathbf{r}_J^\alpha - \mathbf{r}_I^\alpha\right]\left[\mathbf{r}_J^\gamma - \mathbf{r}_I^\gamma\right] + O\left(h^3\right)\end{aligned} \tag{15}$$

$$m(\mathbf{r}_J) = m(\mathbf{r}_I) + m_{,\alpha}(\mathbf{r}_I)\left[\mathbf{r}_J^\alpha - \mathbf{r}_I^\alpha\right] + O\left(h^2\right) \tag{16}$$

$$m_I\left\langle \mathbf{F}(\mathbf{r}_I)\right\rangle_\alpha = m_I \mathbf{F}_{,\alpha}(\mathbf{r}_I) + O\left(h^2\right) \tag{17}$$

$$\sum_{\Omega_{\mathbf{r}_I,h}} V_{\mathbf{r}_J}\left[\mathbf{r}_J^\gamma - \mathbf{r}_I^\gamma\right]\nabla_\alpha^* W(\mathbf{r}_J - \mathbf{r}_I, h) = \delta_{\gamma\alpha}, \quad \forall\gamma,\alpha \tag{18}$$

$$\left[\mathbf{r}_J^\alpha - \mathbf{r}_I^\alpha\right]\frac{\left(\mathbf{r}_J^\gamma - \mathbf{r}_I^\gamma\right)\nabla_\gamma W(\mathbf{r}_J - \mathbf{r}_I, h)}{\|\mathbf{r}_J - \mathbf{r}_I\|^2} = \nabla_\alpha W(\mathbf{r}_J - \mathbf{r}_I, h), \quad \forall\alpha \tag{19}$$

Substituting relations ((15)–(17)) into the scheme (3) and taking into account the relations (18) and (19), it leads to the following relations:

$$\begin{aligned}&\sum_{\Omega_{\mathbf{r}_I,h}} V_{\mathbf{r}_J}\left[\mathbf{F}(\mathbf{r}_J) - \mathbf{F}(\mathbf{r}_I)\right]\frac{(\mathbf{r}_J - \mathbf{r}_I)\cdot(m_J + m_I)\cdot\nabla W(\mathbf{r}_J - \mathbf{r}_I, h)}{\|\mathbf{r}_J - \mathbf{r}_I\|^2} = \\ &= 2m(\mathbf{r}_I)\mathbf{F}_{,\alpha}(\mathbf{r}_I)\sum_{\Omega_{\mathbf{r}_I,h}} V_{\mathbf{r}_J}\left[\mathbf{r}_J^\alpha - \mathbf{r}_I^\alpha\right]\frac{(\mathbf{r}_J - \mathbf{r}_I)\cdot\nabla W(\mathbf{r}_J - \mathbf{r}_I, h)}{\|\mathbf{r}_J - \mathbf{r}_I\|^2} + \\ &+ m(\mathbf{r}_I)\mathbf{F}_{,\alpha\gamma}(\mathbf{r}_I)\sum_{\Omega_{\mathbf{r},h}} V_{\mathbf{r}_J}\left[\mathbf{r}_J^\alpha - \mathbf{r}_I^\alpha\right]\left[\mathbf{r}_J^\gamma - \mathbf{r}_I^\gamma\right]\frac{(\mathbf{r}_J - \mathbf{r}_I)\cdot\nabla W(\mathbf{r}_J - \mathbf{r}_I, h)}{\|\mathbf{r}_J - \mathbf{r}_I\|^2} \\ &+ m_{,\alpha}(\mathbf{r}_I)\mathbf{F}_{,\gamma}(\mathbf{r}_I)\sum_{\Omega_{\mathbf{r}_I,h}} V_{\mathbf{r}_J}\left[\mathbf{r}_J^\alpha - \mathbf{r}_I^\alpha\right]\left[\mathbf{r}_J^\gamma - \mathbf{r}_I^\gamma\right]\frac{(\mathbf{r}_J - \mathbf{r}_I)\cdot\nabla W(\mathbf{r}_J - \mathbf{r}_I, h)}{\|\mathbf{r}_J - \mathbf{r}_I\|^2} \\ &+ O\left(h^2\right)\end{aligned} \tag{20}$$

$$\begin{aligned}&\left[\left\langle\nabla_\alpha\left(m(\mathbf{r}_I)\mathbf{F}(\mathbf{r}_I)\right)\right\rangle - \mathbf{F}(\mathbf{r}_I)\left\langle\nabla_\alpha m(\mathbf{r}_I)\right\rangle + m(\mathbf{r}_I)\left\langle\nabla_\alpha\mathbf{F}(\mathbf{r}_I)\right\rangle\right] = \\ &= 2m(\mathbf{r}_I)\mathbf{F}_{,\alpha}(\mathbf{r}_I) + O\left(h^2\right)\end{aligned} \tag{21}$$

The claim of the theorem can be seen from the comparison of relations (20) and (21) and the fact is that

$$\sum_{\Omega_{\mathbf{r}_I,h}} V_{\mathbf{r}_J}\left[\mathbf{r}_J^\alpha - \mathbf{r}_I^\alpha\right]\frac{\left(\mathbf{r}_J^\gamma - \mathbf{r}_I^\gamma\right)\nabla_\gamma W(\mathbf{r}_J - \mathbf{r}_I, h)}{\|\mathbf{r}_J - \mathbf{r}_I\|^2} = \mathbf{N}^\alpha, \ \forall\alpha \tag{22}$$

and

$$\frac{\Gamma_{\beta\beta}^{-1}}{n} \times \left(\sum_{\Omega_{\mathbf{r}_I,h}} V_{\mathbf{r}_J} \left[\mathbf{r}_J^{\gamma} - \mathbf{r}_I^{\gamma} \right] \left[\mathbf{r}_J^{\alpha} - \mathbf{r}_I^{\alpha} \right] \frac{\left(\mathbf{r}_J^{\beta} - \mathbf{r}_I^{\beta} \right) \nabla_{\beta} W \left(\mathbf{r}_J - \mathbf{r}_I, h \right)}{\left\| \mathbf{r}_J - \mathbf{r}_I \right\|^2} \right) = \tag{23}$$

$$= \delta_{\gamma\alpha} + O\left(h^{\omega}\right), \ 1 \leq \omega < 2, \ \forall \gamma, \alpha$$

for all points $\mathbf{r}_I$ located sufficiently far away from the boundary $\partial\Omega$. The order of accuracy $O\left(h^{\omega}\right)$ has to be understood in statistical average sense with some dispersion around the average value. The scheme ((3)–(7)) does not require exact expressions for the gradient (i.e., spatial derivatives) of the mobility field $\nabla_{\gamma} m\left(\mathbf{r}, p\right)$ to keep a higher order of accuracy for any mobility field. Hence, this scheme can be used with the a discontinuous (or piece-wise) mobility field $m\left(\mathbf{r}, p\left(\mathbf{r}\right)\right) \in L_2\left(\Omega\right)$. The case of a discontinuous mobility field is considered below.

4 Discontinuous Mobility Case

Using the same idea behind the expression (13) and heterogeneous discontinues mobility field $m\left(\mathbf{r}\right)$, it can be shown that the effect of requiring the flux into the adjoining region to be continuous leads to the equivalent to the expression (13) in terms of the effective mobility between particle $\mathbf{r}_I$ and $\mathbf{r}_J$ (clearly and Monaghan [8]):

$$m^{eff} = \left(\frac{m\left(\mathbf{r}_J\right) \cdot m\left(\mathbf{r}_I\right)}{m\left(\mathbf{r}_J\right) + m\left(\mathbf{r}_I\right)} \right) \tag{24}$$

It can be seen that the effective mobility $\left(m\left(\mathbf{r}_J\right) + m\left(\mathbf{r}_I\right)\right)$ does not guarantee the continuity of the flux between the particles with discontinuous mobilities. Taking this into account and applying the relation (24), the final discretization scheme for the discontinuous scalar mobility field can be written as

$$\begin{aligned} &\left\langle \nabla_{\alpha} \left(m\left(\mathbf{r}_I\right) \nabla_{\alpha} \mathbf{F}\left(\mathbf{r}_I\right) \right) \right\rangle = \\ &\frac{4 \cdot \Gamma_{\beta\beta}^{-1}}{n} \left\{ \sum_{\Omega_{\mathbf{r}_I,h}} V_{\mathbf{r}_J} \cdot m^{eff} \cdot \left[\mathbf{F}\left(\mathbf{r}_J\right) - \mathbf{F}\left(\mathbf{r}_I\right) \right] \frac{\left(\mathbf{r}_J^{\alpha} - \mathbf{r}_I^{\alpha} \right) \cdot \nabla_{\alpha} W\left(\mathbf{r}_J - \mathbf{r}_I, h\right)}{\left\| \mathbf{r}_J - \mathbf{r}_I \right\|^2} \right\} \\ &- \frac{2 \cdot \Gamma_{\beta\beta}^{-1}}{n} \left\{ \left(\sum_{\Omega_{\mathbf{r}_I,h}} V_{\mathbf{r}_J} \cdot m_I \cdot \left[\mathbf{F}\left(\mathbf{r}_J\right) - \mathbf{F}\left(\mathbf{r}_I\right) \right] \nabla_{\alpha}^{*} W\left(\mathbf{r}_J - \mathbf{r}_I, h\right) \right) \mathbf{N}^{\alpha} \right\} \end{aligned} \tag{25}$$

This numerical scheme is the final one for the heterogeneous discontinuous isotropic scalar mobility field, which is used for numerical tests throughout this paper. It ensures that the flux is automatically continuous between particles with the reasonable accuracy. Multiple regions with substantially different fluid properties and specific mobilities can then be simulated.

The analytical analysis of the aforementioned scheme for a fully anisotropic mobility tensor field is complicated. However, the numerical analysis reveals that these schemes do not produce a reasonable approximation for a linear pressure field for the anisotropic mobility tensor field. The following section describes a scheme applicable to a fully anisotropic mobility tensor field.

4.1 *Anisotropic Case*

Generally speaking, any second order tensor can be decomposed into the spherical and deviatorical parts. In the case of continuum mechanics, the decomposition of the second order tensor (e.g., stress tensor or strain tensor) into their volumetric and deviatoric components have certain physical justifications. This step is done in order to distinguish between volumetric and shear responses. Hence, any mobility field $\mathbf{M}(\mathbf{r}, p(\mathbf{r}))$ can also be split as:

$$\mathbf{M}(\mathbf{r}, p(\mathbf{r})) = M^S(\mathbf{r}, p(\mathbf{r}))\cdot\mathbf{I}+\mathbf{M}^D(\mathbf{r}, p(\mathbf{r})), \; M^S(\mathbf{r}, p(\mathbf{r})) = \frac{1}{3}tr(\mathbf{M}(\mathbf{r})) \tag{26}$$

where $M^S(\mathbf{r}, p(\mathbf{r}))\cdot\mathbf{I}$ is the spherical part of the mobility tensor, $\mathbf{M}^D(\mathbf{r}, p(\mathbf{r}))$ is the deviatoric part of the mobility tensor. In addition, the Darcy velocity can be written as

$$\begin{aligned} \mathbf{v}(\mathbf{r}) &= \mathbf{v}^S(\mathbf{r}) + \mathbf{v}^D(\mathbf{r}), \\ \mathbf{v}^S(\mathbf{r}) &= -M^S(\mathbf{r}, p(\mathbf{r}))\,\nabla p(\mathbf{r}), \\ \mathbf{v}^D(\mathbf{r}) &= -\mathbf{M}^D(\mathbf{r}, p(\mathbf{r}))\,\nabla p(\mathbf{r}) \end{aligned} \tag{27}$$

where $\mathbf{v}^S(\mathbf{r})$ is the volumetric velocity, $\mathbf{v}^D(\mathbf{r})$ is the deviatoric velocity. The problem discussed in this paper is the discretization of the elliptic operator:

$$\nabla\mathbf{v}(\mathbf{r}) = \nabla\mathbf{v}^S(\mathbf{r}) + \nabla\mathbf{v}^D(\mathbf{r}) = g(\mathbf{r}), \; \forall\mathbf{r}\in\Omega\subset\mathbb{R}^n \tag{28}$$

Hence, the discretization scheme can be constructed in two steps. The first step is to discretize the volumetric term $\nabla\mathbf{v}^S(\mathbf{r})$ following the scheme (25):

$$\begin{aligned} &-\langle\nabla\mathbf{v}^S(\mathbf{r})\rangle = \\ &\frac{4\cdot\Gamma^{-1}_{\beta\beta}}{n}\left\{\sum_{\Omega_{\mathbf{r}_I,h}} V_{\mathbf{r}_J}\cdot M^S_{eff}\cdot[\mathbf{F}(\mathbf{r}_J)-\mathbf{F}(\mathbf{r}_I)]\frac{(\mathbf{r}^\alpha_J-\mathbf{r}^\alpha_I)\cdot\nabla_\alpha W(\mathbf{r}_J-\mathbf{r}_I,h)}{\|\mathbf{r}_J-\mathbf{r}_I\|^2}\right\} \\ &-\frac{2\cdot\Gamma^{-1}_{\beta\beta}}{n}\left\{\left(\sum_{\Omega_{\mathbf{r}_I,h}} V_{\mathbf{r}_J}\cdot M^S(\mathbf{r}_I)\cdot[\mathbf{F}(\mathbf{r}_J)-\mathbf{F}(\mathbf{r}_I)]\,\nabla^*_\alpha W(\mathbf{r}_J-\mathbf{r}_I,h)\right)\mathbf{N}^\alpha\right\} \end{aligned} \tag{29}$$

where M_{eff}^S is defined as

$$M_{eff}^S = \left(\frac{M^S(\mathbf{r}_J) \cdot M^S(\mathbf{r}_I)}{M^S(\mathbf{r}_J) + M^S(\mathbf{r}_I)} \right) \tag{30}$$

The second step is to discretize the deviatoric term $\nabla \mathbf{v}^D(\mathbf{r})$ as follows:

$$\langle \nabla \mathbf{v}^D(\mathbf{r}) \rangle = \sum_{\Omega_{\mathbf{r}_I,h}} V_{\mathbf{r}_J} \left[\langle \mathbf{v}^D(\mathbf{r}_J) \rangle - \langle \mathbf{v}^D(\mathbf{r}_I) \rangle \right] \nabla^* W(\mathbf{r}_J - \mathbf{r}_I, h) \tag{31}$$

$$\begin{aligned} \langle \mathbf{v}^D(\mathbf{r}_I) \rangle &= -\mathbf{M}^D \langle \nabla p(\mathbf{r}_I) \rangle, \\ \langle \nabla p(\mathbf{r}_I) \rangle &= \sum_{\Omega_{\mathbf{r}_I,h}} V_{\mathbf{r}_J} \left[p(\mathbf{r}_J) - p(\mathbf{r}_I) \right] \nabla^* W(\mathbf{r}_J - \mathbf{r}_I, h) \end{aligned} \tag{32}$$

The numerical scheme (31)–(32) can be directly applied to discretize the original Laplace operator (1) with the anisotropic mobility tensor $\mathbf{M}(\mathbf{r}, p(\mathbf{r})) \in L_2(\Omega)$. This scheme provides an exact answer for the linear pressure distribution in both homogeneous and heterogeneous (linear) mobility fields. Figure 1 shows the comparison between numerical and analytical values of the generalized Laplace operator for the linear pressure distribution $p(\mathbf{r}) = 11 \cdot x + 5 \cdot y + 17$ and for mobility tensors defined as

$$\begin{aligned} &(a) \quad \mathbf{M}(\mathbf{r}) = \begin{pmatrix} 6 & 3 \\ 3 & 5 \end{pmatrix}, \\ &(b) \quad \mathbf{M}(\mathbf{r}) = \begin{pmatrix} 10 + 10 \cdot x + 6 \cdot y & 2 + 2 \cdot x + 2 \cdot y \\ 2 + 2 \cdot x + 2 \cdot y & 4 + 4 \cdot x + y \end{pmatrix} \end{aligned} \tag{33}$$

The observed error is of a machine tolerance, which confirms theoretical claims by numerical experiments. Incompleteness of the kernel support combined with the

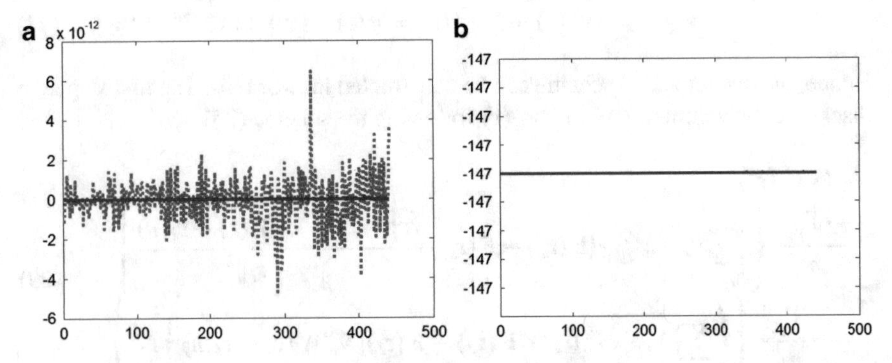

Fig. 1 Comparison between analytical and numerical values for the generalized Laplace operator for the linear pressure and (a) homogenous mobility field, and (b) heterogeneous mobility field defined in (33)

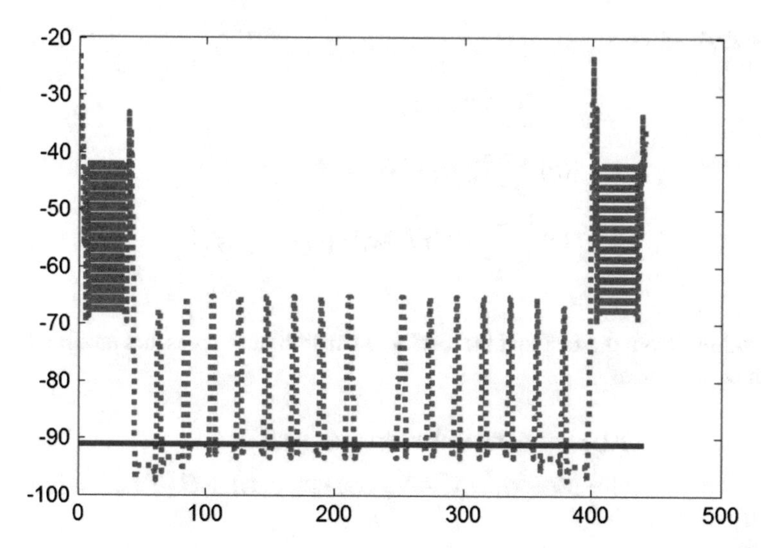

Fig. 2 Comparison between analytical and numerical values for the generalized Laplace operator for the quadratic pressure and homogenous mobility field defined in (33) case (a)

lack of consistency of the kernel interpolation in conventional meshless method results in fuzzy boundaries. For the scheme (31)–(32), the error starts occurring at the boundary particles for quadratic and higher polynomials of the pressure distribution. Figure 2 demonstrates the values of the generalized Laplace operator for the homogeneous mobility field in (33) case (a) for the quadratic pressure distribution $p(\mathbf{r}) = \frac{1}{2}\left(11 \cdot x^2 + 5 \cdot y^2 + 17\right)$. In spite of perfectly adequate general discretization properties, the numerical scheme (31)–(32) is not unconditionally monotone. Knowledge of the capabilities and limitations of these different numerical schemes leads to a better understanding of their impact on various applications and future research on improving and extending modeling capabilities. Hence, it is important to make here a few remarks.

Remark 2 The aforementioned schemes (25) and (31)–(32) can be written in the form:

$$\langle \mathbf{L}(p(\mathbf{r})) \rangle = \sum_S \bar{T}_{SI}^M p_S, \; \sum_S \bar{T}_{SI}^M = 0 \tag{34}$$

where operator $\mathbf{L}$ is defined by either M^S or $\mathbf{M}^D$ and $\bar{T}_{SI}^M$ is the meshless transmissibilities. In the case of the conventional Laplace operator $\nabla^2 p$ (i.e., $m(\mathbf{r}_I) \equiv 1$), it can be derived $\forall \mathbf{r}_I$:

$$\begin{aligned} &(a) \sum_S \bar{T}_{SI}^M = 0, \\ &(b) \sum_S \bar{T}_{SI}^M [\mathbf{r}_S - \mathbf{r}_I] = \mathbf{0}, \\ &(c) \frac{1}{2} \sum_S \bar{T}_{SI}^M [\mathbf{r}_S - \mathbf{r}_I] \cdot [\mathbf{r}_S - \mathbf{r}_I]^T \neq \mathbf{I} \end{aligned} \tag{35}$$

This follows from the fact that the Taylor expansion of the pressure around the point $\mathbf{r}_I$ can be written as

$$\begin{aligned} p(\mathbf{r}_S) &= p(\mathbf{r}_I) + \nabla p(\mathbf{r}_I) \cdot [\mathbf{r}_S - \mathbf{r}_I] + \\ &+ \frac{1}{2} [\mathbf{r}_S - \mathbf{r}_I]^T \cdot \nabla \otimes \nabla p(\mathbf{r}_I) \cdot [\mathbf{r}_S - \mathbf{r}_I] + O\left(h^3\right) \end{aligned} \tag{36}$$

The constraints (35) lead to significant differences between proposed meshless multi-point flux approximation schemes and meshfree finite difference approximation schemes (see, Seibold [32]). It is important to recall here that meshfree finite difference approximation schemes, which satisfy the constraints (a)–(b) and

$$\frac{1}{2} \sum_S \bar{T}_{SI}^M [\mathbf{r}_S - \mathbf{r}_I] \cdot [\mathbf{r}_S - \mathbf{r}_I]^T = \mathbf{I}$$

are based on the following steps: (1) to define the neighbours list for each point (it is important to choose more neighbours than constraints); (2) to select unique stencil which can be satisfied addition requirements (e.g., monotonicity Seibold [32]).

Remark 3 The scheme (31)–(32) can be applied directly to the Darcy velocity (27) with the full mobility tensor. Furthermore, the following theorem is valid for the full mobility tensor:

Theorem 2 *The discretization scheme (31)–(32) is at least of $O\left(h^2\right)$ order of accuracy for any differentiable heterogeneous full mobility tensor field everywhere within the numerical domain Ω.*

The proof of *Theorem* 2 can be seen from the construction of the scheme (31)–(32). However, the scheme (31)–(32) is not unconditionally monotone but as was shown by Seibold [32] in case of meshfree finite difference methods, it is possible to have positive stencils in the scheme (31)–(32), i.e. all neighbor entries are of the same sign.

Remark 4 The schemes (25) and (31)–(32) do not require exact expressions for the gradient (i.e., spatial derivatives) of the mobility field $\nabla_\gamma \mathbf{M}_{\alpha\beta}(\mathbf{r}, p(\mathbf{r})) = \mathbf{M}_{\alpha\beta,\gamma}(\mathbf{r}, p(\mathbf{r}))$ to keep $O(h^\alpha)$, $1 \leq \alpha < 2$ order of accuracy) for any mobility field $\mathbf{M}(\mathbf{r}, p(\mathbf{r})) \in L_2(\Omega)$. Hence, an important feature of reservoir simulations that the mobility field $\mathbf{M}(\mathbf{r}, p(\mathbf{r}))$ is to be a discontinuous (or piece-wise function) is allowed in this scheme.

Remark 5 Applying aforementioned meshless discretization schemes to the elliptic problem $\mathbf{L}(p) = 0$ results in formulating a general non-linear system which can be solved by iterative Newton-Raphson or Picard methods leading to the sequence of linear systems with the matrix $\mathbf{A} = (a_{ij})_{1 \le i \le n, 1 \le j \le n} \in \mathbb{R}^{n \times n}$:

$$\mathbf{A}p = b \tag{37}$$

where the pressure vector p contains approximations to the pressure $p(\mathbf{r})$. It is assumed that $\mathbf{L}(p(\mathbf{r})) = 0$ admits a unique solution with a discontinuity permeability tensor. The I-th row of the matrix $\mathbf{A}$ consists of the stencil corresponding to the point $\mathbf{r}_I$. Let the unknowns be labeled by an index set N same as particles labels. We consider square matrices $\mathbf{A} \in \mathbb{R}^{n \times n}$.

Definition 1 A matrix $\mathbf{A}$ is called essentially irreducible if every point is connected to a Dirichlet boundary point.

The matrix $\mathbf{A}$ resulting from the meshless discretization is a essentially irreducible, which is guaranteed by selecting Kernel supports and the *Heine-Borel theorem.*

Remark 6 Meshless multi-point flux approximation matrices are in general non-symmetric. Consider two points $\mathbf{r}_I$ and $\mathbf{r}_J$ with the corresponding smoothing lengths h_I and h_J. Since each stencil entry depends on the smoothing length, the point $\mathbf{r}_J$ influences the matrix entry a_{ij} if $\|\mathbf{r}_J - \mathbf{r}_I\| \le \lambda \cdot h_I$ whereas $\mathbf{r}_I$ does not influence the matrix entry a_{ji} if $\|\mathbf{r}_J - \mathbf{r}_I\| > \lambda \cdot h_J$, where λ is the scaling factor defined by the shape of the Kernel function. A number of symmetrization methods can be used to overcome this problem. In this paper, the homogeneous smoothing length $h_I = h_J,\ \forall I, J$ is used.

This ends the derivation of a meshless multi-point flux approximation method that can be used to solve different boundary value problems.

5 Numerical Experiments

The verification process is intended to provide, and quantify, the confidence in numerical modelling and the results from the corresponding simulations. Therefore, in order to be confident that the proposed meshless multi-point flux approximation provides the announced accuracy of the elliptic operator (1), it was tested for several functions and different media (diagonal and non-diagonal mobility tensors).

In this section, the results of the numerical experiments using the proposed scheme in Sect. 2.1 are presented, which confirm some of the theoretical results from the previous sections. The problems are solved using 2D (i.e., $n = 2$) square domains (see Fig. 3) with Dirichlet boundary conditions. Following the work by

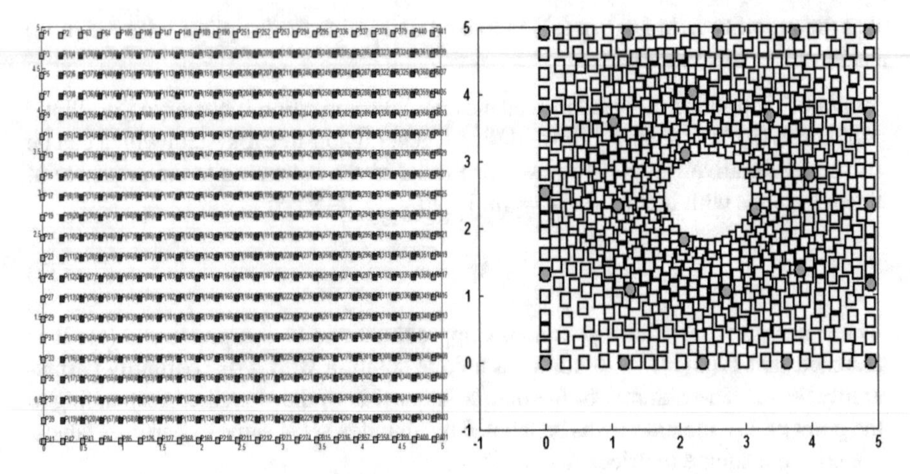

Fig. 3 Numerical domains with different particle distributions

Lukyanov [24–26], the inhomogeneous Dirichlet test cases are considered for the verification purpose in this paper subject to the assumption that $g(\mathbf{r}) = 0,\ \forall \mathbf{r} \in \Omega \subset \mathbb{R}^2$, linear and quadratic pressure boundary conditions:

$$\begin{aligned} p(\mathbf{r}) &= 10 \cdot x + 12 \cdot y + 1, \forall \mathbf{r} = (x, y) \in \partial\Omega \subset \mathbb{R}^2, \\ p(\mathbf{r}) &= \frac{1}{2} \cdot \left(11 \cdot x^2 + 12 \cdot y^2 + 1\right), \forall \mathbf{r} = (x, y) \in \partial\Omega \subset \mathbb{R}^2 \end{aligned} \tag{38}$$

The rectangular 2D (i.e., $n = 2$) domain $\Omega = \{(x, y) \in [0; L] \times [0; H]\} \subset \mathbb{R}^n$ of width $L = 4.9$ m and height $H = 4.9$ m with and without a circle inclusion are considered (see, Fig. 3 cases (a) and (b), respectively).

The components of the heterogeneous mobility field in SI units are defined using the normal distribution with the mean mobility tensor $\mathbf{M}(\mathbf{r})$ and standard deviation matrix $\mathbf{D}(\mathbf{r})$:

$$\mathbf{M}(\mathbf{r}) = \begin{pmatrix} 12 & 5 \\ 5 & 12 \end{pmatrix}, \quad \mathbf{D}(\mathbf{r}) = \begin{pmatrix} 0.1 & 0.2 \\ 0.2 & 0.1 \end{pmatrix} \tag{39}$$

It is clear that the pressure field depicted at Figs. 4 and 5 does not have any spurious oscillations and, hence, satisfies a discrete maximum principle. This suggests that the meshless multi-point flux approximation provides a good approximation of the pressure gradient in the flux computation at least for this study.

Convergence rates are established by running for five levels of particles refinement, starting with the particle distance $h = 0.245$ m on level 1 and refining by a factor of 2 for each successive level. Assuming that the error takes the form $C_p h^{\alpha_p}$, where C_p and α_p are determined to give the best least square fit the data. We consider two types of particle distributions: (a) uniform particle distribution and

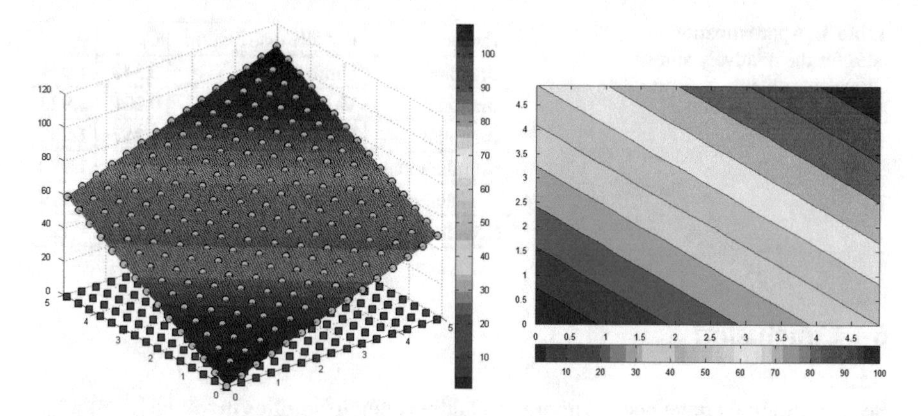

Fig. 4 Single-phase incompressible problem with full-permeability tensor (Cartesian grid)

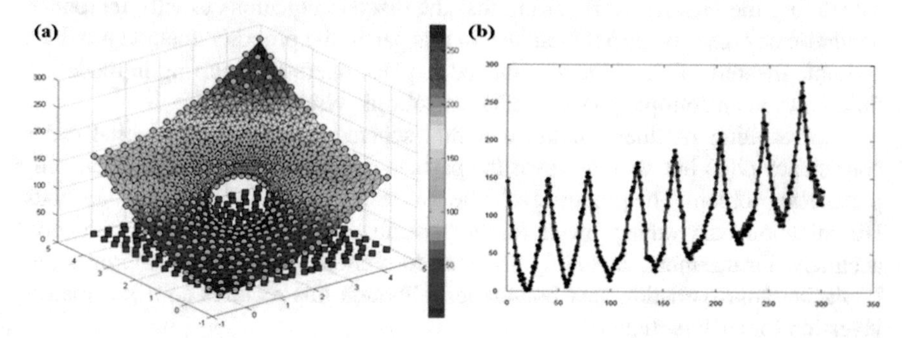

Fig. 5 Different approximate solutions of Dirichlet boundary value problems with the Laplace operator (1) and nonlinear boundary conditions: (**a**) boundary and internal pressure distribution, (**b**) comparison of solutions for different particle distributions

(b) non-uniform particle distribution that is a random perturbation of the uniform particles. The results for the Dirichlet problems using the numerical domain Fig. 3 cases (a) are presented in Table 1. Quadrature rules are used for calculating the error:

$$\|p - p_h\|^2 = \sum_{\xi_K} V_{\xi_K} \left(p\left(\xi_K\right) - p_h\left(\xi_K\right)\right)^2 \tag{40}$$

that is, the results presented for the pressure. The approximation rate for the pressure is between $O(h)$ and $O\left(h^2\right)$. Although, there is no solid proof that the proposed scheme is unconditionally monotone. Numerical results indicate that a relatively small spacing between particles leads to the unconditional monotonicity condition.

Table 1 Approximation rates for the relatively simple Dirichlet problem $\|p - p_h\| \leq C_p h^{\alpha_p}$

Tensor	Particle distribution	C_p	α_p
Diagonal	Uniform	0.348	1.991
Diagonal	Weakly distorted	0.231	1.923
Diagonal	Highly distorted	0.257	1.732
Non-diagonal	Uniform	0.391	1.990
Non-diagonal	Weakly distorted	0.272	1.919
Non-diagonal	Highly distorted	0.293	1.727

6 Conclusion

Several methods have been proposed to address the difficulties involved in calculating second-order derivatives with SPH for heterogeneous scalar mobility fields by calculating the Hessian or requiring that the discrete equations exactly reproduce quadratic or higher order polynomials. In this paper, the proposed method provides a simple discretization of the generalized Laplace operator occurring in modeling fluid flows in anisotropic porous media, anisotropic viscous fluids.

The resulting meshless multi-point flux scheme not only ensures first order consistency $O(h)$ but also improves the particle deficiency (kernel support incompleteness) problem. The proposed scheme was tested by solving an inhomogeneous Dirichlet boundary value problem for the generalized Laplacian equation with good accuracy. Furthermore, including gradient corrections significantly improves the Laplacian approximation near boundaries, although this requires an $n \times n$ matrix inversion for each particle.

The discretization was tested for several boundary value problems using a variety of boundary conditions. Approximation rates of the discretization scheme is smaller with particle disorder; however, the solution remains robust. It is possible that these rates may be improved with different approximations of the spherical part of the Darcy velocity.

References

1. I. Aavatsmark, An introduction to multipoint flux approximations for quadrilateral grids. Comput. Geosci. **6**, 405–432 (2002)
2. I. Aavatsmark, T. Barkve, O. Bøe, T. Mannseth, Discretization on unstructured grids for inhomogeneous, anisotropic media. Part I: derivation of the methods. Part II: discussion and numerical results. SIAM J. Sci. Comput. **19**, 1700–1716 (1998)
3. I. Aavatsmark, T. Barkve, T. Mannseth, Control-volume discretization methods for 3d quadrilateral grids in inhomogeneous, anisotropic reservoirs. SPE J. **3**, 146–154 (1998)
4. J. Bonet, S. Kulasegaram, Correction and stabilization of smooth particle hydrodynamics methods with applications in metal forming simulations. Int. J. Numer. Methods Eng. **47**, 1189–1214 (2000)
5. J. Bonet, T.S.L. Lok, Variational and momentum preservation aspects of smooth particle hydrodynamic formulations. Comput. Methods Appl. Mech. Eng. **180**, 97–115 (1999)

6. L. Brookshaw, A method of calculating radiative heat diffusion in particle simulations. Proc. Astron. Soc. Aust. **6**, 207–210 (1985)
7. J.K. Chen, J.E. Beraun, C.J. Jih, Completeness of corrective smoothed particle method for linear elastodynamics. Comput. Mech. **24**, 273–285 (1999)
8. P.W. Cleary, J.J. Monaghan, Conduction modelling using smoothed particle hydrodynamics. J. Comput. Phys. **148**, 227–264 (1999)
9. F. Colin, R. Egli, F.Y. Lin, Computing a null divergence velocity field using smoothed particle hydrodynamics. J. Comput. Phys. **217**, 680–692 (2006)
10. S.J. Cummins, M. Rudman, An SPH projection method. J. Comput. Phys. **152**, 584–607 (1999)
11. M.G. Edwards, Unstructured, control-volume distributed, full-tensor finite volume schemes with flow based grids. Comput. Geosci. **6**, 433–452 (2002)
12. M.G. Edwards, C.F. Rogers, Finite volume discretization with imposed flux continuity for the general tensor pressure equation. Comput. Geosci. **2**, 259–290 (1998)
13. J. Fang, A. Parriaux, A regularized lagrangian finite point method for the simulation of incompressible viscous flows. J. Comput. Phys. **227**, 8894–8908 (2008)
14. D.A. Fulk, D.W. Quinn, An analysis of 1-d smoothed particle hydrodynamics kernels. J. Comput. Phys. **126**, 165–180 (1996)
15. X.Y. Hu, N.A. Adams, A multi-phase SPH method for macroscopic and mesoscopic flows. J. Comput. Phys. **213**, 844–861 (2006)
16. M. Jubelgas, V. Springel, K. Dolag, Thermal conduction in cosmological SPH simulations. Mon. Not. R.. Astron. Soc. **351**, 423–435 (2004)
17. R.A. Klausen, T.F. Russell, Relationships among some locally conservative discretization methods which handle discontinuous coefficients. Comput. Geosci. **8**, 341–377 (2004)
18. S.H. Lee, L.J. Durlofsky, M.F. Lough, W.H. Chen, Finite difference simulation of geologically complex reservoirs with tensor permeabilities. SPE Reservoir Eval. Eng. **1**, 567–574 (1998)
19. S.H. Lee, P. Jenny, H.A. Tchelepi, A finite-volume method with hexahedral multiblock grids for modeling flow in porous media. Comput. Geosci. **6**, 353–379 (2002)
20. E.-S. Lee, C. Moulinec, R. Xu, D. Violeau, D. Laurence, P. Stansby, Comparison of weakly compressible and truly incompressible algorithms for the SPH mesh free particle method. J. Comput. Phys. **227**, 8417–8436 (2008)
21. W.K. Liu, S. Jun, Multiple-scale reproducing kernel particle method for large deformation problems. Int. J. Numer. Methods Eng. **41**, 1339–1362 (1998)
22. G.R. Liu, M.B. Liu, *Particle Hydrodynamics: A Meshfree Particle Method* (World Scientific Publishing, Singapore, 2003), 449 p.
23. A.A. Lukyanov, Numerical modelling of the material failure under shock loading using particle method, Izvetiya Tula State University. Estestvennonauchn. Ser. **1**, 54–65 (2007)
24. A.A. Lukyanov, Meshless upscaling method and its application to a fluid flow in porous media, in *ECMOR XII Conference Proceedings (2010)*, Oxford (2010)
25. A.A. Lukyanov, Meshless multi-point flux approximation of fluid flow in porous media, in *SPE Reservoir Simulation Symposium (2011), SPE* (2011), 141617 pp.
26. A.A. Lukyanov, Adaptive fully implicit multi-scale meshless multi-point flux method for fluid flow in heterogeneous porous media, in *ECMOR XIII Conference Proceedings (2012)*, Biarritz (2012)
27. A.A. Lukyanov, V.B. Pen'kov, Numerical simulation of solids deformation by a meshless method. Vestn. Samar. Gos. Univ. Estestvennonauchn. Ser. **6**, 62–70 (2007)
28. M.J. Mlacnik, L.J. Durlofsky, Unstructured grid optimization for improved monotonicity of discrete solutions of elliptic equations with highly anisotropic coefficients. J. Comput. Phys. **216**, 337–361 (2006)
29. J.P. Morris, P.J. Fox, Y. Zhu, Modeling low Reynolds number incompressible flows using SPH. J. Comput. Phys. **136**, 214–226 (1997)
30. P. Randles, L. Libersky, Smoothed particle hydrodynamics: some recent improvements and applications. Comput. Methods Appl. Mech. Eng. **139**, 375–408 (1996)
31. H.F. Schwaiger, An implicit corrected SPH formulation for thermal diffusion with linear free surface boundary conditions. Int. J. Numer. Methods Eng. **75**, 647–671 (2008)

32. B. Seibold, M-matrices in meshless finite difference methods, Dissertation, Department of Mathematics, University of Kaiserslautern, 2006
33. S. Shao, E.Y.M. Lo, Incompressible SPH method for simulating newtonian and non-newtonian flows with a free surface. Adv. Water Resour. **26**, 787–800 (2003)
34. S. Zhang, S. Kuwabara, T. Suzuki, Y. Kawano, K. Morita, K. Fukuda, Simulation of solid-fluid mixture flow using moving particle methods. J. Comput. Phys. **228**, 2552–2565 (2009)

Multiscale Petrov-Galerkin Method for High-Frequency Heterogeneous Helmholtz Equations

Donald L. Brown, Dietmar Gallistl, and Daniel Peterseim

Abstract This paper presents a multiscale Petrov-Galerkin finite element method for time-harmonic acoustic scattering problems with heterogeneous coefficients in the high-frequency regime. We show that the method is pollution-free also in the case of heterogeneous media provided that the stability bound of the continuous problem grows at most polynomially with the wave number k. By generalizing classical estimates of Melenk (Ph.D. Thesis, 1995) and Hetmaniuk (Commun. Math. Sci. 5, 2007) for homogeneous medium, we show that this assumption of polynomially wave number growth holds true for a particular class of smooth heterogeneous material coefficients. Further, we present numerical examples to verify our stability estimates and implement an example in the wider class of discontinuous coefficients to show computational applicability beyond our limited class of coefficients.

1 Introduction

The time-harmonic acoustic wave-propagation is customarily described by the Helmholtz equation, which is of second-order, elliptic, but indefinite. Its numerical solution therefore exhibits severe difficulties especially in the regime of high wave numbers k. It is well-known that the mesh size h required for the stability of a standard finite element method must be much smaller than a mesh size H which would be sufficient for a reasonable representation of the solution. The phenomenon

D.L. Brown (✉)
School of Mathematical Sciences, The University of Nottingham, University Park, Nottingham, UK
e-mail: donald.brown@nottingham.ac.uk

D. Gallistl
Dietmar Gallistl, Institut für Angewandte und Numerische Mathematik, Karlsruher Institut für Technologie, Englerstr. 2, 76131 Karlsruhe, Germany
e-mail: gallistl@kit.edu

D. Peterseim
Institut für Numerische Simulation, Universität Bonn, Wegelerstr. 6, 53115 Bonn, Germany
e-mail: peterseim@ins.uni-bonn.de

M. Griebel, M.A. Schweitzer (eds.), *Meshfree Methods for Partial Differential Equations VIII*, Lecture Notes in Computational Science and Engineering 115,
DOI 10.1007/978-3-319-51954-8_6

that the ratio H/h tends to infinity as k grows, is known as the *pollution effect* [1]. A method is referred to as pollution-free, if h and H have the same order of magnitude and so proper resolution of the solution—usually a certain fixed number of grid points per wave length—implies quasi-optimality of the method.

When studying acoustic wave-propagation, it is often assumed to have constant material properties such as density and speed of sound, while in real complex materials, such as composites, these may be heterogeneous. Therefore, in this paper we study a multiscale Petrov-Galerkin method for the Helmholtz equation with large wave numbers k and possibly heterogeneous material coefficients as a generalization of [7, 16]. Standard first-order piecewise polynomials on the scale H serve as trial functions in this method, whereas the test functions involve a correction by solutions to coercive cell problems on the scale h. The size of the cells is proportional to H, where the proportionality constant m—the oversampling parameter—can be adjusted. Typically $m \approx \log k$, depending on the stability of the problem, leads to a quasi-optimal method. These local problems are translation invariant. Therefore, in periodic media only a small number of corrector problems must be solved depending on the number of local mesh configurations.

The stability of the method requires that the stability constant of the continuous operator depends polynomially on k. Such results are very rare in the literature even for the case of homogeneous media. We shall emphasize that such an assumption does not hold true in general [2]. The first positive estimates of this type go back to [14] for convex planar domains with pure Robin boundary. They were later generalized to other settings and three spatial dimensions in [4, 11]. For instance, in the particular case of pure impedance boundary conditions with $\partial\Omega = \Gamma_R$, it was proved in [4, 6, 14], by employing a technique of Makridakis et al. [12], that the inf-sup constant is bounded, i.e. $\gamma(k, \Omega, A, V^2) \lesssim k$. Further setups allowing for polynomially well-posedness in the presence of a single star-shaped sound-soft scatterer are described in [11]. For multiple scattering and, in particular, for scattering in heterogeneous media, the situation is completely open. To show that the assumption is satisfiable for non-trivial heterogeneous media, in this work we determine a class of smooth heterogeneous coefficients that allow for explicit-in-k stability estimates.

1.1 Heterogeneous Helmholtz Problem

We begin with some standard notation on complex-valued Lebesgue and Sobolev spaces that applies throughout this paper. The bar indicates complex conjugation and i is the imaginary unit. The L^2 inner product is denoted by $(v, w)_{L^2(\Omega)} := \int_\Omega v\bar{w}\,dx$. The Sobolev space of complex-valued L^p functions over a domain ω whose generalized derivatives up to order l belong to L^p is denoted by $W^{l,p}(\omega)$ and $H^l(\omega) := W^{l,2}(\omega)$. Further, the notation $A \lesssim B$ abbreviates $A \leq CB$ for some constant C that is independent of the mesh-size, the wave number k, and all further parameters in the method like the oversampling parameter m or the fine-scale mesh-size h; $A \approx B$ abbreviates $A \lesssim B \lesssim A$.

We now begin with some notation and problem setting. Let $\Omega \subset \mathbb{R}^d$ be an open bounded Lipschitz domain with polyhedral boundary for $d \in \{1, 2, 3\}$. We wish to find a solution u that satisfies

$$-\operatorname{div} A(x)\nabla u - k^2V^2(x)u = f \text{ in } \Omega, \tag{1}$$

along with the boundary conditions

$$u = 0 \text{ on } \Gamma_D, \tag{2a}$$

$$A(x)\nabla u \cdot \nu = 0 \text{ on } \Gamma_N, \tag{2b}$$

$$A(x)\nabla u \cdot \nu - ik\beta(x)u = g \text{ on } \Gamma_R. \tag{2c}$$

Here, ν denotes the outer normal to $\partial\Omega = \overline{\Gamma_D \cup \Gamma_N \cup \Gamma_R}$, where the boundary sections are assumed disjoint. We suppose that $|\Gamma_R| > 0$, but allow the other portions of the boundary to have measure zero. Although the results in this paper hold for a weaker dual space here we suppose $f \in L^2(\Omega)$ and $g \in L^2(\Gamma_R)$. For the coefficients, we suppose $A(x), V^2(x) \in W^{1,\infty}(\Omega)$, and $\beta(x) \in L^\infty(\Omega)$ are real valued. Moreover, we suppose there exist positive constants $A_{min}, A_{max}, \beta_{min}, \beta_{max}, V_{min}$, and V_{max} independent of k such that for almost all $x \in \Omega$ we have

$$A_{min} \leq A(x) \leq A_{max}, \tag{3a}$$

$$\beta_{min} \leq \beta(x) \leq \beta_{max}, \tag{3b}$$

$$V^2_{min} \leq V^2(x) \leq V^2_{max}. \tag{3c}$$

We denote the space

$$V := \{u \in H^1(\Omega) \mid u = 0 \text{ on } \Gamma_D\}$$

and denote the norm weighted with $A(x)$, $V(x)$, and k to be for $\omega \subset \Omega$

$$\|u\|_{V,\omega} := \sqrt{\|kVu\|^2_{L^2(\omega)} + \left\|A^{\frac{1}{2}}\nabla u\right\|^2_{L^2(\omega)}}, \tag{4}$$

where if $\omega = \Omega$, we simply write $\|u\|_V$. We have the following variational form corresponding to (1): Find $u \in V$ such that

$$a(u, v) = (f, v)_{L^2(\Omega)} + (g, v)_{L^2(\Gamma_R)} \text{ for all } v \in V, \tag{5}$$

where the complex-valued sesquilinear form $a : V \times V \to \mathbb{C}$ is given by

$$a(u, v) = (A(x)\nabla u, \nabla v)_{L^2(\Omega)} - (k^2V^2(x)u, v)_{L^2(\Omega)} - (ik\beta(x)u, v)_{L^2(\Gamma_R)}. \tag{6}$$

Here we write $(u, v)_{L^2(\Omega)} = \int_\Omega u\bar{v}dx$ and similarly $(u, v)_{L^2(\Gamma_R)} = \int_{\Gamma_R} u\bar{v}ds$.

1.2 Motivation for a Multiscale Method and Stability Analysis

It is well known [1] that the pollution effect cannot be avoided in standard methods. However, it may be overcome by coupling the polynomial degree of the method with the wave number k [15]. Therefore, multiscale methods appear to be a natural tool to incorporate fine-scale features in a low-order discretization. Moreover, the parameters of this method must be coupled logarithmically with the wave number and therefore require the stability constant of the continuous problem to be polynomially dependent of k to arrive at a computationally efficient method. Hence, the stability of the continuous heterogeneous problem (1) is critical to the analysis of the related algorithms. In general, it is often shown (or possibly assumed) that there exists some constant $C_{stab}(k, \Omega, A, V^2) > 0$, which depends on k, the geometry, and the coefficients, such that

$$\|u\|_V \leq C_{stab}(k, \Omega, A, V^2) \left(\|f\|_{L^2(\Omega)} + \|g\|_{L^2(\Gamma_R)} \right). \tag{7}$$

Further, turning to the inf-sup type lower bound, it is often shown, or possibly assumed, that there exists some constant $\gamma(k, \Omega, A, V^2)$, related to $C_{stab}(k, \Omega, A, V^2)$, such that

$$\gamma(k, \Omega, A, V^2)^{-1} \leq \inf_{v \in V \setminus \{0\}} \sup_{w \in V \setminus \{0\}} \frac{\operatorname{Re} a(v, w)}{\|v\|_V \|w\|_V}. \tag{8}$$

As noted, it is often the case that these constants depend merely polynomially on k. However, it has been demonstrated that there are special instances of exponential k dependence on $C_{stab}(k, \Omega, A, V^2)$ [2], and thus, highly unstable inf-sup constants $\gamma(k, \Omega, A, V^2)$.

2 Stability of the Heterogeneous Helmholtz Model

As discussed in Sect. 1, the stability and regularity of the continuous problem has been investigated for constant coefficients in various contexts. In this section, we shall investigate the stability of the continuous problem with respect to wave number in the case of heterogeneous coefficients. We proceed using the variational techniques with geometric constraints [11].

As noted in Sect. 1, in the case of constant coefficients, there exist various methods to bound $\gamma(k, \Omega, A, V^2)$ from (8) in terms of k. Most importantly, the possible exponential dependence discussion in [2], will be excluded here. We will show in this section, that for certain classes of coefficients, we are able to obtain a favorable polynomial bound for $\gamma(k, \Omega, A, V^2)$. To this end, we will employ variational techniques and so-called Rellich type identities with restrictions on the types of geometries similar to the work of Hetmaniuk [11] and references therein.

As we use the variational techniques we will make the geometric assumptions made by Hetmaniuk [11]. That is we suppose that there exists a $x_0 \in \mathbb{R}^d$ and a $\eta > 0$ such that

$$(x - x_0) \cdot \nu \leq 0 \text{ on } \Gamma_D, \tag{9a}$$

$$(x - x_0) \cdot \nu = 0 \text{ on } \Gamma_N, \tag{9b}$$

$$(x - x_0) \cdot \nu \geq \eta \text{ on } \Gamma_R. \tag{9c}$$

For a summary of such possible domains, we refer the reader to [11]. However, to get some sense of a geometry the reader may envision a convex domain with pure impedance boundary conditions. This of course may be weakened.

2.1 *Statement of Stability, Connections to Inf-Sup Constants, and Boundedness*

In this section we present our main stability result. The variational techniques employed require assumptions on the class of coefficients to remain valid. We outline these constraints and obtain a bounded-in-k result. We further relate these to the inf-sup constants and explore the boundedness of the non-constant coefficient case.

We assume throughout that (5) has a unique solution for any L^2 right-hand side f and focus on quantified stability.

Theorem 1 *Suppose $\Omega \subset \mathbb{R}^d$, is a bounded connected Lipschitz domain and satisfies the geometric assumptions* (9)*. Let u be a solution of* (1) *with the boundary conditions* (2)*, coefficients satisfying the bounds* (3)*, and $k \geq k_0 > 0$, for some k_0. Further, we suppose the regularity $u \in H^{3/2+\delta}(\Omega)$ for some $\delta > 0$.*

Define the following function

$$S(x) := \operatorname{div}\left(\left(\frac{V^2(x)}{A(x)}\right)(x - x_0)\right) \tag{10}$$

and further, we will denote C_G to be the minimal constant so that

$$2\left|\int_\Omega \left(\frac{\nabla A}{A}\right) \nabla u((x - x_0) \cdot \nabla \bar{u}) dx\right| \leq C_G \left\|\left(\frac{\nabla A}{A}\right)\right\|_{L^\infty(\Omega)} \|\nabla u\|^2_{L^2(\Omega)}. \tag{11}$$

We suppose that

$$S_{min} = \min_{x \in \Omega} S(x) > 0, \tag{12a}$$

$$S_{min} - \left((d - 2) + C_G \left\|\left(\frac{\nabla A}{A}\right)\right\|_{L^\infty(\Omega)}\right) \frac{V_{max}^2}{A_{min}} > 0. \tag{12b}$$

We then have the following estimate

$$\|u\|_V^2 \le C^* \left(1+\frac{1}{k^2}\right)\left(\|f\|^2_{L^2(\Omega)} + \|g\|^2_{L^2(\Gamma_R)}\right), \tag{13}$$

where C^* depends only on the (3) *and Ω, but not on k.*

Proof See Appendix below. □

Remark 1 The assumption from Theorem 1 that u satisfy the regularity $u \in H^{3/2+\delta}(\Omega)$ is an assumption on the configuration of the boundary decomposition into Γ_D, Γ_N, Γ_R. It is not a further restriction on the coefficients A or V^2.

Now that we have an explicit bound for a class of constant variable coefficients, we now will relate the constant $C_{stab}(k,\Omega,A,V^2) := C^*\left(1+\frac{1}{k^2}\right)$ to $\gamma(k,\Omega,A,V^2)$ given by (8).

Theorem 2 *Supposing the assumptions in Theorem 1, we have the following estimate*

$$k^{-1} \lesssim \widetilde{\gamma}^{-1} \lesssim \inf_{v\in V\setminus\{0\}} \sup_{w\in V\setminus\{0\}} \frac{\operatorname{Re} a(v,w)}{\|v\|_V \|w\|_V}. \tag{14}$$

Where, $\widetilde{\gamma} := (1 + C^*\left(k+\frac{1}{k}\right) V^2_{max})$.

Proof We proceed by a standard argument from [6], adapted to the heterogeneous case. Given $u \in H^1(\Omega)$, define $z \in H^1(\Omega)$ as the solution of

$$2k^2(v, V^2 u)_{L^2(\Omega)} = a(v,z), \text{ for all } v \in V. \tag{15}$$

Then, from the estimate (13), we have

$$\|z\|_V \le C^*\left(1+\frac{1}{k^2}\right) V^2_{max} k^2 \|u\|_{L^2(\Omega)}. \tag{16}$$

Note that

$$\operatorname{Re} a(u,u) = (A(x)\nabla u, \nabla u)_{L^2(\Omega)} - (k^2 V^2(x) u, u)_{L^2(\Omega)}$$

and using (15) and taking $v = u + z$ implies

$$\operatorname{Re} a(u,v) = \operatorname{Re} a(u,u) + \operatorname{Re} a(u,z) = \|u\|_V^2. \tag{17}$$

Using (16) we obtain

$$\begin{aligned}\|v\|_V \le \|u\|_V + \|z\|_V &\le \|u\|_V + C^*\left(1+\frac{1}{k^2}\right) V^2_{max} k^2 \|u\|_{L^2(\Omega)} \\ &\le (1 + C^*\left(k+\frac{1}{k}\right) V^2_{max}) \|u\|_V.\end{aligned}$$

Hence, $\operatorname{Re} a(u, v) = \|u\|_V^2 \geq (1 + C^* \left(k + \frac{1}{k}\right) V_{max}^2)^{-1} \|v\|_V \|u\|_V$, taking

$$\widetilde{\gamma} := (1 + C^* \left(k + \frac{1}{k}\right) V_{max}^2) \approx k$$

yields the result. □

Finally, for completeness, we include a brief proof of the boundedness of the variational from.

Theorem 3 *Supposing the assumptions in Theorem 1, the variational form* (6) *has the following boundedness property*

$$|a(u, v)| \leq C_a \|u\|_V \|v\|_V . \tag{18}$$

Here C_a may depend on the bounds (3)*, multiplicative trace constants, and Ω, but not k.*

Proof From the variational form we have

$$\begin{aligned} |a(u, v)| &\leq \left|(A^{\frac{1}{2}}\nabla u, A^{\frac{1}{2}}\nabla v)_{L^2(\Omega)}\right| + \left|(kVu, kVv)_{L^2(\Omega)}\right| \\ &\quad + \left|((\beta k)^{\frac{1}{2}} u, (\beta k)^{\frac{1}{2}} v)_{L^2(\Gamma_R)}\right| \\ &\leq \left\|A^{\frac{1}{2}}\nabla u\right\|_{L^2(\Omega)} \left\|A^{\frac{1}{2}}\nabla v\right\|_{L^2(\Omega)} + \|kVu\|_{L^2(\Omega)} \|kVv\|_{L^2(\Omega)} \\ &\quad + \left\|(\beta k)^{\frac{1}{2}} u\right\|_{L^2(\Gamma_R)} \left\|(\beta k)^{\frac{1}{2}} v\right\|_{L^2(\Gamma_R)} \\ &\lesssim \|u\|_V \|v\|_V + \left\|(\beta k)^{\frac{1}{2}} u\right\|_{L^2(\Gamma_R)} \left\|(\beta k)^{\frac{1}{2}} v\right\|_{L^2(\Gamma_R)}. \end{aligned}$$

We have from the multiplicative trace inequality

$$\begin{aligned} \left\|k^{\frac{1}{2}} u\right\|_{L^2(\Gamma_R)}^2 &\leq C_M \left(\left\|k^{\frac{1}{2}} u\right\|_{L^2(\Omega)} \left|k^{\frac{1}{2}} u\right|_{H^1(\Omega)} + \operatorname{diam}(\Omega)^{-1} \left\|k^{\frac{1}{2}} u\right\|_{L^2(\Omega)}^2\right) \\ &\leq C_M \left(\|ku\|_{L^2(\Omega)}^2 + |u|_{H^1(\Omega)}^2 + \operatorname{diam}(\Omega)^{-1} \left\|k^{\frac{1}{2}} u\right\|_{L^2(\Omega)}^2\right) \\ &\lesssim C_M \left(\|u\|_V + \operatorname{diam}(\Omega)^{-1} \|ku\|_{L^2(\Omega)}^2\right) \lesssim C_M \|u\|_V^2 \end{aligned}$$

since $k \geq 1$. Applying this to the Γ_R terms we arrive at (18). □

2.2 Example Coefficients

In this subsection, we will provide a few examples that satisfy the assumptions on the coefficients (12). Hence, the set of bounded smooth coefficients that yields polynomial-in-k bounds is non-trivial. We show that for some coefficients, as the oscillations become more frequent we violate the conditions (12). In particular, it appears that the restriction on the amplitude of the coefficients is related to the restrictions on the frequency of oscillations.

To simplify things, yet provide non-trivial coefficients, we will only consider radially symmetric conditions in $\mathbb{R}^2$. Indeed, even with this symmetry, we are able to highlight the complexities and restrictiveness in these conditions. We will see that the frequency of oscillations play a considerable role in violation of these conditions, as well as the amplitude.

We take $\Omega \subset \mathbb{R}^2$ to be given by the unit circle $\Omega := \{(x,y) \in \mathbb{R}^2 \mid x^2 + y^2 \leq 1\}$ and $\partial\Omega = \{(x,y) \in \mathbb{R}^2 \mid x^2 + y^2 = 1\}$. Further, we will take $\Gamma_N = \Gamma_D = \emptyset$, so that $\Gamma_R = \partial\Omega$. We take $x_0 = (0,0) \in \Omega$, and so $m = (x - x_0) = r\hat{r}$, where $r^2 = x^2 + y^2$ and $\hat{r}$ is the standard unit normal in radial coordinates. Then, clearly, $m \cdot \nu = 1$ on Γ_R and so the geometric assumptions (9) are satisfied with this domain. We will take $\beta(x) = 1$, $g(x) = 0$, and suppose that $f := f(r)$, is a given radially symmetric forcing. We finally suppose that the heterogeneities are radially symmetric, $V^2(x) = V^2(r)$, and $A(x) = A(r)$. We briefly recall in radial coordinates that for a function A and a vector field $\sigma = (\sigma_r, \sigma_\theta)$

$$\mathrm{div}(\sigma) = \frac{1}{r}\frac{\partial}{\partial r}(r\sigma_r) + \frac{1}{r}\frac{\partial \sigma_\theta}{\partial \theta}.$$

$$\nabla A = \frac{\partial A}{\partial r}\hat{r} + \frac{1}{r}\frac{\partial A}{\partial \theta}\hat{\theta}.$$

$$\int_\Omega A dx dy = \int_0^{2\pi}\int_0^1 A r dr d\theta,$$

where $\hat{\theta}$ is the standard angular coordinate. By examining the conditions (12), we are able to produce a few interesting examples.

Case 1: $A = 1$ Note that from condition (12b), that if $A = 1$ (or constant), we see that the conditions simplify slightly since the gradient terms in A will vanish. Indeed, now we see that only condition (12a) must be satisfied. In this setting, we must have that $\mathrm{div}(V^2 m) > 0$ for our estimates to hold, or rewritten in radial coordinates as

$$\frac{1}{r}\frac{\partial}{\partial r}\left(V^2(r)r^2\right) > 0. \tag{19}$$

From this condition we may choose a few possible coefficients for $V(r)$. A trivial example is when $V^2(r) = r + 1$. Clearly,

$$\frac{1}{r}\frac{\partial}{\partial r}\left(r^3 + r^2\right) = \frac{1}{r}(3r^2 + 2r) = 3r + 2 > 0.$$

Many such polynomial in r choices exist as long as they do not violate boundedness and positivity.

More interesting examples come from oscillatory coefficients. Suppose, for $\epsilon > 0$, we take now the innocent looking example

$$V^2(r) = \frac{1}{2}\sin\left(\frac{2\pi r}{\epsilon}\right) + 5, \tag{20}$$

and so

$$\frac{1}{r}\frac{\partial}{\partial r}\left(\frac{r^2}{2}\sin\left(\frac{2\pi r}{\epsilon}\right) + 5r^2\right) = \sin\left(\frac{2\pi r}{\epsilon}\right) + \frac{r\pi}{\epsilon}\cos\left(\frac{2\pi r}{\epsilon}\right) + 10. \tag{21}$$

A quick investigation shows that if $\epsilon = 1$, then (19) is satisfied, however, when $\epsilon = 0.1$ it is violated. Hence, if the coefficient becomes highly oscillatory, the stability condition is not satisfied. Also note that if we fix $\epsilon = 1$, but extend the domain from a unit circle to one of radius R, we will eventually enter a negative region. Hence, the domain size also may have an effect on stability from the viewpoint of conditions (12).

Case 2: $A = V^2$ Turning to the definition of $S(x)$ in (10), we see that if $A = V^2$, the functions simplifies to $S(x) = d$. Thus, condition (12a) is always satisfied. For $d = 2$, (12b) becomes

$$2 - \left(C_G \left\| \left(\frac{\nabla A}{A}\right) \right\|_{L^\infty(\Omega)}\right) \frac{A_{max}}{A_{min}} > 0. \tag{22}$$

Taking a closer look at the terms related to C_G from Theorem 1, we have in radial coordinates

$$\begin{aligned} &2\left|\int_\Omega \left(\frac{\nabla A}{A}\right) \nabla u((x - x_0) \cdot \nabla \bar{u}) dx\right| \\ &= 2\left|\int_0^{2\pi}\int_0^1 \left(\frac{r^2}{A(r)}\frac{\partial A(r)}{\partial r}\right)\left|\frac{\partial u(r)}{\partial r}\right|^2 drd\theta\right| \\ &\leq 2\left\|\frac{1}{A(r)}\frac{\partial A(r)}{\partial r}\right\|_{L^\infty(\Omega)} \|\nabla u\|^2_{L^2(\Omega)} . \end{aligned}$$

Hence, we may take here $C_G = 2$. Noting that

$$\frac{\partial}{\partial r}\ln(A) = \frac{1}{A(r)}\frac{\partial A(r)}{\partial r},$$

then the condition (22) becomes

$$1 - \left(\left\|\frac{\partial}{\partial r}\ln(A)\right\|\right)\frac{A_{max}}{A_{min}} > 0. \tag{23}$$

Taking

$$V^2(r) = A(r) = \exp\left(\alpha\left(\sin\left(\frac{r}{\epsilon}\right) + \delta\right)\right), \tag{24}$$

for ϵ, α, and δ positive, then

$$\left\|\frac{\partial}{\partial r}\ln(A)\right\|_{L^\infty(\Omega)} = \left\|\frac{\alpha}{\epsilon}\cos\left(\frac{r}{\epsilon}\right)\right\|_{L^\infty(\Omega)} = \frac{\alpha}{\epsilon}.$$

Note further that $A_{max} = \exp(\alpha(\delta + 1))$ and $A_{min} = \exp(\alpha(\delta - 1))$, and so $\frac{A_{max}}{A_{min}} = \exp(2\alpha)$. Hence,

$$1 - \left(\left\|\frac{\partial}{\partial r}\ln(A)\right\|\right)\frac{A_{max}}{A_{min}} = 1 - \frac{\alpha}{\epsilon}\exp(2\alpha) > 0 \tag{25}$$

or $\alpha\exp(2\alpha) < \epsilon$. We see from this calculation that the frequency of oscillation in the coefficients is related to the amplitude as far as the conditions (12) are concerned. The more oscillatory the function, the smaller the amplitude must be in this example.

3 The Multiscale Method

In this section, we will introduce the notation on finite element spaces and meshes that define the multiscale Petrov-Galerkin method (msPGFEM) for the heterogeneous Helmholtz problem. This method is based on ideas in an algorithm developed for homogenization problems in [3, 9, 13] also known as Localized Orthogonal Decomposition. The ideas have been adapted to the Helmholtz problem for homogeneous coefficients in [16], and later presented in the Petrov-Galerkin framework [7, 17]. We will stay in line with the notation and presentation of [7], as this is the basis for the algorithm applied to a heterogeneous medium. We begin by defining the basic components needed, then define the multiscale method as well as some computational aspects. Finally, we will briefly discuss the error analysis for the method, however, this will not differ too far from the homogeneous coefficient algorithm and as thus, will refer the reader to technical proofs in [7].

3.1 Meshes and Data Structures

We begin with the basic notation needed regarding the relevant mesh and data structures. For the sake of clarity and completeness, we will briefly recall the notation used in [7]. Let $\mathcal{G}_H$ be a regular partition of Ω into intervals, parallelograms, parallelepipeds for $d = 1, 2, 3$, respectively, such that $\bigcup \mathcal{G}_H = \overline{\Omega}$ and any two distinct $T, T' \in \mathcal{G}_H$ are either disjoint or share exactly one lower-dimensional hyper-face (that is a vertex or an edge for $d \in \{2, 3\}$ or a face for $d = 3$). We suppose the mesh is quasi-uniform. For simplicity, we are considering quadrilaterals (resp. hexahedra) with parallel faces, this guarantees the non-degeneracy of the elements in $\mathcal{G}_H$. Again, the theory of this paper carries over to partitions satisfying suitable non-degeneracy conditions or even to meshless methods based on proper partitions of unity [10].

Given any subdomain $S \subseteq \overline{\Omega}$, we define its neighborhood to be

$$\mathsf{N}(S) := \operatorname{int}\Big(\cup \{T \in \mathcal{G}_H : T \cap \overline{S} \neq \emptyset\} \Big).$$

Furthermore, we introduce for any $m \geq 2$ the patch extensions

$$\mathsf{N}^1(S) := \mathsf{N}(S) \qquad \text{and} \qquad \mathsf{N}^m(S) := \mathsf{N}(\mathsf{N}^{m-1}(S)).$$

Note that the shape-regularity implies that there is a uniform bound denoted $C_{\mathrm{ol},m}$, on the number of elements in the mth-order patch, $\#\{K \in \mathcal{G}_H : K \subseteq \overline{\mathsf{N}^m(T)}\} \leq C_{\mathrm{ol},m}$ for all $T \in \mathcal{G}_H$. We will abbreviate $C_{\mathrm{ol}} := C_{\mathrm{ol},1}$. The assumption that the coarse-scale mesh $\mathcal{G}_H$ is quasi-uniform implies that $C_{\mathrm{ol},m}$ depends polynomially on m. The global mesh-size is $H := \max\{\operatorname{diam}(T)\}$ for all $T \in \mathcal{G}_H$.

We will denote $Q_p(\mathcal{G}_H)$ to be the space of piecewise polynomials of partial degree less than or equal to p. The space of globally continuous piecewise first-order polynomials is given by $\mathcal{S}^1(\mathcal{G}_H) := C^0(\Omega) \cap Q_1(\mathcal{G}_H)$, and by incorporating the Dirichlet condition we arrive at the standard Q_1 finite element space denoted here as

$$V_H := \mathcal{S}^1(\mathcal{G}_H) \cap V.$$

The set of free vertices, or the degrees of freedom, is denoted by

$$\mathcal{N}_H := \{z \in \overline{\Omega} : z \text{ is a vertex of } \mathcal{G}_H \text{ and } z \notin \Gamma_D\}.$$

To construct our fine-scale and, thus, multiscale spaces we will need to define a coarse-grid quasi-interpolation operator. For simplicity of presentation,we suppose here that this quasi-interpolation is also projective. This assumption may be lifted c.f. [10] and references therein. We let $I_H : V \to V_H$ be a surjective quasi-interpolation operator that acts as a stable quasi-local projection in the sense that $I_H^2 = I_H$ and that for any $T \in \mathcal{G}_H$ and all $v \in V$ the following local stability result holds

$$H^{-1}\|v - I_H v\|_{L^2(T)} + \|\nabla I_H v\|_{L^2(T)} \leq C_{I_H}\|\nabla v\|_{L^2(\mathsf{N}(T))}. \quad (26)$$

Under the mesh condition that

$$kH \lesssim 1$$

is bounded by a generic constant, this implies stability in the $\|\cdot\|_V$ norm

$$\|I_H v\|_V \leq C_{I_H,V}\|v\|_V \quad \text{for all } v \in V, \quad (27)$$

with a k-independent constant $C_{I_H,V}$. However, $C_{I_H,V}$, will depend on the constants in (3).

One possible choice and which we use in our implementation of the method, is to define $I_H := E_H \circ \Pi_H$, where Π_H is the piecewise L^2 projection onto $Q_1(\mathcal{G}_H)$ and E_H is the averaging operator that maps $Q_1(\mathcal{G}_H)$ to V_H by assigning to each free vertex the arithmetic mean of the corresponding function values of the neighbouring cells, that is, for any $v \in Q_1(\mathcal{G}_H)$ and any free vertex $z \in \mathcal{N}_H$,

$$(E_H(v))(z) = \sum_{\substack{T \in \mathcal{G}_H \\ \text{with } z \in T}} v|_T(z) \Big/ \#\{K \in \mathcal{G}_H \, : \, z \in K\}.$$

Note that with this choice of quasi-interpolation, $E_H(v)|_{\Gamma_D} = 0$ by construction. For this choice, the proof of (26) follows from combining the well-established approximation and stability properties of Π_H and E_H shown in [5].

3.2 Definition of the Method

The multiscale method is determined by three parameters, namely the coarse-scale mesh-size H, the fine-scale mesh-size h, and the oversampling parameter m. We assign to any $T \in \mathcal{G}_H$ its m-th order patch $\Omega_T := \mathsf{N}^m(T)$, $m \in \mathbb{N}$, and define for any $v, w \in V$ the localized sesquilinear forms of (6) to Ω_T as

$$a_{\Omega_T}(u, v) \\ = (A(x)\nabla u, \nabla v)_{L^2(\Omega_T)} - (k^2 V^2(x)u, v)_{L^2(\Omega_T)} - (ik\beta(x)u, v)_{L^2(\Gamma_R \cap \partial\Omega_T)}.$$

and to T, we have

$$a_{\Omega_T}(u, v) = (A(x)\nabla u, \nabla v)_{L^2(T)} - (k^2 V^2(x)u, v)_{L^2(T)} - (ik\beta(x)u, v)_{L^2(\Gamma_R \cap \partial T)}.$$

Let the fine-scale mesh $\mathcal{G}_h$, be a global uniform refinement of the mesh $\mathcal{G}_H$ over Ω and define

$$V_h(\Omega_T) := \{v \in Q_1(\mathcal{G}_h) \cap V : v = 0 \text{ outside } \Omega_T\}.$$

Define the null space

$$W_h(\Omega_T) := \{v_h \in V_h(\Omega_T) : I_H(v_h) = 0\}$$

of the quasi-interpolation operator I_H defined in the previous section. This is the space often referred to as the fine-scale or small-scale space. Given any nodal basis function $\Lambda_z \in V_H$, let $\lambda_{z,T} \in W_h(\Omega_T)$ solve the subscale corrector problem

$$a_{\Omega_T}(w, \lambda_{z,T}) = a_T(w, \Lambda_z) \quad \text{for all } w \in W_h(\Omega_T). \tag{28}$$

Let $\lambda_z := \sum_{T \in \mathcal{G}_H} \lambda_{z,T}$ and define the multiscale test function

$$\widetilde{\Lambda}_z := \Lambda_z - \lambda_z.$$

The space of multiscale test functions then reads

$$\widetilde{V}_H := \operatorname{span}\{\widetilde{\Lambda}_z : z \in \mathcal{N}_H\}.$$

We emphasize that the dimension of the multiscale space is the same as the original coarse space, $\dim V_H = \dim \widetilde{V}_H$. Moreover, it is independent of the parameters m and h. Finally, the multiscale Petrov-Galerkin FEM seeks to find $u_H \in V_H$ such that

$$a(u_H, \tilde{v}_H) = (f, \tilde{v}_H)_{L^2(\Omega)} + (g, \tilde{v}_H)_{L^2(\Gamma_R)} \quad \text{for all } \tilde{v}_H \in \widetilde{V}_H. \tag{29}$$

As in [7], the error analysis and the numerical experiments will show that the choice $H \lesssim k^{-1}$, $m \approx \log(k)$ will be sufficient to guarantee stability and quasi-optimality properties, provided that $k^\alpha h \lesssim 1$ where α depends on the stability and regularity of the continuous problem. This constant α was the subject of Sect. 2. The conditions on h are the same as for the standard Q_1 FEM on the global fine scale. For example, in two dimensions, in the case of pure Robin boundary conditions on a convex domain, it is required that $k^{3/2}h \lesssim 1$ for stability [18] and $k^2 h \lesssim 1$ for quasi-optimality [14] is satisfied.

4 Error Analysis

The error analysis for the algorithm presented in Sect. 3, is very similar to that developed in [16] and references therein, and in particular for the Petrov-Galerkin formulation we discuss now in [7]. It is clear the proofs are unaffected by the coefficients as the arguments rely on very general constants being bounded such as C_a, $C_{stab}(k, \Omega, A, V^2)$, and $\gamma(k, \Omega, A, V^2)$, for example. This is primarily due to the upper and lower boundedness on the coefficients (3). However, we will highlight the main themes of the analysis here as this will be useful to refer to in our discussion on Numerical Examples in Sect. 5 as well as general completeness of the discussion.

We begin the error analysis with some notation. We denote the global finite element space on the fine scale by $V_h := V_h(\Omega) = \mathcal{S}^1(\mathcal{G}_h) \cap V$. We denote the solution operator of the truncated element corrector problem (28) by $\mathcal{C}_{T,m}$. Then, any $z \in \mathcal{N}_H$ and any $T \in \mathcal{G}_H$ satisfy $\lambda_{z,T} = \mathcal{C}_{T,m}(\Lambda_z)$ and we refer to $\mathcal{C}_{T,m}$ as the truncated element correction operator. The map $\Lambda_z \mapsto \lambda_z$ described in Sect. 3.2 defines a linear operator $\mathcal{C}_m$ via $\mathcal{C}_m(\Lambda_z) = \lambda_z$ for any $z \in \mathcal{N}_H$, referred to as correction operator.

For the analysis we introduce idealized counterparts of these correction operators where the patch Ω_T equals Ω. These global corrections are never computed and are merely used in the analysis. We define the null space

$$W_h := \{v \in V_h : I_H(v) = 0\},$$

also referred to as the fine-scale space on the global domain. For any $v \in V$, the idealized element corrector problem seeks $\mathcal{C}_T v \in W_h$ such that

$$a(w, \mathcal{C}_T v) = a_T(w, v) \quad \text{for all } w \in W_h. \tag{30}$$

Furthermore, define

$$\mathcal{C}v := \sum_{T \in \mathcal{G}_H} \mathcal{C}_T v. \tag{31}$$

Recall, we proved in Sect. 2 that the form a with heterogeneous coefficients given by (6), is continuous and there is a constant C_a such that

$$a(v, w) \leq C_a \|v\|_V \|w\|_V \quad \text{for all } v, w \in V.$$

The following result implies the well-posedness of the idealized corrector problems.

Lemma 1 (Well-Posedness for Idealized Corrector Problems) *Provided*

$$C_{I_H} \sqrt{C_{\mathrm{ol}}} Hk \leq 1, \tag{32}$$

we have for all $w \in W_h$ equivalence of norms

$$A_{min}^{\frac{1}{2}} \|\nabla w\|_{L^2(\Omega)} \leq \|w\|_V \leq \left(V_{max}^2 + A_{max}\right)^{\frac{1}{2}} \|\nabla w\|_{L^2(\Omega)},$$

and coercivity

$$\left(V_{max}^2 + A_{max}\right) \|\nabla w\|_{L^2(\Omega)}^2 \leq \operatorname{Re} a(w, w).$$

Proof The lower bound is trivial, indeed we have that

$$\|w\|_V^2 = \|kVw\|_{L^2(\Omega)}^2 + \|A^{\frac{1}{2}} \nabla w\|_{L^2(\Omega)}^2 \geq A_{min} \|\nabla w\|_{L^2(\Omega)}^2.$$

For the upper bound, we note for any $w \in W_h$ the property (26) implies

$$k^2 \|Vw\|_{L^2(\Omega)}^2 = k^2 \|V(1 - I_H)w\|_{L^2(\Omega)}^2 \leq V_{max}^2 C_{I_H}^2 C_{\mathrm{ol}} H^2 k^2 \|\nabla w\|_{L^2(\Omega)}^2.$$

Thus, using (32) we arrive at

$$\begin{aligned}
\|w\|_V^2 &= \|kVw\|_{L^2(\Omega)}^2 + \|A^{\frac{1}{2}} \nabla w\|_{L^2(\Omega)}^2 \\
&\leq V_{max}^2 C_{I_H}^2 C_{\mathrm{ol}} H^2 k^2 \|\nabla w\|_{L^2(\Omega)}^2 + A_{max} \|\nabla w\|_{L^2(\Omega)}^2 \\
&\leq \left(V_{max}^2 + A_{max}\right) \|\nabla w\|_{L^2(\Omega)}^2.
\end{aligned}$$

Note from this we have

$$\begin{aligned}
\|kVw\|_{L^2(\Omega)}^2 &\leq \left(V_{max}^2 + A_{max}\right) \|\nabla w\|_{L^2(\Omega)}^2 - \|A^{\frac{1}{2}} \nabla w\|_{L^2(\Omega)}^2 \\
&\leq \left(V_{max}^2 + A_{max} - A_{min}\right) \|\nabla w\|_{L^2(\Omega)}^2,
\end{aligned}$$

and so

$$\begin{aligned}
\operatorname{Re} a(w, w) &= \|A^{\frac{1}{2}} \nabla w\|_{L^2(\Omega)}^2 - \|kVw\|_{L^2(\Omega)}^2 \\
&\geq \left(V_{max}^2 + A_{max}\right) \|\nabla w\|_{L^2(\Omega)}^2.
\end{aligned}$$

Thus, equivalence and coercivity is proven. □

Lemma 1 implies that the idealized corrector problems (31) are well-posed and the correction operator C is continuous in the sense that

$$\|Cv_H\|_V \leq C_C\|v_H\|_V \quad \text{for all } v_H \in V_H$$

for some constant $C_C \approx 1$. Since the inclusion $W_h(\Omega_T) \subseteq W_h$ holds, the well-posedness result of Lemma 1 carries over to the corrector problems (28) in the subspace $W_h(\Omega_T)$ with the sesquilinear form a_{Ω_T}.

Again as with the homogeneous coefficient case [7], the proof of well-posedness of the Petrov-Galerkin method (29) is based on the fact that the difference $(C - C_m)(v)$ decays exponentially with the distance from $\text{supp}(v)$. In the next theorem, we quantify the difference between the idealized and the discrete correctors. As the proof is a bit technical and does not differ fundamentally from the homogeneous case, we refer the reader to Appendix of [7] and references therein. The proof is based on the exponential decay of the corrector $C\Lambda_z$ and requires the resolution condition (32), namely $kH \lesssim 1$.

Theorem 4 *Under the resolution condition* (32), *there exist constants* $C_1 \approx 1 \approx C_2$ *and* $0 < \theta < 1$ *such that any* $v \in V_H$, *any* $T \in \mathcal{G}_H$ *and any* $m \in \mathbb{N}$ *satisfy*

$$\|\nabla(C_T v - C_{T,m} v)\|_{L^2(\Omega)} \leq C_1 \theta^m \|\nabla v\|_{L^2(T)}, \tag{33}$$

$$\|\nabla(C v - C_m v)\|_{L^2(\Omega)} \leq C_2 \sqrt{C_{\text{ol},m}} \theta^m \|\nabla v\|_{L^2(\Omega)}. \tag{34}$$

Proof See Appendix of [7]. □

Provided we choose the fine-mesh h small enough, the standard finite element over the mesh $\mathcal{G}_h$ is stable in the sense that there exists a constant C_{FEM} such that with $\gamma(k, \Omega, A, V^2)$ from (8) it holds that

$$\left(C_{\text{FEM}}\gamma(k, \Omega, A, V^2)\right)^{-1} \leq \inf_{v \in V_h \setminus \{0\}} \sup_{w \in V_h \setminus \{0\}} \frac{\operatorname{Re} a(v, w)}{\|v\|_V \|w\|_V}. \tag{35}$$

Recall, this is actually a condition on the fine-scale parameter h. In general, the requirements on h depend on the stability of the continuous problem [14]. We now recall the conditions on the oversampling parameter for the well-posedness of the discrete problem. Again, the proof here does not rely heavily on the coefficients, just the general boundedness and ellipticity constants etc. Thus, we again refer the reader to [7].

Theorem 5 (Well-Posedness of the Discrete Problem) *Under the resolution conditions* (32) *and* (35) *and the following oversampling condition*

$$m \gtrsim |\log\left(C_{\text{FEM}}\gamma(k, \Omega, A, V^2)\right)| \big/ |\log(\theta)|, \tag{36}$$

problem (29) *is well-posed and the constant* $C_{\mathrm{PG}} := 2C_{I_H,V}C_C C_{\mathrm{FEM}}$ *satisfies*

$$\left(C_{\mathrm{PG}}\gamma(k,\Omega,A,V^2)\right)^{-1} \leq \inf_{v_H\in V_H\setminus\{0\}} \sup_{\tilde{v}_H\in\widetilde{V}_H\setminus\{0\}} \frac{\operatorname{Re} a(v_H,\tilde{v}_H)}{\|v_H\|_V\|\tilde{v}_H\|_V}.$$

Proof See [7]. □

The quasi-optimality requires the following additional condition on the oversampling parameter m,

$$m \gtrsim |\log\left(C_{\mathrm{PG}}\gamma(k,\Omega,A,V^2)\right)| \Big/ |\log(\theta)|. \tag{37}$$

Theorem 6 (Quasi-Optimality) *The resolution conditions* (32) *and* (35) *and the oversampling conditions* (36) *and* (37) *imply that the solution* u_H *to* (29) *with parameters* H, h, *and* m *and the solution* u_h *of the standard Galerkin FEM on the mesh* $\mathcal{G}_h$ *satisfy*

$$\|u_h - u_H\|_V \lesssim \|(1-I_H)u_h\|_V \approx \min_{v_H\in V_H}\|u_h - v_H\|_V.$$

Proof See [7]. □

The following consequence of Theorem 6 states an estimate for the error $u - u_H$.

Corollary 1 *Under the conditions of Theorem 6, the discrete solution* u_H *to* (29) *satisfies with some constant* $C \approx 1$ *that*

$$\|u-u_H\|_V \leq \|u-u_h\|_V + C\min_{v_H\in V_H}\|u_h - v_H\|_V.$$

For the class of coefficients described in Theorem 1, this leads to the following convergence rates. Provided that the geometry allows for H^2 regularity of the solution and that h is sufficiently small such that the standard FEM is quasi-optimal on the fine scale h and the error is dominated by the coarse-scale part, we have

$$\|u-u_H\|_V \leq O(kH).$$

5 Numerical Examples

In this section, we present the results from our numerical experiments on a smooth coefficient for both cases when the conditions are satisfied and when it is violated. Further, we implement the method on discontinuous periodic coefficients to highlight broader applicability of the method. We give three example coefficients; based on (20), (24), and a discontinuous example. In all three experiments we took $\Omega = (-1,1)^2$ to be the unit square. We use triangular meshes and continuous P_1

finite elements as trial functions. We used $k = 2^5$, $g = 0$, and the approximate point source

$$f(x) = \begin{cases} \exp\left(-\frac{1}{1-(20|x|)^2}\right) & \text{for } |x| < 1/20 \\ 0 & \text{else.} \end{cases}$$

The coarse-scale mesh-sizes are $H = 2^{-3}, 2^{-4}, 2^{-5}, 2^{-6}$ and the fine-scale mesh-size is $h = 2^{-8}$.

The convergence history plots display the errors in the $\|\cdot\|_V$ norm as well as L^2 norms. We compare the multiscale Petrov-Galerkin method for oversampling parameters $m = 1, 2, 3$ with the standard P_1 finite element method and the best-approximation. To compute the error quantity we take the standard finite element solution at the fine scale h to be the overkill solution.

For the first example, we take $A = 1$ and V^2 as (20), with $\epsilon = 1$ and refer to this as example 1. Note that this does not violate the stability condition. The coefficient V^2 is displayed in Fig. 1a and the corresponding computational solution is displayed in Fig. 1b. We note the spurious oscillation in Fig. 1b that breaks the rotational symmetry of the problem. However, this is due to the Robin boundary condition on the square domain being a poor choice for an absorbing boundary condition. The normal vector on the square is a crude approximation to $\hat{r}$. Computing on a circular domain would yield radially symmetric results.

Figure 2a, b display the convergence history in the V-norm and the L^2 norm for Example 1. In general, we see that the multiscale method appears to perform much better than the corresponding standard P_1 finite element. However, there appears to be some resonance effects of some sort that is particularly pronounced in the V norm just before the resolution condition is satisfied. This is not in contradiction with the theory. It has been demonstrated in [7] that there is no decay of the

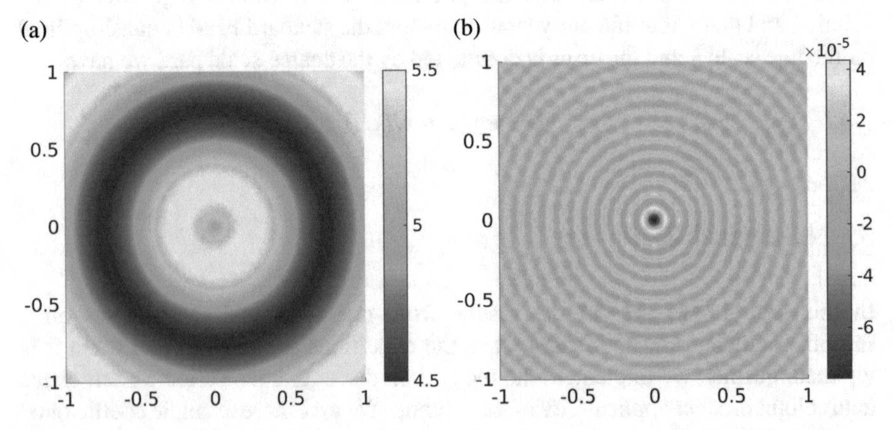

Fig. 1 Plots for Example 1. (**a**) The coefficient V^2 for Example 1. (**b**) Plot of the solution for Example 1

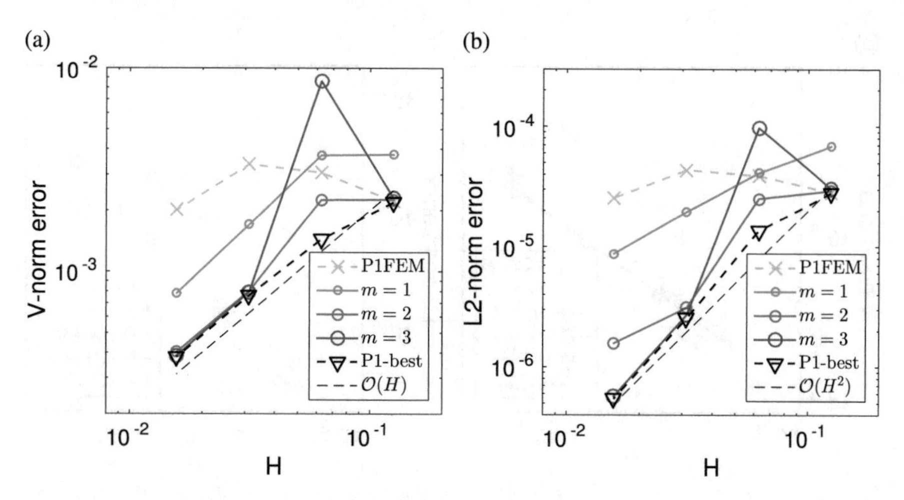

Fig. 2 Convergence history for Example 1. (**a**) Convergence in V norm: Example 1. (**b**) Convergence in L^2 norm: Example 1

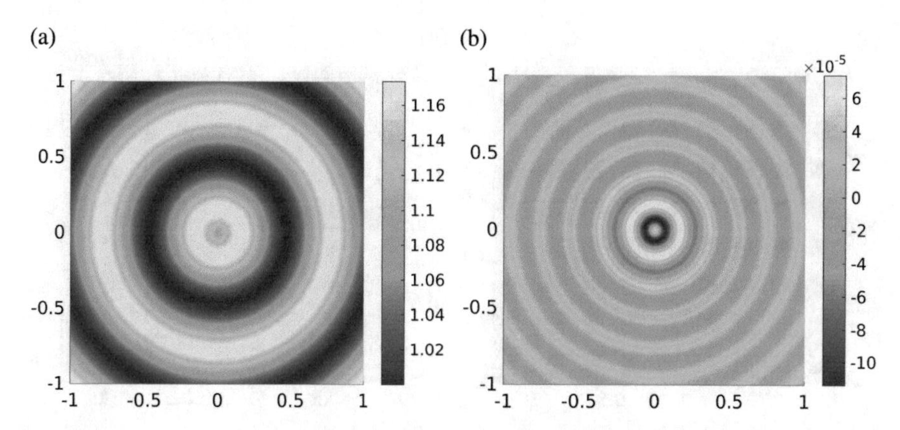

Fig. 3 Plots for Example 2. (**a**) The coefficient V^2 for Example 2. (**b**) Plot of the solution for Example 2

corrector functions if the resolution condition is not satisfied, so that in this regime the localization is not justified and leads to unreliable results.

For the second example, we take $A = V^2$ and V^2 as (24), and refer to this as Example 2. For the parameters we took $\delta = 1$, $\epsilon = 0.1$, $\alpha = 0.08$, and note that the corresponding stability condition $\alpha \exp(2\alpha) < \epsilon$ is narrowly satisfied. The coefficient V^2 is displayed in Fig. 3a and the computational solution is displayed in Fig. 3b. Figure 4a, b display the convergence history in the V-norm and the L^2 norm for Example 2. We see that in this example, we achieve faster convergence and do not see the resonance effects. This is also the case for the standard finite elements.

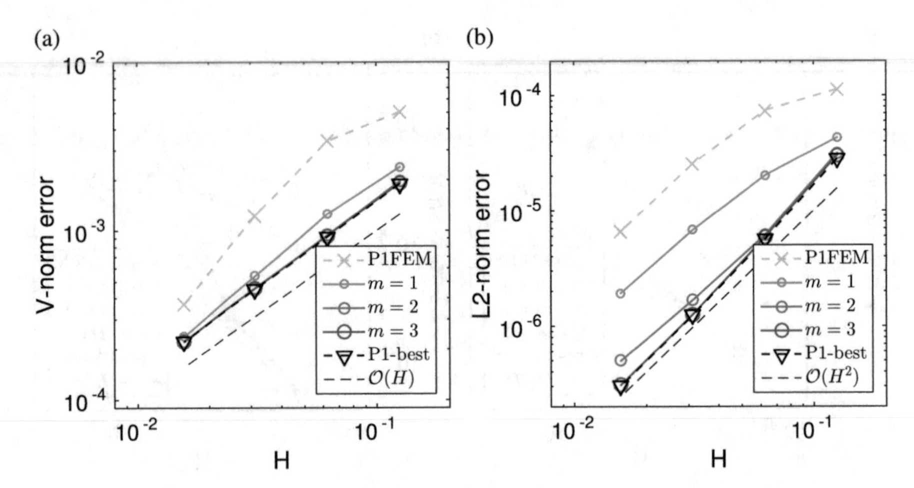

Fig. 4 Convergence history for Example 2. (**a**) Convergence in V norm Example 2. (**b**) Convergence in L^2 norm Example 2

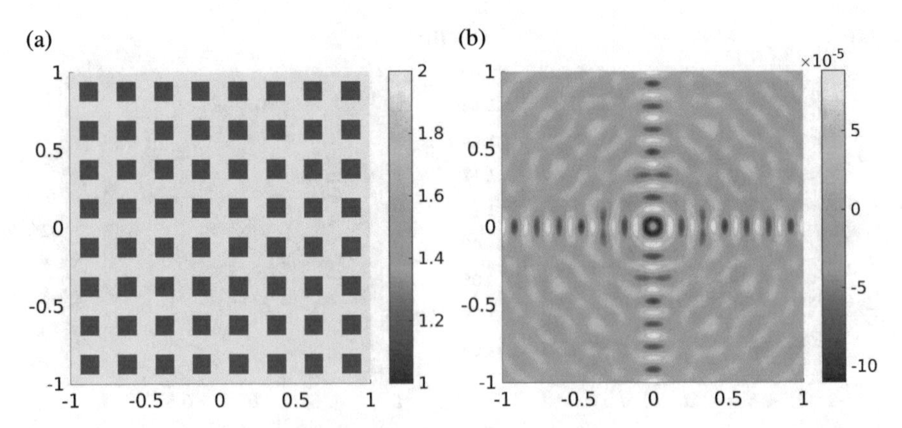

Fig. 5 Plots for Example 3. (**a**) The coefficient V^2 for Example 3. (**b**) Plot of the solution for Example 3

We now present a numerical example outside of our stability theory. We take $V^2 = 2$ except at periodically placed blocks where $V^2 = 1$ and plot the function in Fig. 5a. We refer to this as Example 3. The computational solution is displayed in Fig. 5b. Figure 6a, b display the convergence history in the V-norm and the L^2 norm for Example 3. We observe that the method performs particularly well in this example, especially when compared against the corresponding P_1 finite element. We do not see the resonances as with Example 1.

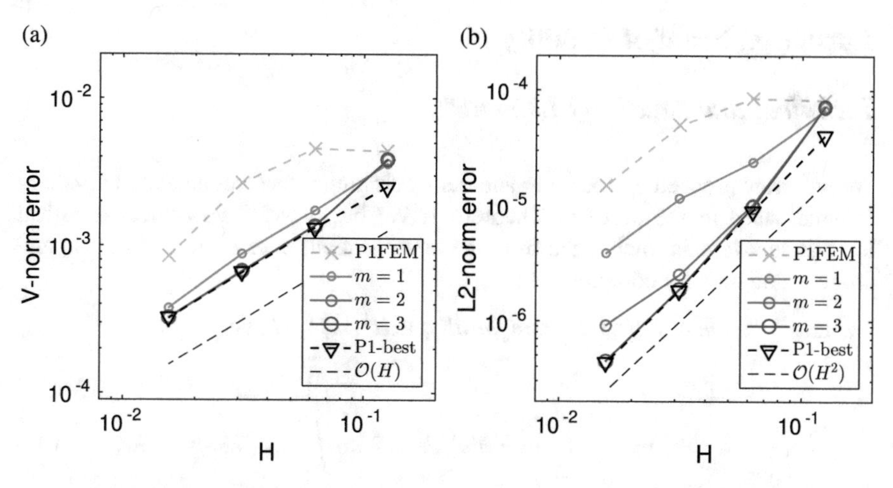

Fig. 6 Convergence history for Example 3. (**a**) Convergence in V norm Example 3. (**b**) Convergence in L^2 norm Example 3

6 Conclusions

ptIn this work, we developed a multiscale method to efficiently solve the heterogeneous Helmholtz equation at high frequency. The primary challenge was establishing k-explicit bounds for the continuous problem as these are critical in the analysis of the patch truncation parameter. We established these bounds for a class of smooth coefficients given some restrictions that appear to depend heavily on the frequency of oscillations and the amplitude of the coefficients. We then presented our multiscale method whose error analysis is not significantly modified by the heterogeneities assuming standard upper and lower boundedness. Finally, we implemented the algorithm on two coefficients that fit inside the class of coefficients in our main theorem and one that is discontinuous. We see that the method performs well in these cases. Future work includes exploring if these stability estimates apply to a greater class of more heterogeneous coefficients with less smoothness.

Acknowledgements The authors acknowledge the support given by the Hausdorff Center for Mathematics Bonn. D. Peterseim is supported by Deutsche Forschungsgemeinschaft in the Priority Program 1748 "Reliable simulation techniques in solid mechanics. Development of non-standard discretization methods, mechanical and mathematical analysis" under the project "Adaptive isogeometric modeling of propagating strong discontinuities in heterogeneous materials".

We thank Professor S. Sauter for helpful discussions on the stability analysis of Helmholtz problems.

Appendix: Proof of Stability

Technical and Auxiliary Lemmas

We will now proceed by recalling and demonstrating a few technical and auxiliary Lemmas used in the proof of Theorem 1. We begin with two critical technical lemmas that remain unchanged from the homogeneous case examined in [11] and are repeated here for completeness.

Lemma 2 *Let $m \in W^{1,\infty}(\Omega)^d$ and for all $q \in H^1(\Omega)$ we have*

$$\int_{\partial\Omega} |q|^2 m \cdot \nu ds = \int_{\Omega} \operatorname{div}(m)|q|^2 dx + 2\operatorname{Re}\int_{\Omega} qm \cdot \nabla\bar{q} dx. \tag{38}$$

Proof See [11], Lemma 3.1. □

Lemma 3 *Let $m \in W^{1,\infty}(\Omega)^d$ and for all $q \in H^1_{\Gamma_D}(\Omega) \cap H^{3/2+\delta}$, $\delta > 0$, we have*

$$\begin{aligned}\int_{\partial\Omega\setminus\Gamma_D} |\nabla q|^2 m \cdot \nu ds - \int_{\Gamma_D} |\partial_\nu q|^2 m \cdot \nu ds \\ = \int_{\Omega} \operatorname{div}(m)|\nabla q|^2 dx - 2\operatorname{Re}\int_{\Omega} \nabla q \cdot (\nabla\bar{q}\nabla) m dx \\ - 2\operatorname{Re}\int_{\Omega} \Delta q(m \cdot \nabla\bar{q}) dx + 2\operatorname{Re}\int_{\partial\Omega\setminus\Gamma_D} \partial_\nu q(m \cdot \nabla\bar{q}) ds\end{aligned} \tag{39}$$

Proof See [8]. □

Here we will present a few auxiliary Lemmas.

Lemma 4 *Let $\Omega \subset \mathbb{R}^d$ be a bounded connected Lipschitz domain. Let $u \in H^1(\Omega)$ be a weak solution of* (1), *with $f \in L^2(\Omega)$ and $g \in L^2(\Gamma_R)$. Then, we have for any $\epsilon > 0$*

$$k^2\|u\|^2_{L^2(\Gamma_R)} \leq \frac{1}{\beta_{min}}\left(\frac{1}{\epsilon}\|f\|^2_{L^2(\Omega)} + k^2\epsilon\|u\|^2_{L^2(\Omega)} + \frac{1}{\beta_{min}}\|g\|^2_{L^2(\Gamma_R)}\right). \tag{40}$$

Proof Taking $v = u$ into the variational form (5) and looking at the imaginary part we have

$$\Im(a(u,u)) = -(k\beta(x)u,u) = \Im((g,u)_{L^2(\Gamma_R)} + (f,u)_{L^2(\Omega)}),$$

and so

$$\begin{aligned}
&k\beta_{min}\|u\|^2_{L^2(\Gamma_R)} \\
&\leq \|u\|_{L^2(\Omega)}\|f\|_{L^2(\Omega)} + \|u\|_{L^2(\Gamma_R)}\|g\|_{L^2(\Gamma_R)} \\
&\leq \frac{1}{2k\xi_1}\|f\|^2_{L^2(\Omega)} + \frac{k\xi_1}{2}\|u\|^2_{L^2(\Omega)} + \frac{1}{2\xi_2}\|g\|^2_{L^2(\Gamma_R)} + \frac{\xi_2}{2}\|u\|^2_{L^2(\Gamma_R)}.
\end{aligned}$$

Multiplying by k, dividing by β_{min}, and setting $\xi_2 = \beta_{min}k$ we obtain

$$\begin{aligned}
k^2\|u\|^2_{L^2(\Gamma_R)} \leq \frac{1}{\beta_{min}}\bigg(&\frac{1}{2\xi_1}\|f\|^2_{L^2(\Omega)} + \frac{k^2\xi_1}{2}\|u\|^2_{L^2(\Omega)} \\
&+ \frac{1}{2\beta_{min}}\|g\|^2_{L^2(\Gamma_R)} + \frac{k^2\beta_{min}}{2}\|u\|^2_{L^2(\Gamma_R)}\bigg),
\end{aligned}$$

and we obtain

$$\frac{k^2}{2}\|u\|^2_{L^2(\Gamma_R)} \leq \frac{1}{\beta_{min}}\left(\frac{1}{2\xi_1}\|f\|^2_{L^2(\Omega)} + \frac{k^2\xi_1}{2}\|u\|^2_{L^2(\Omega)} + \frac{1}{2\beta_{min}}\|g\|^2_{L^2(\Gamma_R)}\right).$$

Taking $\xi_1 = \epsilon > 0$ we arrive at the estimate. □

We will also need the estimate below.

Lemma 5 *Let $\Omega \subset \mathbb{R}^d$ be a bounded connected Lipschitz domain. Let $u \in H^1(\Omega)$ be a weak solution of* (1) *with $f \in L^2(\Omega)$ and $g \in L^2(\Gamma_R)$. Then, we have*

$$\begin{aligned}
&\|\nabla u\|^2_{L^2(\Omega)} \\
&\leq \frac{1}{A_{min}}\bigg[k^2\left(V^2_{max} + \frac{\xi_4}{\beta_{min}} + \frac{\xi_3}{2}\right)\|u\|^2_{L^2(\Omega)} \\
&\quad + \left(\frac{1}{2k^2\xi_3} + \frac{1}{\beta_{min}\xi_4}\right)\|f\|^2_{L^2(\Omega)} + \left(\frac{1}{\beta^2_{min}} + \frac{1}{4k^2}\right)\|g\|^2_{L^2(\Gamma_R)}\bigg].
\end{aligned} \tag{41}$$

for any $\xi_3, \xi_4 > 0$.

Proof Taking $v = u$ into the variational form (5) and looking at the real part we have

$$\begin{aligned}
\mathrm{Re}(a(u,u)) &= (A(x)\nabla u, \nabla u)_{L^2(\Omega)} - (k^2V^2(x)u, u)_{L^2(\Omega)} \\
&= \mathrm{Re}((g,u)_{L^2(\Gamma_R)} + (f,u)_{L^2(\Omega)}),
\end{aligned}$$

and so we have

$$\left\|A^{\frac{1}{2}}\nabla u\right\|^2_{L^2(\Omega)} \leq k^2\|Vu\|^2_{L^2(\Omega)} + \|u\|_{L^2(\Omega)}\|f\|_{L^2(\Omega)} + \|u\|_{L^2(\Gamma_R)}\|g\|_{L^2(\Gamma_R)}.$$

Using the maximal and minimal values we have for any $\xi_3 > 0$ that

$$\begin{aligned} A_{min}\|\nabla u\|^2_{L^2(\Omega)} &\leq k^2\|Vu\|^2_{L^2(\Omega)} + \|u\|_{L^2(\Omega)}\|f\|_{L^2(\Omega)} + \|u\|_{L^2(\Gamma_R)}\|g\|_{L^2(\Gamma_R)} \\ &\leq \left(k^2V^2_{max} + \frac{k^2\xi_3}{2}\right)\|u\|^2_{L^2(\Omega)} + \frac{1}{2k^2\xi_3}\|f\|^2_{L^2(\Omega)} \\ &+ \frac{1}{4k^2}\|g\|^2_{L^2(\Gamma_R)} + k^2\|u\|^2_{L^2(\Gamma_R)}. \end{aligned} \tag{42}$$

Using estimate (40) we may write for any $\epsilon > 0$

$$k^2\|u\|^2_{L^2(\Gamma_R)} \leq \frac{1}{\beta_{min}}\left(k^2\epsilon\|u\|^2_{L^2(\Omega)} + \frac{1}{\epsilon}\|f\|^2_{L^2(\Omega)} + \frac{1}{\beta_{min}}\|g\|^2_{L^2(\Gamma_R)}\right). \tag{43}$$

Inserting the above inequality into (42) we obtain

$$\begin{aligned} &A_{min}\|\nabla u\|^2_{L^2(\Omega)} \\ &\leq \left(k^2V^2_{max} + \frac{k^2\xi_3}{2}\right)\|u\|^2_{L^2(\Omega)} + \frac{1}{2k^2\xi_3}\|f\|^2_{L^2(\Omega)} + \frac{1}{4k^2}\|g\|^2_{L^2(\Gamma_R)} \\ &+ \frac{1}{\beta_{min}}\left(k^2\epsilon\|u\|^2_{L^2(\Omega)} + \frac{1}{\epsilon}\|f\|^2_{L^2(\Omega)} + \frac{1}{\beta_{min}}\|g\|^2_{L^2(\Gamma_R)}\right). \end{aligned}$$

Taking $\epsilon = \xi_4$ the above inequality becomes

$$\begin{aligned} A_{min}\|\nabla u\|^2_{L^2(\Omega)} &\leq k^2\left(V^2_{max} + \frac{\xi_4}{\beta_{min}} + \frac{\xi_3}{2}\right)\|u\|^2_{L^2(\Omega)} \\ &+ \left(\frac{1}{2k^2\xi_3} + \frac{1}{\beta_{min}\xi_4}\right)\|f\|^2_{L^2(\Omega)} + \left(\frac{1}{\beta^2_{min}} + \frac{1}{4k^2}\right)\|g\|^2_{L^2(\Gamma_R)}. \end{aligned}$$

Thus, we obtained our estimate. □

Proof of Main Stability Result

We are now in a position to prove Theorem 1. The key observation is that the Laplacian may be rewritten using (1) and combined with the technical and auxiliary lemmas. This leads to the conditions on the coefficients (12).

Proof (Proof of Theorem 1)
Using (39) where we write

$$-\Delta u = \frac{1}{A}(f + k^2V^2u + \nabla A \cdot \nabla u),$$

$\partial_\nu u = 0$ on Γ_N, and $\partial_\nu u = ik\beta u + g$ on Γ_R, we obtain

$$\begin{aligned}\int_{\partial\Omega\setminus\Gamma_D} &|\nabla u|^2 m\cdot\nu ds - \int_{\Gamma_D} |\partial_\nu u|^2 m\cdot\nu ds\\ &= \int_\Omega \mathrm{div}(m)|\nabla u|^2 dx - 2\,\mathrm{Re}\int_\Omega \nabla u\cdot(\nabla\bar{u}\nabla)m dx\\ &\quad + 2\,\mathrm{Re}\int_\Omega \frac{1}{A}(f + k^2V^2u + \nabla A\cdot\nabla u)(m\cdot\nabla\bar{u})dx\\ &\quad + 2\,\mathrm{Re}\int_{\Gamma_R}(ik\beta u + g)(m\cdot\nabla\bar{u})ds.\end{aligned} \tag{44}$$

Using (38) with the transform $m \to \frac{V^2}{A}m$, we have

$$\begin{aligned}k^2\int_{\partial\Omega}&|u|^2\left(\frac{V^2}{A}\right)m\cdot\nu ds\\ &= k^2\int_\Omega \mathrm{div}\left(\frac{V^2}{A}m\right)|u|^2dx + 2k^2\,\mathrm{Re}\int_\Omega u\left(\frac{V^2}{A}\right)m\cdot\nabla\bar{u}dx.\end{aligned}$$

Using this to replace the term $\mathrm{Re}\int_\Omega\left(\frac{V^2}{A}\right)u(m\cdot\nabla\bar{u})dx$, we have

$$\begin{aligned}\int_{\partial\Omega\setminus\Gamma_D}&|\nabla u|^2m\cdot\nu ds - \int_{\Gamma_D}|\partial_\nu u|^2 m\cdot\nu ds\\ &= \int_\Omega \mathrm{div}(m)|\nabla u|^2dx - 2\,\mathrm{Re}\int_\Omega\nabla u\cdot(\nabla\bar{u}\nabla)mdx\\ &+ 2\,\mathrm{Re}\int_\Omega\left(\frac{f}{A}\right)(m\cdot\nabla\bar{u})dx + 2\,\mathrm{Re}\int_\Omega\left(\frac{\nabla A}{A}\right)\cdot\nabla u(m\cdot\nabla\bar{u})dx\\ &+ 2\,\mathrm{Re}\int_{\Gamma_R}(ik\beta u + g)(m\cdot\nabla\bar{u})ds\\ &- k^2\int_\Omega\mathrm{div}\left(\frac{V^2}{A}m\right)|u|^2dx + k^2\int_{\partial\Omega}|u|^2\left(\frac{V^2}{A}\right)m\cdot\nu ds.\end{aligned}$$

Expanding out the boundary terms in each of the portions we have

$$
\begin{aligned}
&-\int_{\Gamma_D} |\partial_\nu u|^2 m \cdot \nu ds + \int_{\Gamma_N} |\nabla u|^2 m \cdot \nu ds \\
&\quad + \int_{\Gamma_R} |\nabla u|^2 m \cdot \nu ds + k^2 \int_\Omega \operatorname{div}\left(\frac{V^2}{A} m\right) |u|^2 dx \\
&= \int_\Omega \operatorname{div}(m) |\nabla u|^2 dx - 2\,\mathrm{Re} \int_\Omega \nabla u \cdot (\nabla \bar{u} \nabla) m dx \\
&\quad + 2\,\mathrm{Re} \int_\Omega \left(\frac{f}{A}\right) (m \cdot \nabla \bar{u}) dx + 2\,\mathrm{Re} \int_\Omega \left(\frac{\nabla A}{A}\right) \cdot \nabla u (m \cdot \nabla \bar{u}) dx \\
&\quad + k^2 \int_{\Gamma_N} |u|^2 \left(\frac{V^2}{A}\right) m \cdot \nu ds + k^2 \int_{\Gamma_R} |u|^2 \left(\frac{V^2}{A}\right) m \cdot \nu ds \\
&\quad + 2\,\mathrm{Re} \int_{\Gamma_R} (ik\beta u + g)(m \cdot \nabla \bar{u}) ds.
\end{aligned}
\tag{45}
$$

Now we suppose we make the geometric assumptions made by Hetmaniuk [11] outlined in (9). Recall, we have for $m = x - x_0$, thus we compute

$$
\begin{aligned}
\operatorname{div}(x - x_0) &= d \text{ in } \Omega, \\
\nabla u \cdot (\nabla \bar{u} \nabla)(x - x_0) &= |\nabla u|^2 \text{ in } \Omega, \\
(x - x_0) \cdot \nu &\leq 0 \text{ on } \Gamma_D, \\
(x - x_0) \cdot \nu &= 0 \text{ on } \Gamma_N, \\
(x - x_0) \cdot \nu &\geq \eta \text{ on } \Gamma_R.
\end{aligned}
$$

Using the above relations in (45) we obtain

$$
\begin{aligned}
&\eta \int_{\Gamma_R} |\nabla u|^2 ds + k^2 \int_\Omega \operatorname{div}\left(\frac{V^2}{A}(x - x_0)\right) |u|^2 dx \\
&\leq (d-2) \int_\Omega |\nabla u|^2 dx + 2\,\mathrm{Re} \int_\Omega \left(\frac{f}{A}\right) ((x - x_0) \cdot \nabla \bar{u}) dx \\
&+ 2\,\mathrm{Re} \int_\Omega \left(\frac{\nabla A}{A}\right) \nabla u ((x - x_0) \cdot \nabla \bar{u}) dx \\
&+ k^2 \int_{\Gamma_R} |u|^2 \left(\frac{V^2}{A}\right) (x - x_0) \cdot \nu ds + 2\,\mathrm{Re} \int_{\Gamma_R} (ik\beta u + g)(m \cdot \nabla \bar{u}) ds.
\end{aligned}
\tag{46}
$$

Recall, (10), where we define the following function

$$\begin{aligned} S(x) &:= \operatorname{div}\left(\left(\frac{V^2(x)}{A(x)}\right)(x-x_0)\right) \\ &= d\left(\frac{V^2(x)}{A(x)}\right) + \left(2\frac{V(x)\nabla V(x)}{A(x)} - \frac{V^2(x)\nabla A(x)}{A^2(x)}\right)\cdot(x-x_0), \end{aligned} \tag{47}$$

and from (12), we have a minimum for $S(x)$ exists and is positive

$$S_{min} = \min_{x\in\Omega} S(x) > 0.$$

Further, from (12), we have C_G to be the minimal constant so that

$$2\left|\int_\Omega \left(\frac{\nabla A}{A}\right)\nabla u((x-x_0)\cdot\nabla\bar{u})dx\right| \le C_G\left\|\left(\frac{\nabla A}{A}\right)\right\|_{L^\infty(\Omega)}\|\nabla u\|^2_{L^2(\Omega)}. \tag{48}$$

Returning to inequality (46), we obtain

$$\begin{aligned} &\eta\|\nabla u\|^2_{L^2(\Gamma_R)} + k^2 S_{min}\|u\|^2_{L^2(\Omega)} \\ &\le (d-2)\|\nabla u\|^2_{L^2(\Omega)} + C_G\left\|\left(\frac{\nabla A}{A}\right)\right\|_{L^\infty(\Omega)}\|\nabla u\|^2_{L^2(\Omega)} \\ &+ C_1\left(\frac{1}{A_{min}}\|f\|_{L^2(\Omega)}\|\nabla u\|_{L^2(\Omega)} + \|g\|_{L^2(\Gamma_R)}\|\nabla u\|_{L^2(\Gamma_R)}\right) \\ &+ C_1\left(k^2\left(\frac{V^2_{max}}{A_{min}}\right)\|u\|^2_{L^2(\Gamma_R)} + k\,\|\beta\|_{L^\infty(\Gamma_R)}\,\|u\|_{L^2(\Gamma_R)}\|\nabla u\|_{L^2(\Gamma_R)}\right), \end{aligned} \tag{49}$$

where C_1 is independent of k and the bounds (3). Note that on the right hand side we have for any $\xi_5, \xi_6 > 0$ the terms

$$\begin{aligned} k\,\|\beta\|_{L^\infty(\Gamma_R)}\,\|u\|_{L^2(\Gamma_R)}\|\nabla u\|_{L^2(\Gamma_R)} &\le \frac{k^2}{2\xi_5}\|u\|^2_{L^2(\Gamma_R)} + \frac{\xi_5}{2}\,\|\beta\|^2_{L^\infty(\Gamma_R)}\,\|\nabla u\|^2_{L^2(\Gamma_R)} \\ \|g\|_{L^2(\Gamma_R)}\|\nabla u\|_{L^2(\Gamma_R)} &\le \frac{1}{2\xi_6}\|g\|^2_{L^2(\Gamma_R)} + \frac{\xi_6}{2}\|\nabla u\|^2_{L^2(\Gamma_R)}. \end{aligned}$$

We choose ξ_5, ξ_6 so that

$$\frac{\eta}{2} = C_1\frac{\xi_5}{2}\,\|\beta\|^2_{L^\infty(\Gamma_R)} = C_1\frac{\xi_6}{2},$$

and so

$$\frac{k^2}{2\xi_5} \leq \frac{C_1}{2\eta} \|\beta\|^2_{L^\infty(\Gamma_R)} k^2.$$

We then obtain

$$\begin{aligned} k^2 S_{min}\|u\|^2_{L^2(\Omega)} &\leq C_1 \left(\left(\frac{C_1}{2\eta} \|\beta\|^2_{L^\infty(\Gamma_R)} + \frac{V^2_{max}}{A_{min}} \right) k^2 \|u\|^2_{L^2(\Gamma_R)} \right) \\ &+ C_1 \left(\frac{1}{A_{min}} \|f\|_{L^2(\Omega)} \|\nabla u\|_{L^2(\Omega)} + \frac{C_1}{2\eta} \|g\|^2_{L^2(\Gamma_R)} \right) \\ &+ (d-2)\|\nabla u\|^2_{L^2(\Omega)} + C_G \left\| \left(\frac{\nabla A}{A} \right) \right\|_{L^\infty(\Omega)} \|\nabla u\|^2_{L^2(\Omega)}. \end{aligned} \tag{50}$$

Taking $C_2^{bd} = C_1 \left(\frac{C_1}{2\eta} \|\beta\|^2_{L^\infty(\Gamma_R)} + \frac{V^2_{max}}{A_{min}} \right)$ and letting $\epsilon = \beta_{min}\xi_7 / C_2^{bd}$ in the inequality (40) we have the relation

$$C_2^{bd} k^2 \|u\|^2_{L^2(\Gamma_R)} \leq \frac{(C_2^{bd})^2}{\beta^2_{min}\xi_7} \|f\|^2_{L^2(\Omega)} + k^2 \xi_7 \|u\|^2_{L^2(\Omega)} + \frac{C_2^{bd}}{\beta^2_{min}} \|g\|^2_{L^2(\Gamma_R)}. \tag{51}$$

Applying this above inequality to (50), we obtain

$$\begin{aligned} &k^2 (S_{min} - \xi_7) \|u\|^2_{L^2(\Omega)} \\ &\quad \leq C_1 \left(\frac{1}{A_{min}} \|f\|_{L^2(\Omega)} \|\nabla u\|_{L^2(\Omega)} + \frac{C_1}{2\eta} \|g\|^2_{L^2(\Gamma_R)} \right) \\ &\quad + \left((d-2) + C_G \left\| \left(\frac{\nabla A}{A} \right) \right\|_{L^\infty(\Omega)} \right) \|\nabla u\|^2_{L^2(\Omega)} \\ &\quad + \frac{(C_2^{bd})^2}{\beta^2_{min}\xi_7} \|f\|^2_{L^2(\Omega)} + \frac{C_2^{bd}}{\beta^2_{min}} \|g\|^2_{L^2(\Gamma_R)}. \end{aligned} \tag{52}$$

Recall the estimate (41), with $C_3^{bd} = \left((d-2) + C_G \left\| \left(\frac{\nabla A}{A} \right) \right\|_{L^\infty(\Omega)} \right)$, and taking $\xi_4 = \frac{\xi_3}{2} = \xi_8$

$$\begin{aligned} &C_3^{bd} \|\nabla u\|^2_{L^2(\Omega)} \\ &\leq \frac{C_3^{bd} k^2}{A_{min}} \left(V^2_{max} + \frac{\xi_8}{\beta_{min}} + \xi_8 \right) \|u\|^2_{L^2(\Omega)} \\ &+ \frac{C_3^{bd}}{A_{min}} \left(\frac{1}{4k^2 \xi_8} + \frac{1}{\beta_{min}\xi_8} \right) \|f\|^2_{L^2(\Omega)} + \frac{C_3^{bd}}{A_{min}} \left(\frac{1}{\beta^2_{min}} + \frac{1}{4k^2} \right) \|g\|^2_{L^2(\Gamma_R)}. \end{aligned}$$

and so, using the above estimate (52)we obtain

$$
\begin{aligned}
&k^2(S_{min} - \xi_7 - \frac{C_3^{bd}}{A_{min}}\left(V_{max}^2 + \frac{\xi_8}{\beta_{min}} + \xi_8\right))\|u\|^2_{L^2(\Omega)} \\
&\le C_1\left(\frac{1}{A_{min}}\|f\|_{L^2(\Omega)}\|\nabla u\|_{L^2(\Omega)} + \frac{C_1}{2\eta}\|g\|^2_{L^2(\Gamma_R)}\right) \\
&+ \frac{C_3^{bd}}{A_{min}}\left(\frac{1}{4k^2\xi_8} + \frac{1}{\beta_{min}\xi_8}\right)\|f\|^2_{L^2(\Omega)} + \frac{C_3^{bd}}{A_{min}}\left(\frac{1}{\beta_{min}^2} + \frac{1}{4k^2}\right)\|g\|^2_{L^2(\Gamma_R)} \\
&+ \frac{(C_2^{bd})^2}{\beta_{min}^2\xi_7}\|f\|^2_{L^2(\Omega)} + \frac{C_2^{bd}}{\beta_{min}^2}\|g\|^2_{L^2(\Gamma_R)}.
\end{aligned}
\tag{53}
$$

Finally to deal with the remaining term on the right hand side that contains ∇u, we note using (41), letting $\frac{\xi_4}{\beta_{min}} = \frac{\xi_3}{2} = \frac{V_{max}^2}{2}$, and multiplying by $\xi_9/(2A_{min}), \xi_9 > 0$, we obtain

$$
\begin{aligned}
&\frac{\xi_9}{2A_{min}}\|\nabla u\|^2_{L^2(\Omega)} \\
&\quad \le \frac{\xi_9}{2A_{min}^2}\Bigg[2V_{max}^2k^2\|u\|^2_{L^2(\Omega)} + \left(\frac{2}{\beta_{min}^2V_{max}^2} + \frac{1}{2k^2V_{max}^2}\right)\|f\|^2_{L^2(\Omega)} \\
&\qquad + \left(\frac{1}{\beta_{min}^2} + \frac{1}{4k^2}\right)\|g\|^2_{L^2(\Gamma_R)}\Bigg],
\end{aligned}
$$

and so

$$
\begin{aligned}
&\frac{1}{A_{min}}\|f\|_{L^2(\Omega)}\|\nabla u\|_{L^2(\Omega)} \\
&\quad \le \frac{1}{2\xi_9A_{min}}\|f\|^2_{L^2(\Omega)} + \frac{\xi_9}{2A_{min}}\|\nabla u\|^2_{L^2(\Omega)} \\
&\quad \le \frac{\xi_9V_{max}^2}{A_{min}^2}k^2\|u\|^2_{L^2(\Omega)} \\
&\qquad + \left(\frac{1}{2A_{min}\xi_9} + \frac{\xi_9}{2A_{min}^2}\left(\frac{2}{\beta_{min}^2V_{max}^2} + \frac{1}{2k^2V_{max}^2}\right)\right)\|f\|^2_{L^2(\Omega)} \\
&\qquad + \frac{\xi_9}{2A_{min}^2}\left(\frac{1}{\beta_{min}^2} + \frac{1}{4k^2}\right)\|g\|^2_{L^2(\Gamma_R)}.
\end{aligned}
$$

Applying this into (53), we obtain

$$
\begin{aligned}
&k^2(S_{min} - \xi_7 - \frac{C_3^{bd}}{A_{min}}\left(V_{max}^2 + \frac{\xi_8}{\beta_{min}} + \xi_8\right) - \frac{C_1\xi_9 V_{max}^2}{A_{min}^2})\|u\|_{L^2(\Omega)}^2 \\
&\leq C_1\left(\frac{1}{2A_{min}\xi_9} + \frac{\xi_9}{2A_{min}^2}\left(\frac{2}{\beta_{min}^2 V_{max}^2} + \frac{1}{2k^2 V_{max}^2}\right)\right)\|f\|_{L^2(\Omega)}^2 \\
&+ C_1\left(\frac{C_1}{2\eta} + \frac{\xi_9}{2A_{min}^2}\left(\frac{1}{\beta_{min}^2} + \frac{1}{4k^2}\right)\right)\|g\|_{L^2(\Gamma_R)}^2 \\
&+ \frac{C_3^{bd}}{A_{min}}\left(\frac{1}{4k^2\xi_8} + \frac{1}{\beta_{min}\xi_8}\right)\|f\|_{L^2(\Omega)}^2 \\
&+ \frac{C_3^{bd}}{A_{min}}\left(\frac{1}{\beta_{min}^2} + \frac{1}{4k^2}\right)\|g\|_{L^2(\Gamma_R)}^2 + \frac{(C_2^{bd})^2}{\beta_{min}^2\xi_7}\|f\|_{L^2(\Omega)}^2 + \frac{C_2^{bd}}{\beta_{min}^2}\|g\|_{L^2(\Gamma_R)}^2.
\end{aligned} \tag{54}
$$

Hence, we see that the critical term is $S_{min} - \frac{C_3^{bd} V_{max}^2}{A_{min}}$. Recall,

$$
C_3^{bd} := \left((d-2) + C_G\left\|\left(\frac{\nabla A}{A}\right)\right\|_{L^\infty(\Omega)}\right),
$$

thus, from (12), we have

$$
S_{min} - \left((d-2) + C_G\left\|\left(\frac{\nabla A}{A}\right)\right\|_{L^\infty(\Omega)}\right)\frac{V_{max}^2}{A_{min}} > 0. \tag{55}
$$

Since (55) is assumed to hold, we take ξ_7, ξ_8, and ξ_9, so that

$$
\left(S_{min} - \frac{C_3^{bd} V_{max}^2}{A_{min}} - \xi_7 - \frac{C_3^{bd}\xi_8}{A_{min}}\left(\frac{1}{\beta_{min}} + 1\right) - \frac{C_1\xi_9 V_{max}^2}{A_{min}^2}\right) > \delta
$$

for some $\delta > 0$, and taking C_4^{bd} to be the global constant bound for (54) we obtain

$$
k^2\|u\|_{L^2(\Omega)}^2 \leq \frac{C_4^{bd}}{\delta}\left(1 + \frac{1}{k^2}\right)\left(\|f\|_{L^2(\Omega)}^2 + \|g\|_{L^2(\Gamma_R)}^2\right), \tag{56}
$$

and using (41), and taking C_5^{bd} to be the global constant bound we obtain

$$
\|\nabla u\|_{L^2(\Omega)}^2 \leq C_5^{bd}\left(1 + \frac{1}{k^2}\right)\left(\|f\|_{L^2(\Omega)}^2 + \|g\|_{L^2(\Gamma_R)}^2\right), \tag{57}
$$

as desired. □

References

1. I.M. Babuska, S.A. Sauter, Is the pollution effect of the FEM avoidable for the Helmholtz equation considering high wave numbers? SIAM Rev. **42**(3), 451–484 (2000)
2. T. Betcke, S.N. Chandler-Wilde, I.G. Graham, S. Langdon, M. Lindner, Condition number estimates for combined potential integral operators in acoustics and their boundary element discretisation. Numer. Methods Partial Differ. Equ. **27**(1), 31–69 (2011). MR 2743599 (2012a:65355)
3. D. Brown, D. Peterseim, A multiscale method for porous microstructures. SIAM MMS **14**, 1123–1152 (2016)
4. P. Cummings, X. Feng, Sharp regularity coefficient estimates for complex-valued acoustic and elastic Helmholtz equations. Math. Models Methods Appl. Sci. **16**(1), 139–160 (2006). MR 2194984 (2007d:35030)
5. D.A. Di Pietro, A. Ern, *Mathematical Aspects of Discontinuous Galerkin Methods*. Mathématiques and Applications (Berlin), vol. 69 (Springer, Heidelberg, 2012). MR 2882148
6. S. Esterhazy, J.M. Melenk, On stability of discretizations of the Helmholtz equation, in *Numerical Analysis of Multiscale Problems*. Lecture Notes in Computational Science and Engineering, vol. 83 (Springer, Heidelberg, 2012), pp. 285–324. MR 3050917
7. D. Gallistl, D. Peterseim, Stable multiscale Petrov-Galerkin finite element method for high frequency acoustic scattering. Comput. Methods Appl. Mech. Eng. **295**, 1–17 (2015)
8. P. Grisvard, Contrôlabilité exacte des solutions de l'équation des ondes en présence de singularités. J. Math. Pures Appl. (9) **68**(2), 215–259 (1989). MR 1010769 (90i:49045)
9. P. Henning, D. Peterseim, Oversampling for the multiscale finite element method. Multiscale Model. Simul. **11**(4), 1149–1175 (2013). MR 3123820
10. P. Henning, P. Morgenstern, D. Peterseim, Multiscale Partition of Unity, in *Meshfree Methods for Partial Differential Equations VII*, ed. by M. Griebel, M.A. Schweitzer. Lecture Notes in Computational Science and Engineering, vol. 100 (Springer, Berlin, 2014)
11. U. Hetmaniuk, Stability estimates for a class of Helmholtz problems. Commun. Math. Sci. **5**(3), 665–678 (2007). MR 2352336 (2008m:35050)
12. Ch. Makridakis, F. Ihlenburg, I. Babuška, Analysis and finite element methods for a fluid-solid interaction problem in one dimension. Math. Models Methods Appl. Sci. **06**(08), 1119–1141 (1996)
13. A. Målqvist, D. Peterseim, Localization of elliptic multiscale problems. Math. Comput. **83**(290), 2583–2603 (2014). MR 3246801
14. J.M. Melenk, On generalized finite-element methods. ProQuest LLC, Ann Arbor, MI, 1995. Thesis (Ph.D.), University of Maryland, College Park. MR 2692949
15. J.M. Melenk, S.A. Sauter, Wave-number explicit convergence analysis for Galerkin discretizations of the Helmholtz equation. SIAM J. Numer. Anal. **49**, 1210–1243 (2011)
16. D. Peterseim, Eliminating the pollution effect in Helmholtz problems by local subscale correction. Math. Comp. **86**, 1005–1036 (2017)
17. D. Peterseim, Variational multiscale stabilization and the exponential decay of fine-scale correctors, in *Building Bridges: Connections and Challenges in Modern Approaches to Numerical Partial Differential Equations*. Lecture Notes in Computational Science and Engineering, vol. 114 (Springer International Publishing, 2016). doi:10.1007/978-3-319-41640-3
18. H. Wu, Pre-asymptotic error analysis of CIP-FEM and FEM for the Helmholtz equation with high wave number. Part I: linear version. IMA J. Numer. Anal. **34**(3), 1266–1288 (2014). MR 3232452

Error Analysis of Nodal Meshless Methods

Robert Schaback

Abstract There are many application papers that solve elliptic boundary value problems by meshless methods, and they use various forms of generalized stiffness matrices that approximate derivatives of functions from values at scattered nodes $x_1, \dots, x_M \in \Omega \subset \mathbb{R}^d$. If u^* is the true solution in some Sobolev space S allowing enough smoothness for the problem in question, and if the calculated approximate values at the nodes are denoted by $\tilde{u}_1, \dots, \tilde{u}_M$, the canonical form of error bounds is

$$\max_{1 \le j \le M} |u^*(x_j) - \tilde{u}_j| \le \epsilon \|u^*\|_S$$

where ϵ depends crucially on the problem and the discretization, but not on the solution. This contribution shows how to calculate such ϵ *numerically and explicitly*, for any sort of discretization of strong problems via nodal values, may the discretization use Moving Least Squares, unsymmetric or symmetric RBF collocation, or localized RBF or polynomial stencils. This allows users to compare different discretizations with respect to error bounds of the above form, without knowing exact solutions, and admitting all possible ways to set up generalized stiffness matrices. The error analysis is proven to be sharp under mild additional assumptions. As a byproduct, it allows to construct worst cases that push discretizations to their limits. All of this is illustrated by numerical examples.

1 Introduction

Following the seminal survey [5] by Ted Belytschko et al. in 1996, meshless methods for PDE solving often work "*entirely in terms of values at nodes*". This means that large linear systems are set up that have values $u(x_1), \dots, u(x_M)$ of an unknown function u as unknowns, while the equations model the underlying PDE

R. Schaback (✉)
Institut für Numerische und Angewandte Mathematik, Univ. Göttingen, Lotzestraße 16-18, 37083 Göttingen, Germany
e-mail: schaback@math.uni-goettingen.de

M. Griebel, M.A. Schweitzer (eds.), *Meshfree Methods for Partial Differential Equations VIII*, Lecture Notes in Computational Science and Engineering 115,
DOI 10.1007/978-3-319-51954-8_7

problem in discretized way. Altogether, the discrete problems have the form

$$\sum_{j=1}^{M} a_{kj}u(x_j) \approx f_k,\ 1 \le k \le N \tag{1}$$

with $N \ge M$, whatever the underlying PDE problem is, and the $N \times M$ matrix $\mathbf{A}$ with entries a_{kj} can be called a *generalized stiffness matrix*.

Users solve the system somehow and then get values $\tilde{u}_1, \dots, \tilde{u}_M$ that satisfy

$$\sum_{j=1}^{M} a_{kj}\tilde{u}_j \approx f_k,\ 1 \le k \le N,$$

but they should know how far these values are from the values $u^*(x_j)$ of the true solution of the PDE problem that is supposed to exist.

The main goal of this paper is to provide tools that allow users to assess the quality of their discretization, no matter how the problem was discretized or how the system was actually solved. The computer should tell the user whether the discretization is useful or not. It will turn out that this is possible, and at tolerable computational cost that is proportional to the complexity for setting up the system, not for solving it.

The only additional ingredient is a specification of the smoothness of the true solution u^*, and this is done in terms of a strong norm $\|.\|_S$, e.g. a higher-order Sobolev norm or seminorm. The whole problem will then be implicitly scaled by $\|u^*\|_S$, and we assert an absolute bound of the form

$$\max_{1\le j\le M} |u^*(x_j) - \tilde{u}_j| \le \epsilon \|u^*\|_S$$

or a relative bound

$$\frac{\max_{1\le j\le M} |u^*(x_j) - \tilde{u}_j|}{\|u^*\|_S} \le \epsilon$$

with an entity ϵ that can be calculated. It will be a product of two values caring for *stability* and *consistency*, respectively, and these are calculated and analyzed separately.

Section 2 will set up the large range of PDE or, more generally, operator equation problems we are able to handle, and Sect. 3 provides the backbone of our error analysis. It must naturally contain some versions of *consistency* and *stability*, and we deal with these in Sects. 5 and 7, with an interlude on polyharmonic kernels in Sect. 6. For given Sobolev smoothness order m, these provide stable, sparse, and error-optimal nodal approximations of differential operators [8]. Numerical examples follow in Sect. 8, demonstrating how to work with the tools of this paper. It turns out that the evaluation of stability is easier than expected, while the evaluation of consistency often suffers from severe numerical cancellation that is to be overcome by future research, or that is avoided by using special scale-invariant approximations, e.g. via polyharmonic kernels along the lines of Sect. 6.

2 Problems and Their Discretizations

We have to connect the system (1) back to the original PDE problem, and we do this in an unconventional but useful way that we use successfully since [30] in 1999.

2.1 Analytic Problems

For example, consider a model boundary value problem of the form

$$\begin{aligned} Lu &= f \quad \text{in } \Omega \subset \mathbb{R}^d \\ Bu &= g \quad \text{in } \Gamma := \partial\Omega \end{aligned} \tag{2}$$

where $f,\ g$ are given functions on Ω and Γ, respectively, and $L,\ B$ are linear operators, defined and continuous on some normed linear space U in which the true solution u^* should lie. Looking closer, this is an infinite number of linear constraints

$$\begin{aligned} Lu(y) &= f(y) \quad \text{for all } y \in \Omega \subset \mathbb{R}^d \\ Bu(z) &= g(z) \quad \text{for all } z \in \Gamma := \partial\Omega \end{aligned}$$

and these can be generalized as infinitely many linear functionals acting on the function u, namely

$$\lambda(u) = f_\lambda \quad \text{for all } \lambda \in \Lambda \subset U^* \tag{3}$$

where the set Λ is contained in the topological dual U^* of U, in our example

$$\Lambda = \{\delta_y \circ L,\ y \in \Omega\} \cup \{\delta_z \circ B,\ z \in \Gamma\}. \tag{4}$$

Definition 1 An admissible problem in the sense of this paper consists in finding an u from some normed linear space U such that (3) holds for a fixed set $\Lambda \subset U^*$. Furthermore, solvability via $f_\lambda = \lambda(u^*)$ for all $\lambda \in \Lambda \subset U^*$ for some $u^* \in U$ is always assumed.

Clearly, this allows various classes of differential equations and boundary conditions. in weak or strong form. For examples, see [29]. Here, we just mention that the standard functionals for weak problems with $L = -\Delta$ are of the form

$$\lambda_v(u) := \int_\Omega (\nabla u)^T \nabla v \tag{5}$$

where v is an arbitrary test function from $W_0^1(\Omega)$.

2.2 Discretization

The connection of the problem (3) to the discrete linear system (1) usually starts with specifying a finite subset $\Lambda_N = \{\lambda_1, \dots, \lambda_N\} \subset \Lambda$ of *test* functionals. But then it splits into two essentially different branches.

The *shape function* approach defines functions $u_j : \Omega \to \mathbb{R}$ with the Lagrange property $u_i(x_j) = \delta_{ij},\ 1 \le i,j \le M$ and defines the elements a_{kj} of the stiffness matrix as $a_{kj} := \lambda_k(u_j)$. This means that the application of the functionals λ_k on *trial functions*

$$u(x) = \sum_{j=1}^{M} u(x_j) u_j(x)$$

is exact, and the linear system (1) describes the exact action of the selected test functionals on the trial space. Typical instances of the shape function approach are standard applications of Moving Least Squares (MLS) trial functions [2, 3, 32]. Such applications were surveyed in [5] and incorporate many versions of the Meshless Local Petrov Galerkin (MLPG) technique [4]. Another popular shape function method is unsymmetric or symmetric kernel-based collocation, see [10, 12, 13, 17].

But one can omit shape functions completely, at the cost of sacrificing exactness. Then the selected functionals λ_k are each approximated by linear combinations of the functionals $\delta_{x_1}, \dots, \delta_{x_M}$ by requiring

$$\lambda_k(u) \approx \sum_{j=1}^{M} a_{kj} \delta_{x_j} u = \sum_{j=1}^{M} a_{kj} u(x_j),\ 1 \le k \le N, \quad \text{for all } u \in U. \tag{6}$$

This approach can be called *direct* discretization, because it bypasses shape functions. It is the standard technique for *generalized finite differences* (FD) [23], and it comes up again in meshless methods at many places, starting with [24, 31] and called *RBF-FD* or *local RBF collocation* by various authors, e.g. [11, 34]. The generalized finite difference approximations may be calculated via radial kernels using local selections of nodes only [26, 35, 36], and there are papers on how to calculate such approximations, e.g. [7, 19]. Bypassing Moving Least Squares trial functions, direct methods in the context of Meshless Local Petrov Galerkin techniques are in [21, 22], connected to *diffuse derivatives* [24]. For a mixture of kernel-based and MLS techniques, see [18].

This contribution will work in both cases, with a certain preference for the direct approach. The paper [29] focuses on shape function methods instead. It proves that uniform stability can be achieved for all well-posed problems by choosing a suitable discretization, and then convergence can be inferred from standard convergence rates of approximations of derivatives of the true solution from derivatives of trial functions. The methods of [29] fail for direct methods, and this was the main reason to write this paper.

2.3 Nodal Trial Approximations

In addition to Definition 1 we now assume that U is a space of functions on some set Ω, and that point evaluation is continuous, i.e. $\delta_x \in U^*$ for all $x \in \Omega$. We fix a finite set X_M of M *nodes* $x_1, \dots, x_M$ and denote the span of the functionals δ_{x_j} by D_M.

For each $\lambda \in \Lambda$ we consider a linear approximation $\tilde{\lambda}$ to λ from D_M, i.e.

$$\lambda(u) \approx \sum_{j=1}^{M} a_j(\lambda)u(x_j) =: \tilde{\lambda}(u) \tag{7}$$

Note that there is no trial space of functions, and no *shape functions* at all, just *nodal values* and approximations of functionals from nodal values. It should be clear how the functionals in (4) can be approximated as in (7) via values at nodes.

In the sense of the preceding section, this looks like a *direct* discretization, but it also covers the *shape function* approach, because it is allowed to take $a_j(\lambda) = \lambda(u_j)$ for shape functions u_j with the Lagrange property.

2.4 Testing

Given a nodal trial approximation, consider a finite subset Λ_N of functionals $\lambda_1, \dots, \lambda_N$ and pose the possibly overdetermined linear system

$$\lambda_k(u^*) = f_{\lambda_k} = \sum_{j=1}^{M} a_j(\lambda_k)u_j \tag{8}$$

for unknown *nodal values* $u_1, \dots, u_M$ that may be interpreted as approximations to $u^*(x_1), \dots, u^*(x_M)$. We call Λ_N a *test selection* of functionals, and remark that we have obtained a system of the form (1).

For what follows, we write the linear system (8) in matrix form also

$$\mathbf{f} = \mathbf{A}\mathbf{u} \tag{9}$$

with

$$\begin{aligned} \mathbf{A} &= (a_j(\lambda_k))_{1\le k\le N, 1\le j\le M} &\in \mathbb{R}^{N\times M} \\ \mathbf{f} &= (f_{\lambda_1},\dots,f_{\lambda_N})^T &\in \mathbb{R}^N \\ \mathbf{u} &= (u(x_1),\dots,u(x_M))^T &\in \mathbb{R}^M. \end{aligned}$$

Likewise, we denote the vector of exact nodal values $u^*(x_j)$ by $\mathbf{u}^*$, and $\tilde{\mathbf{u}}$ will be the vector of nodal values $\tilde{u}_j$ that is obtained by some numerical method that solves the system (8) approximately.

It is well-known [15] that square systems of certain meshless methods may be singular, but it is also known [29] that one can bypass that problem by *overtesting*, i.e. choosing N larger than M. This leads to overdetermined systems, but they can be handled by standard methods like the MATLAB backslash in a satisfactory way. Here, we expect that users set up their $N \times M$ stiffness matrix $\mathbf{A}$ by sufficiently thorough testing, i.e. by selecting many test functionals $\lambda_1,\dots,\lambda_N$ so that the matrix has rank $M \le N$. Section 7 will show that users can expect good stability if they handle a well-posed problem with sufficient overtesting. Note further that for cases like the standard Dirichlet problem (2), the set Λ_N has to contain a reasonable mixture of functionals connected to the differential operator and functionals connected to boundary values. Since we focus on general worst-case error estimates here, insufficient overtesting and an unbalanced mixture of boundary and differential equation approximations will result in error bounds that either cannot be calculated due to rank loss or come out large. The computer should reveal whether a discretization is good or not.

3 Error Analysis

The goal of this paper is to derive useful bounds for $\|\mathbf{u}^* - \tilde{\mathbf{u}}\|_\infty$, but we do not care for an error analysis away from the nodes. Instead, we assume a *postprocessing step* that interpolates the elements of $\tilde{\mathbf{u}}$ to generate an approximation $\tilde{u}$ to the solution u^* in the whole domain. Our analysis will accept any numerical solution $\tilde{\mathbf{u}}$ in terms of nodal values and provide an error bound with small additional computational effort.

3.1 Residuals

We start with evaluating the *residual* $\mathbf{r} := \mathbf{f} - \mathbf{A}\tilde{\mathbf{u}} \in \mathbb{R}^N$ no matter how the numerical solution $\tilde{\mathbf{u}}$ was obtained. This can be explicitly done except for roundoff errors, and needs no derivation of upper bounds. Since in general the final error at the nodes will be larger than the observed residuals, users should refine their discretization when they encounter residuals that are very much larger than the expected error in the solution.

3.2 *Stability*

In Sect. 2.4 we postulated that users calculate an $N \times M$ stiffness matrix $\mathbf{A}$ that has no rank loss. Then the *stability constant*

$$C_S(\mathbf{A}) := \sup_{\mathbf{u}\neq 0} \frac{\|\mathbf{u}\|_p}{\|\mathbf{A}\mathbf{u}\|_q} \tag{10}$$

is finite for any choice of discrete norms $\|.\|_p$ and $\|.\|_q$ on $\mathbb{R}^M$ and $\mathbb{R}^N$, respectively, with $1 \leq p, q \leq \infty$ being fixed here, and dropped from the notation. In principle, this constant can be explicitly calculated for standard norms, but we refer to Sect. 7 on how it is treated in theory and practice. We shall mainly focus on well-posed cases where $C_S(\mathbf{A})$ can be expected to be reasonably bounded, while norms of $\mathbf{A}$ get very large. This implies that the ratios $\|\mathbf{u}\|_p/\|\mathbf{A}\mathbf{u}\|_q$ can vary in a wide range limited by

$$\|\mathbf{A}\|_{q,p}^{-1} \leq \frac{\|\mathbf{u}\|_p}{\|\mathbf{A}\mathbf{u}\|_q} \leq C_S(\mathbf{A}). \tag{11}$$

If we assume that we can deal with the stability constant $C_S(\mathbf{A})$, the second step of error analysis is

$$\begin{aligned} \|\mathbf{u}^* - \tilde{\mathbf{u}}\|_p &\leq C_S(\mathbf{A})\|\mathbf{A}(\mathbf{u}^* - \tilde{\mathbf{u}})\|_q \\ &\leq C_S(\mathbf{A})(\|\mathbf{A}\mathbf{u}^* - \mathbf{f}\|_q + \|\mathbf{f} - \mathbf{A}\tilde{\mathbf{u}}\|_q) \\ &\leq C_S(\mathbf{A})(\|\mathbf{A}\mathbf{u}^* - \mathbf{f}\|_q + \|\mathbf{r}\|_q) \end{aligned} \tag{12}$$

and we are left to handle the *consistency term* $\|\mathbf{A}\mathbf{u}^* - \mathbf{f}\|_q$ that still contains the unknown true solution $\mathbf{u}^*$. Note that $\mathbf{f}$ is not necessarily in the range of $\mathbf{A}$, and we cannot expect to get zero residuals $\mathbf{r}$.

3.3 *Consistency*

For all approximations (7) we assume that there is a *consistency* error bound

$$|\lambda(u) - \tilde{\lambda}(u)| \leq c(\lambda)\|u\|_S \tag{13}$$

for all u in some *regularity* subspace U_S of U that carries a strong norm or seminorm $\|.\|_S$. In case of a seminorm, we have to assume that the approximation $\tilde{\lambda}$ is an exact approximation to λ on the nullspace of the seminorm, but we shall use seminorms only in Sect. 6 below. If the solution u^* has plenty of smoothness, one may expect that $c(\lambda)\|u^*\|_S$ is small, provided that the discretization quality keeps up with the smoothness. In Sect. 5, we shall consider cases where the $c(\lambda)$ can be calculated explicitly.

The bound (13) now specializes to

$$\|\mathbf{A}\mathbf{u}^* - \mathbf{f}\|_q \leq \|\mathbf{c}\|_q \|u^*\|_S$$

with the vector

$$\mathbf{c} = (c(\lambda_1), \ldots, c(\lambda_N))^T \in \mathbb{R}^N,$$

and the error in (12) is bounded absolutely by

$$\|\mathbf{u}^* - \tilde{\mathbf{u}}\|_p \leq C_S(\mathbf{A}) \left(\|\mathbf{c}\|_q \|u^*\|_S + \|\mathbf{r}\|_q\right)$$

and relatively by

$$\frac{\|\mathbf{u}^* - \tilde{\mathbf{u}}\|_p}{\|u^*\|_S} \leq C_S(\mathbf{A}) \left(\|\mathbf{c}\|_q + \frac{\|\mathbf{r}\|_q}{\|u^*\|_S}\right). \tag{14}$$

This still contains the unknown solution u^*. But in kernel-based spaces, there are ways to get estimates of $\|u^*\|_S$ via interpolation. A strict but costly way is to interpolate the data vector $\mathbf{f}$ by symmetric kernel collocation to get a function $u^*_{\mathbf{f}}$ with $\|u^*_{\mathbf{f}}\|_S \leq \|u^*\|_S$, and this norm can be plugged into (14). In single applications, users would prefer to take the values of $u^*_{\mathbf{f}}$ in the nodes as results, since they are known to be error-optimal [28]. But if discretizations with certain given matrices $\mathbf{A}$ are to be evaluated or compared, this suggestion makes sense to get the right-hand side of (14) independent of u^*.

3.4 Residual Minimization

To handle the awkward final term in (14) without additional calculations, we impose a rather weak additional condition on the numerical procedure that produces $\tilde{\mathbf{u}}$ as an approximate solution to (9). In particular, we require

$$\|\mathbf{A}\tilde{\mathbf{u}} - \mathbf{f}\|_q \leq K(\mathbf{A}) \|\mathbf{A}\mathbf{u}^* - \mathbf{f}\|_q, \tag{15}$$

which can be obtained with $K(\mathbf{A}) = 1$ if $\tilde{\mathbf{u}}$ is calculated via minimization of the residual $\|\mathbf{A}\mathbf{u} - \mathbf{f}\|_q$ over all $\mathbf{u} \in \mathbb{R}^M$, or with $K(\mathbf{A}) = 0$ if $\mathbf{f}$ is in the range of $\mathbf{A}$.

Anyway, we assume that users have a way to solve the system (9) approximately such that (15) holds with a known and moderate constant $K(\mathbf{A})$.

Then (15) implies

$$\begin{aligned}\|\mathbf{r}\|_q &= \|\mathbf{A}\tilde{\mathbf{u}} - \mathbf{f}\|_q \\ &\leq K(\mathbf{A})\|\mathbf{A}\mathbf{u}^* - \mathbf{f}\|_q \\ &\leq K(\mathbf{A})\|\mathbf{c}\|_q\|u^*\|_S\end{aligned}$$

and bounds $\|\mathbf{r}\|_q$ in terms of $\|u^*\|_S$.

3.5 Final Relative Error Bound

Theorem 1 *Under the above assumptions,*

$$\frac{\|\mathbf{u}^* - \tilde{\mathbf{u}}\|_p}{\|u^*\|_S} \leq (1 + K(\mathbf{A}))C_S(\mathbf{A})\|\mathbf{c}\|_q. \tag{16}$$

Proof We can insert (15) directly into (12) to get

$$\begin{aligned}\|\mathbf{u}^* - \tilde{\mathbf{u}}\|_p &\leq C_S(\mathbf{A})(1 + K(\mathbf{A}))\|\mathbf{A}\mathbf{u}^* - \mathbf{f}\|_q \\ &\leq (1 + K(\mathbf{A}))C_S(\mathbf{A})\|\mathbf{c}\|_q\|u^*\|_S\end{aligned}$$

and finally (16), where now all elements of the right-hand side are accessible.

This is as far as one can go, not having any additional information on how u^* scales. The final form of (16) shows the classical elements of convergence analysis, since the right-hand side consists of a *stability* term $C_S(\mathbf{A})$ and a *consistency* term $\|\mathbf{c}\|_q$. The factor $1 + K(\mathbf{A})$ can be seen as a *computational accuracy* term.

Examples in Sect. 8 will show how these relative error bounds work in practice. Before that, the next sections will demonstrate theoretically why users can expect that the ingredients of the bound in (16) can be expected to be small. For this analysis, we shall assume that users know which regularity the true solution has, because we shall have to express everything in terms of $\|u^*\|_S$.

At this point, some remarks on error bounds should be made, because papers focusing on applications of meshless methods often contain one of the two standard crimes of error assessment.

The first is to take a problem with a known solution u^* that supplies the data, calculate nodal values $\tilde{\mathbf{u}}$ by some hopefully new method and then compare with $\mathbf{u}^*$ to conclude that the method is good because $\|\mathbf{u}^* - \tilde{\mathbf{u}}\|$ is small. But the method may be intolerably unstable. If the input is changed very slightly, it may produce a seriously different numerical solution $\hat{\mathbf{u}}$ that reproduces the data as well as $\tilde{\mathbf{u}}$. The "quality" of the result $\tilde{\mathbf{u}}$ may be just lucky, it does not prove anything about the method used.

The second crime, usually committed when there is no explicit solution known, is to evaluate residuals $\mathbf{r} = \mathbf{A}\tilde{\mathbf{u}} - \mathbf{f}$ and to conclude that $\|\mathbf{u}^* - \tilde{\mathbf{u}}\|$ is small because residuals are small. This also ignores stability. There even are papers that claim convergence of methods by showing that residuals converge to zero when the discretization is refined. This reduces convergence rates of a PDE solver to rates of consistency, again ignoring stability problems that may counteract against good consistency. Section 8 will demonstrate this effect by examples.

This paper will avoid these crimes, but on the downside our error analysis is a worst-case theory that will necessarily overestimate errors of single cases.

3.6 Sharpness

In particular, if users take a specific problem (2) with data functions f and g and a known solution u^*, and if they evaluate the observed error and the bound (16), they will often see quite an overestimation of the error. This is due to the fact that they have a special case that is far away from being worst possible for the given PDE discretization, and this is comparable to a lottery win, as we shall prove now.

Theorem 2 *For all $K(\mathbf{A}) > 1$ there is some $u^* \in U_S$ and an admissible solution vector $\tilde{\mathbf{u}}$ satisfying (15) such that*

$$(K(\mathbf{A}) - 1)C_S(\mathbf{A})\|u^*\|_S\|\mathbf{c}\|_\infty \le \|\mathbf{u}^* - \tilde{\mathbf{u}}\|_\infty \le (K(\mathbf{A}) + 1)C_S(\mathbf{A})\|u^*\|_S\|\mathbf{c}\|_\infty \tag{17}$$

showing that the above worst-case error analysis cannot be improved much.

Proof We first take the worst possible value vector $\mathbf{u}_S$ for stability, satisfying

$$\|\mathbf{u}_S\|_\infty = C_S(\mathbf{A})\|\mathbf{A}\mathbf{u}_S\|_\infty$$

and normalize it to $\|\mathbf{u}_S\|_\infty = 1$. Then we consider the worst case of consistency, and we go into a kernel-based context.

Let the consistency vector $\mathbf{c}$ attain its norm at some index j, $1 \le j \le N$, i.e. $\|\mathbf{c}\|_\infty = c(\lambda_j)$. Then there is a function $u_j \in U_S$ with

$$|\lambda_j(u_j) - \tilde{\lambda}_j(u_j)| = c(\lambda_j)\|u_j\|_S = c(\lambda_j)^2,$$

namely by taking the Riesz representer $u_j := (\lambda_j - \tilde{\lambda}_j)^x K(x, \cdot)$ of the error functional. The values of u_j at the nodes form a vector $\mathbf{u}_j$, and we take the data f as exact values of u_j, i.e. $f_k := \lambda_k(u_j)$, $1 \le k \le N$ to let u_j play the role of the true solution u^*, in particular $\mathbf{u}^* = \mathbf{u}_j$ and $\|u^*\|_S = \|u_j\|_S = c(\lambda_j) = \|\mathbf{c}\|_\infty$.

We then define $\tilde{\mathbf{u}} := \mathbf{u}^* + \alpha C_S(\mathbf{A})\mathbf{u}_S$ as a candidate for a numerical solution and check how well it satisfies the system and what its error bound is. We have

$$\begin{aligned}
\|\mathbf{A}\tilde{\mathbf{u}} - \mathbf{f}\|_\infty &= \|\mathbf{A}\,(\mathbf{u}^* + \alpha C_S(\mathbf{A})\mathbf{u}_S) - \mathbf{f}\|_\infty \\
&\le \|\mathbf{A}\mathbf{u}^* - \mathbf{f}\|_\infty + |\alpha| C_S(\mathbf{A}) \|\mathbf{A}\mathbf{u}_S\|_\infty \\
&= |\alpha| + \|\mathbf{A}\mathbf{u}^* - \mathbf{f}\|_\infty \\
&= K(\mathbf{A})\|\mathbf{A}\mathbf{u}^* - \mathbf{f}\|_\infty
\end{aligned}$$

if we choose

$$\alpha = (K(\mathbf{A}) - 1)\|\mathbf{A}\mathbf{u}^* - \mathbf{f}\|_\infty.$$

Thus $\tilde{\mathbf{u}}$ is a valid candidate for numerical solving. The actual error is

$$\begin{aligned}
\|\mathbf{u}^* - \tilde{\mathbf{u}}\|_\infty &= (K(\mathbf{A}) - 1)\|\mathbf{A}\mathbf{u}^* - \mathbf{f}\|_\infty C_S(\mathbf{A}) \\
&= (K(\mathbf{A}) - 1)C_S(\mathbf{A}) \max_{1\le k\le N} |\lambda_k(u_j) - \tilde{\lambda}_k(u_j)| \\
&\ge (K(\mathbf{A}) - 1)C_S(\mathbf{A})|\lambda_j(u_j) - \tilde{\lambda}_j(u_j)| \\
&= (K(\mathbf{A}) - 1)C_S(\mathbf{A})\|u_j\|_S\|\mathbf{c}\|_\infty
\end{aligned} \tag{18}$$

proving the assertion.

We shall come back to this worst-case construction in the examples of Sect. 8.

4 Dirichlet Problems

The above error analysis simplifies for problems where Dirichlet values are given on boundary nodes, and where approximations of differential operators are only needed in interior points. Then we have N approximations of functionals that are based on M_I interior nodes and M_B boundary nodes, with $M = M_I + M_B$. We now use subscripts I and B to indicate vectors of values on interior and boundary nodes, respectively. The linear system now is

$$\mathbf{B}\mathbf{u}_I = \mathbf{f}_I - \mathbf{C}\mathbf{g}_B$$

while the previous section dealt with the full system

$$\mathbf{A}\begin{pmatrix}\mathbf{u}_I\\ \mathbf{u}_B\end{pmatrix} = \begin{pmatrix}\mathbf{f}_I\\ \mathbf{g}_B\end{pmatrix} \quad \text{with } \mathbf{A} = \begin{pmatrix}\mathbf{B} & \mathbf{C}\\ 0 & \mathbf{I}_B\end{pmatrix}$$

that has trivial approximations on the boundary. Note that this splitting is standard practice in classical finite elements when nonzero Dirichlet boundary conditions are given. We now use the stability constant $C_S(\mathbf{B})$ for $\mathbf{B}$, not for $\mathbf{A}$, and examples will show that it often comes out much smaller than $C_S(\mathbf{A})$. The consistency bounds (13) stay the same, but they now take the form

$$\|\mathbf{B}\mathbf{u}_I^* + \mathbf{C}\mathbf{u}_B^* - \mathbf{f}_I\|_q = \|\mathbf{B}\mathbf{u}_I^* + \mathbf{C}\mathbf{g}_B - \mathbf{f}_I\|_q \le \|\mathbf{c}_I\|_q \|u^*\|_S.$$

The numerical method should now guarantee

$$\|\mathbf{B}\tilde{\mathbf{u}}_I + \mathbf{C}\mathbf{g}_B - \mathbf{f}_I\|_q \le K(\mathbf{B})\|\mathbf{B}\mathbf{u}_I^* + \mathbf{C}\mathbf{g}_B - \mathbf{f}_I\|_q$$

with a reasonable $K(\mathbf{B}) \ge 1$. Then the same error analysis applies, namely

$$\begin{aligned}
\|\mathbf{u}_I^* - \tilde{\mathbf{u}}_I\|_p &\le C_S(\mathbf{B})\|\mathbf{B}(\mathbf{u}_I^* - \tilde{\mathbf{u}}_I)\|_q \\
&\le C_S(\mathbf{B})\|\mathbf{B}\mathbf{u}_I^* - \mathbf{C}\mathbf{g}_B - \mathbf{f}_I\|_q + C_S(\mathbf{B})\|\mathbf{B}\tilde{\mathbf{u}}_I - \mathbf{C}\mathbf{g}_B - \mathbf{f}_I)\|_q \\
&\le C_S(\mathbf{B})(1 + K(\mathbf{B}))\|\mathbf{B}\mathbf{u}_I^* - \mathbf{C}\mathbf{g}_B - \mathbf{f}_I\|_q \\
&\le C_S(\mathbf{B})(1 + K(\mathbf{B}))\|\mathbf{c}_I\|_q \|u^*\|_S.
\end{aligned}$$

5 Consistency Analysis

There are many ways to determine the *stiffness matrix elements* $a_j(\lambda_k)$ arising in (9) and (7), but they are either based on *trial/shape functions* or on *direct discretizations* as described in Sect. 2.2. We do not care here which technique is used. As a by-product, our method will allow to compare different approaches on a fair basis.

To make the constants $c(\lambda)$ in (13) numerically accessible, we assume that the norm $\|.\|_S$ comes from a Hilbert subspace U_S of U that has a reproducing kernel

$$K : \Omega \times \Omega \to \mathbb{R}.$$

The squared norm of the error functional $\lambda - \tilde{\lambda}$ of the approximation $\tilde{\lambda}$ in (7) then is the value of the quadratic form

$$\begin{aligned}
Q^2(\lambda, \tilde{\lambda}) &:= \|\lambda - \tilde{\lambda}\|_{U_S^*}^2 \\
&= \lambda^x \lambda^y K(x, y) - 2\sum_{j=1}^{M} a_j(\lambda)\lambda_j^x \lambda^y K(x, y) \\
&\quad + \sum_{j,k=1}^{M} a_j(\lambda)a_k(\lambda)\lambda_j^x \lambda_k^y K(x, y)
\end{aligned} \tag{19}$$

which can be explicitly evaluated, though there will be serious numerical cancellations because the result is small while the input is not. It provides the explicit error bound

$$|\lambda(u^*) - \tilde{\lambda}(u^*)|^2 \le Q^2(\lambda, \tilde{\lambda})\|u^*\|_S^2$$

such that we can work with

$$c(\lambda) = Q(\lambda, \tilde{\lambda}).$$

As mentioned already, the quadratic form (19) in its naïve form has an unstable evaluation due to serious cancellation. In [7], these problems were partly overcome by variable precision arithmetic, while the paper [19] provides a very nice stabilization technique, but unfortunately confined to approximations based on the Gaussian kernel. We hope to be able to deal with stabilization of the evaluation of the quadratic form in a forthcoming paper.

On the positive side, there are cases where these instabilities do not occur, namely for *polyharmonic kernels*. We shall come back to this in Sect. 6.

Of course, there are many *theoretical* results bounding the consistency error (13), e.g. [7, 22] in terms of $\|u^*\|_S$, with explicit convergence orders in terms of powers of *fill distances*

$$h := \sup_{y \in \Omega} \min_{x_j} \|y - x_j\|_2.$$

We call these orders *consistency orders* in what follows. Except for Sect. 6, we do not survey such results here, but users can be sure that a sufficiently fine fill distance and sufficient smoothness of the solution will always lead to a high consistency order. Since rates increase when more nodes are used, we target p-methods, not h-methods in the language of the finite element literature, and we assume sufficient regularity for this.

Minimizing the quadratic form (19) over the weights $a_j(\lambda)$ yields discretizations with *optimal* consistency with respect to the choice of the space U_S [7]. But their calculation may be unstable [19] and they usually lead to non-sparse matrices unless users restrict the used nodes for each single functional. If they are combined with a best possible choice of trial functions, namely the Riesz representers $v_j(x) = \lambda_j^y K(x, y)$ of the test functionals, the resulting linear system is symmetric and positive definite, provided that the functionals are linearly independent. This method is *symmetric collocation* [10, 12, 13], and it is an *optimal recovery* method in the space U_S [28]. It leads to non-sparse matrices and suffers from severe instability, but it is error-optimal. Here, we focus on non-optimal methods that allow sparsity.

Again, the instability of optimal approximations can be avoided using polyharmonic kernels, and the next section will describe how this works.

6 Approximations by Polyharmonic Kernels

Assume that we are working in a context where we know that the true solution u^* lies in Sobolev space $W_2^m(\Omega)$ for $\Omega \subset \mathbb{R}^d$, or, by Whitney extension also in $W_2^m(\mathbb{R}^d)$. Then the consistency error (13) of any given approximation should be evaluated in that space, and taking an optimal approximation in that space would yield a system with optimal consistency.

But since the evaluation and calculation of approximations in $W_2^m(\mathbb{R}^d)$ is rather unstable, a workaround is appropriate. Instead of the full norm in $W_2^m(\mathbb{R}^d)$ one takes the seminorm involving only the order m derivatives. This originates from early work of Duchon [9] and leads to Beppo-Levi spaces instead of Sobolev spaces (see e.g. [33]), but we take a summarizing shortcut here. Instead of the Whittle-Matérn kernel reproducing $W_2^m(\mathbb{R}^d)$, the radial *polyharmonic* kernel

$$H_{m,d}(r) := \left\{ \begin{array}{ll} (-1)^{\lceil m-d/2 \rceil} r^{2m-d}, & 2m-d \text{ odd} \\ (-1)^{1+m-d/2} r^{2m-d} \log r, & 2m-d \text{ even} \end{array} \right\} \tag{20}$$

is taken, up to a scalar multiple

$$\left\{ \begin{array}{ll} \dfrac{\Gamma(m-d/2)}{2^{2m}\pi^{d/2}(m-1)!} & 2m-d \text{ odd} \\ \dfrac{1}{2^{2m-1}\pi^{d/2}(m-1)!(m-d/2)!} & 2m-d \text{ even} \end{array} \right\} \tag{21}$$

that is used to match the seminorm in Sobolev space $W^m(\mathbb{R}^d)$. We allow m to be integer or half-integer. This kernel is *conditionally positive definite* of order $k = \lfloor m-d/2 \rfloor + 1$, and this has the consequence that approximations working in that space must be exact on polynomials of al least that order (= degree plus one). In some sense, this is the price to be paid for omitting the lower order derivatives in the Sobolev norm, but polynomial exactness will turn out to be a good feature, not a bug.

As an illustration for the connection between the polyharmonic kernel $H_{m,d}(r)$ and the Whittle-Matérn kernel $K_{m-d/2}(r)r^{m-d/2}$ reproducing $W_2^m(\mathbb{R}^d)$, we state the observation that (up to constants) the polyharmonic kernel arises as the first term in the expansion of the Whittle-Matérn kernel that is not an even power of r. For instance, up to higher-order terms,

$$K_3(r)r^3 = 16 - 2r^2 + \frac{1}{4}r^4 + \frac{1}{24}r^6 \log(r)$$

containing $H_{4,2}(r) = r^6 \log(r)$ up to a constant. This seems to hold in general for $K_n(r)r^n$ and $n = m - d/2$ for integer n and even dimension d. Similarly,

$$\frac{1}{\sqrt{2\pi}} K_{5/2}(r)r^{5/2} = 3 - \frac{1}{2}r^2 + \frac{1}{8}r^4 - \frac{1}{15}r^5$$

contains $H_{4,3}(r) = r^5$ up to a constant, and this generalizes to half-integer n with $n = m - d/2$. The upshot is that the polyharmonic kernel, if written with $r = \|x-y\|_2$, differs from the Whittle-Matérn kernel only by lower-order polynomials and higher-order terms, being simpler to evaluate. A rigid proof is in [8].

If we have an arbitrary approximation (7) that is exact on polynomials of order k, we can insert its coefficients a_j into the usual quadratic form (19) using the polyharmonic kernel there, and evaluate the error. Clearly, the error is not smaller than the error of the optimal approximation using the polyharmonic kernel, and let us denote the coefficients of the latter by a_j^*.

We now consider *scaling*. Due to shift-invariance, we can assume that we have a homogeneous differential operator of order p that is to be evaluated at the origin, and we use scaled points hx_j for its nodal approximation. It then turns out [8] that the optimal coefficients $a_j^*(h)$ scale like $a_j^*(h) = h^{-p}a_j^*(1)$, and the quadratic form Q of (19) written in terms of coefficients as

$$\begin{aligned} Q^2(a) &= \lambda^x\lambda^y K(x,y) - 2\sum_{j=1}^M a_j(\lambda)\lambda_j^x\lambda^y K(x,y) \\ &+ \sum_{j,k=1}^M a_j(\lambda)a_k(\lambda)\lambda_j^x\lambda_k^y K(x,y) \end{aligned}$$

scales *exactly* like

$$Q(a^*(h)) = h^{2m-d-2p}Q(a^*(1)),$$

proving that *there is no approximation of better order* in that space, no matter how users calculate their approximation. Note that strong methods (i.e. collocation) for second-order PDE problems (2) using functionals (4) have $p = 2$ while the weak functionals of (5) have $p = 1$. This is a fundamental difference between weak and strong formulations, but note that it is easy to have methods of arbitrarily high consistency order.

In practice, any set of given and centralized nodes x_j can be blown up to points Hx_j of average pairwise distance 1. Then the error and the weights can be calculated for the blown-up situation, and then the scaling laws for the coefficients and the error are applied using $h = 1/H$. This works for all scalings, without serious instabilities.

Now that we know an optimal approximation with a simple and stable scaling, why bother with other approximations? They will not have a smaller worst-case consistency error, and they will not always have the scaling property $a_j(h) = h^{-p}a_j(1)$, causing instabilities when evaluating the quadratic form. If they do have that scaling law, then

$$Q(a(h)) = h^{2m-d-2p}Q(a(1)) \geq h^{2m-d-2p}Q(a^*(1)) = Q(a^*(h))$$

can easily be proven, leading to stable calculation for an error that is not smaller than the optimal one. In contrast to standard results on the error of kernel-based approximations, we have no restriction like $h \leq h_0$ here, since the scaling law is exact and holds for all h.

If the smoothness m for error evaluation is *fixed*, it will not pay off to use approximations with higher orders of polynomial exactness, or using kernels with higher smoothness. They cannot beat the optimal approximations for that smoothness class, and the error bounds of these are sharp. Special approximations can be better in a single case, but this paper deals with worst-case bounds, and then the optimal approximations are always superior.

The optimal approximations can be calculated for small numbers of nodes, leading to sparse stiffness matrices. One needs enough points to guarantee polynomial exactness of order $k = \lfloor m - d/2 \rfloor + 1$. The minimal number of points actually needed will depend on their geometric placement. The five-point star is an extremely symmetric example with exactness of order 4 in $d = 2$, but this order will normally need 15 points in general position because the dimension of the space of third-degree polynomials in $\mathbb{R}^2$ is 15.

The upshot of all of this is that, given a fixed smoothness m and a dimension d, polyharmonic stencils yield sparse optimal approximations that can be stably calculated and evaluated. Examples are in [8] and in Sect. 8 below. See [16] for an early work on stability of interpolation by polyharmonic kernels, and [1] for an example of an advanced application.

7 Stability Analysis

We now take a closer look at the stability constant $C_S(\mathbf{A})$ from (10). It can be rewritten as

$$C_S(\mathbf{A}) = \sup\{\|\mathbf{u}\|_p \ : \ \|\mathbf{A}\mathbf{u}\|_q \leq 1\} \tag{22}$$

and thus $2C_S(\mathbf{A})$ is the p-norm diameter of the convex set $\{\mathbf{u} \in \mathbb{R}^M \ : \ \|\mathbf{A}\mathbf{u}\|_q \leq 1\}$. In the case $p = q = \infty$ that will be particularly important below, this set is a polyhedron, and the constant $C_S(\mathbf{A})$ can be calculated via linear optimization. We omit details here, but note that the calculation tends to be computationally unstable and complicated. It is left to future research to provide a good estimation technique for the stability constant $C_S(\mathbf{A})$ like MATLAB's `condest` for estimating the L_1 condition number of a square matrix.

In case $p = q = 2$ we get

$$C_S(\mathbf{A})^{-1} = \min_{1 \leq j \leq M} \sigma_j$$

for the M positive singular values $\sigma_1, \ldots, \sigma_M$ of A, and these are obtainable by *singular value decomposition*.

To simplify the computation, one might calculate the pseudoinverse $\mathbf{A}^\dagger$ of $\mathbf{A}$ and then take the standard (p,q)-norm of it, namely

$$\|\mathbf{A}^\dagger\|_{p,q} := \sup_{\mathbf{v}\neq 0} \frac{\|\mathbf{A}^\dagger \mathbf{v}\|_p}{\|\mathbf{v}\|_q}.$$

This overestimates $C_S(\mathbf{A})$ due to

$$\|\mathbf{A}^\dagger\|_{p,q} \geq \sup_{\mathbf{v}=\mathbf{Au}\neq 0} \frac{\|\mathbf{A}^\dagger \mathbf{Au}\|_p}{\|\mathbf{Au}\|_q} = \sup_{\mathbf{u}\neq 0} \frac{\|\mathbf{u}\|_p}{\|\mathbf{Au}\|_q} = C_S(\mathbf{A})$$

since $C_S(\mathbf{A})$ is the norm of the pseudoinverse not on all of $\mathbb{R}^N$, but restricted to the M-dimensional range of $\mathbf{A}$ in $\mathbb{R}^N$. Here, we again used that $\mathbf{A}$ has full rank, thus $\mathbf{A}^\dagger\mathbf{A} = I_{M\times M}$.

Calculating the pseudoinverse may be as expensive as the numerical solution if the system (8) itself, but if a user wants to have a close grip on the error, it is worth while. It assures stability of the numerical process, if not intolerably large, as we shall see. Again, we hope for future research to produce an efficient estimator.

A simple possibility, restricted to square systems, is to use the fact that MATLAB's `condest` estimates the 1-norm-condition number, which is the L_∞ condition number of the transpose. Thus

$$\tilde{C}_S(\mathbf{A}) := \frac{\texttt{condest}(\mathbf{A}')}{\|\mathbf{A}\|_\infty} \tag{23}$$

is an estimate of the L_∞ norm of $\mathbf{A}^{-1}$. This is computationally very cheap for sparse matrices and turns out to work fine on the examples in Sect. 8, but an extension to non-square matrices is missing.

We now switch to theory and want to show that users can expect $C_S(\mathbf{A})$ to be bounded above independent of the discretization details, if the underlying problem is well-posed. To this end, we use the approach of [29] in what follows.

Well-posed analytic problems of the form (3) allow a stable reconstruction of $u \in U$ from their full set of data $f_\lambda(u)$, $\lambda \in \Lambda$. This *analytic stability* can often be described as

$$\|u\|_{WP} \leq C_{WP} \sup_{\lambda\in\Lambda} |\lambda(u)| \text{ for all } u \in U, \tag{24}$$

where the *well-posedness norm* $\|.\|_{WP}$ usually is weaker than the norm $\|.\|_U$. For instance, elliptic second-order Dirichlet boundary value problems written in strong form satisfy

$$\|u\|_{\infty,\Omega} \leq \|u\|_{\infty,\partial\Omega} + C\|Lu\|_{\infty,\Omega} \text{ for all } u \in U := C^2(\Omega) \cap C(\overline{\Omega}) \tag{25}$$

see e.g. [6, (2.3), p. 14], and this is (24) for $\|.\|_{WP} = \|.\|_\infty$.

The results of [29] then show that for each trial space $U_M \subset U$ one can find a test set Λ_N such that (24) takes a discretized form

$$\|u\|_\infty \leq 2C_{WP} \sup_{\lambda_k \in \Lambda_N} |\lambda_k(u)| \text{ for all } u \in U_M,$$

and this implies

$$|u(x_j)| \leq 2C_{WP} \sup_{\lambda_k \in \Lambda_N} |\lambda_k(u)| \text{ for all } u \in U_M$$

for all nodal values. This proves a uniform stability property of the stiffness matrix with entries $\lambda_k(u_i)$. The functional approximations in [29] were of the form $a_j(\lambda) = \lambda(u_j)$, and then

$$\begin{aligned} \|\mathbf{u}\|_\infty &\leq 2C_{WP} \sup_{\lambda_k \in \Lambda_N} |\lambda_k(u)| \\ &= 2C_{WP} \sup_{\lambda_k \in \Lambda_N} |\lambda_k \left(\sum_{i=1}^M u(x_i) u_i\right)| \\ &= 2C_{WP} \sup_{\lambda_k \in \Lambda_N} |\sum_{i=1}^M u(x_i) \lambda_k(u_i)| \\ &= 2C_{WP} \|\mathbf{Au}\|_\infty \end{aligned}$$

and thus

$$C_S(\mathbf{A}) \leq 2C_{WP}.$$

This is a prototype situation encouraging users to expect reasonably bounded norms of the pseudoinverse, provided that the norms are properly chosen.

However, the situation of [29] is much more special than here, because it is confined to the trial function approach. While we do not even specify trial spaces here, the paper [29] relies on the condition $a_j(\lambda) = \lambda(u_j)$ for a Lagrange basis of a trial space, i.e. exactness of the approximations on a chosen trial space. This is satisfied in nodal methods based on trial spaces, but not in direct nodal methods. In particular, it works for Kansa-type collocation and MLS-based nodal meshless methods, but not for localized kernel approximations and direct MLPG techniques in nodal form.

For general choices of $a_j(\lambda)$, the stability problem is a challenging research area that is not addressed here. Instead, users are asked to monitor the row-sum norm of the pseudoinverse numerically and apply error bounds like (16) for $p = q = \infty$. Note that the choice of discrete L_∞ norms is dictated by the well-posedness inequality (25). As pointed out above, chances are good to observe numerical stability for well-posed problems, provided that test functionals are chosen properly. We shall see this in the examples of Sect. 8. In case of square stiffness matrices, users can apply (23) to get a cheap and fairly accurate estimate of the stability constant.

For problems in weak form, the well-posedness norm usually is not $\|.\|_{\infty,\Omega}$ but $\|.\|_{L_2(\Omega)}$, and then we might get into problems using a nodal basis. In such cases, an L_2-orthonormal basis would be needed for uniform stability, but we refrain from considering weak formulations here.

8 Examples

In all examples to follow, the nodal points are $x_1, \dots, x_M$ in the domain $\Omega = [-1,+1]^2 \subset \mathbb{R}^2$, and parts of them are placed on the boundary. We consider the standard Dirichlet problem for the Laplacian throughout, and use testing points $y_1, \dots, y_n \in \Omega$ for the Laplacian and $z_1, \dots, z_k \in \partial\Omega$ for the Dirichlet boundary data in the sense of (4). Note that in our error bound (16) the right-hand sides of problems like (2) do not occur at all. This means that everything is only dependent on how the discretization works, it does not depend on any specific choice of f and g.

We omit detailed examples that show how the stability constant $C_S(\mathbf{A})$ decreases when increasing the number N of test functionals. An example is in [29], and (22) shows that stability must improve if rows are added to $\mathbf{A}$. Users are urged to make sure that their approximations (6), making up the rows of the stiffness matrix, have roughly the same consistency order, because adding equations will then improve stability without serious change of the consistency error.

We first take regular points on a 2D grid of sidelength h in $\Omega = [-1,+1]^2 \subset \mathbb{R}^2$ and interpret all points as nodes. On interior nodes, we approximate the Laplacian by the usual five-point star which is exact on polynomials up to degree 3 or order 4. On boundary nodes, we take the boundary values as given. This yields a square linear system. Since the coefficients of the five-point star blow up like $O(h^2)$ for $h \to 0$, the row-sum norm of $\mathbf{A}$ and the condition must blow up like $O(h^{-2})$, which can easily be observed. The pseudoinverse does not blow up since the Laplacian part of $\mathbf{A}$ just takes means and the boundary part is the identity. For the values of h we computed, its norm was bounded by roughly 1.3. This settles the stability issue from a practical point of view. Theorems on stability are not needed.

Consistency depends on the regularity space U_S chosen. We have a fixed classical discretization strategy via the five-point star, but we can evaluate the consistency error in different spaces. Table 1 shows the results for Sobolev space $W_2^4(\mathbb{R}^d)$. It clearly shows linear convergence, and its last column has the major part of the worst-case relative error bound (16). The estimate $\tilde{C}_S(\mathbf{A})$ from (23) agrees with $C_S(\mathbf{A})$ to all digits shown. Note that for all methods that need continuous point evaluations of the Laplacian in 2D, one cannot work with less smoothness, because the Sobolev inequality requires $W_2^m(\mathbb{R}^2)$ with $m > 2 + d/2 = 3$. The arguments in Sect. 6 show that the consistency order then is at most $m - d/2 - p = m - 3 = 1$, as observed. Table 2 shows the improvement if one uses the partial matrix $\mathbf{B}$ of Sect. 4.

We now demonstrate the sharpness of our error bounds. We implemented the construction of Sect. 3.6 for $K(\mathbf{A}) = 2$ and the situation in the final row of Table 1. This means that, given $\mathbf{A}$, we picked values of f and g to realize worst-case stability and consistency, with known value vectors $\mathbf{u}^*$ and $\tilde{\mathbf{u}}$. Figure 1 shows the values of $\mathbf{u}_S$ and $\mathbf{u}_j = \mathbf{u}^*$ in the notation of the proof of Theorem 2, while Fig. 2 displays $\tilde{\mathbf{u}}$.

Table 1 Results for five-point star on the unit square, for $W_2^4(\mathbb{R}^2)$ and the full matrix **A**

$M = N$	h	$C_S(\mathbf{A})$	$\|\mathbf{c}\|_\infty$	$C_S(\mathbf{A})\,\|\mathbf{c}\|_\infty$
25	0.5000	1.281250	0.099045	0.126901
81	0.2500	1.291131	0.051766	0.066837
289	0.1250	1.293783	0.026303	0.034030
1089	0.0625	1.294459	0.013222	0.017116

Table 2 Results for five-point star on the unit square, for $W_2^4(\mathbb{R}^2)$ and the partial matrix **B**

$M_I = N_I$	h	$C_S(\mathbf{B})$	$\|\mathbf{c}_I\|_\infty$	$C_S(\mathbf{B})\,\|\mathbf{c}_I\|_\infty$
9	0.5000	0.281250	0.099045	0.027856
49	0.2500	0.291131	0.051766	0.015071
225	0.1250	0.293783	0.026303	0.007727
961	0.0625	0.294459	0.013222	0.003893

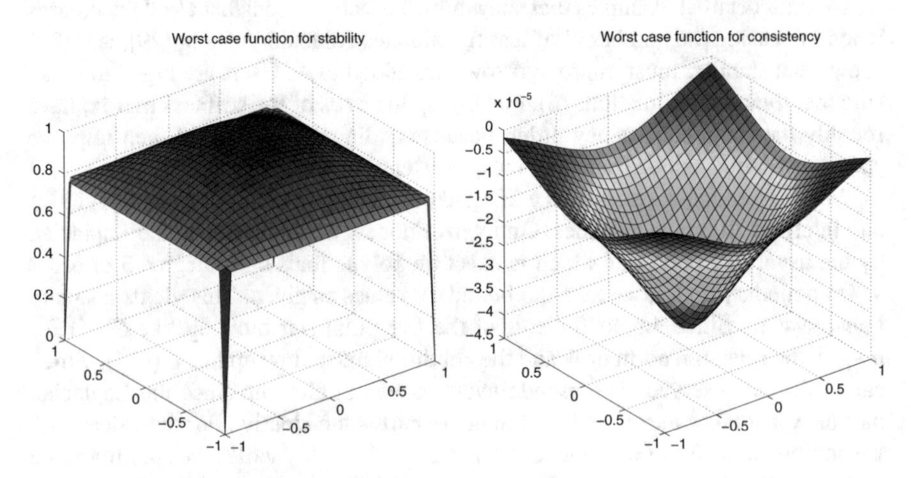

Fig. 1 Stability and consistency worst case

The inequality (17) is in this case

$$0.000226 = C_S(\mathbf{A})\|u^*\|_S\|\mathbf{c}\|_\infty \le \|\mathbf{u}^* - \tilde{\mathbf{u}}\|_\infty = 0.000226 \le 3C_S(\mathbf{A})\|u^*\|_S\|\mathbf{c}\|_\infty = 0.000679$$

and the admissibility inequality (15) is exactly satisfied with $K(\mathbf{A}) = 2$. Even though this example is worst-case, the residuals and the error $\|\mathbf{u}^* - \tilde{\mathbf{u}}\|_\infty$ are small compared to the last line of Table 1, and users might suspect that the table has a useless overestimation of the error. But the explanation is that the above bounds are absolute, not relative, while the norm of the true solution is $\|u^*\|_S = \|\mathbf{c}\|_\infty = 0.0132$. The relative form of the above bound is

$$0.0171 = \frac{\|\mathbf{u}^* - \tilde{\mathbf{u}}\|_\infty}{\|u^*\|_S} \le 0.0513,$$

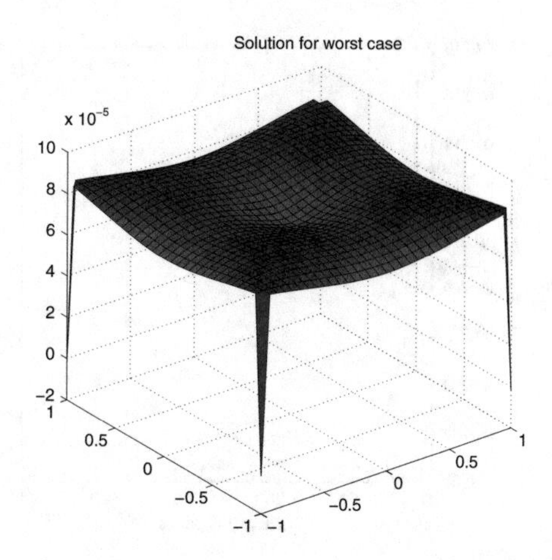

Fig. 2 Solution for joint worst case

showing that the relative error bound 0.0171 in Table 1 is attained by a specific example. Thus our error estimation technique covers this situation well. The lower bound in the worst-case construction is attained because this example has equality in (18).

Note that our constructed case combines worst-case consistency with worst-case stability, but in practical situations these two worst cases will rarely happen at the same time. Figure 1 shows that the worst case for stability seems to be a discretization of a discontinuous function, and therefore it may be that practical situations are systematically far away from the worst case. This calls for a redefinition of the stability constant by restricting the range of **A** in an appropriate way. The worst case for stability arises for vectors of nodal values that are close to the eigenvector of the smallest eigenvalue of **A**, but the worst case for consistency might systematically have small inner products with eigenvectors for small eigenvalues.

If we take the polyharmonic kernel $H_{4,2}(r) = r^6 \log r$ (up to a constant), the five-point star is unique and therefore optimal, with consistency order 1, see Sect. 6. This means that for given smoothness order $m = 4$ and gridded nodes, the five-point star already has the optimal convergence order. Taking approximations of the Laplacian using larger subsets of nodes might be exact on higher-order polynomials, and will have smaller factors if front of the scaling law, but the consistency and convergence *order* will not be better, at the expense of losing sparsity.

To see how much consistency can be gained by using non-sparse optimal approximations by polyharmonic kernels, we worked at $h = 1$, approximating the error of the Laplacian at the origin by data in the integer nodes (m, n) with $-1 \leq m, n \leq K$ for increasing K. This models the case where the Laplacian is approximated in a near-corner point of the square. Smaller h can be handled by the scaling law. The consistency error in $W_2^4(\mathbb{R}^2)$ goes down from 0.07165 to 0.070035 when going from 25 to 225 neighbors (see Fig. 3), while 0.08461 is the error of the five-point star at the origin. The gain is not worth the effort. The optimal stencils decay extremely quickly away from the origin. This is predicted by results

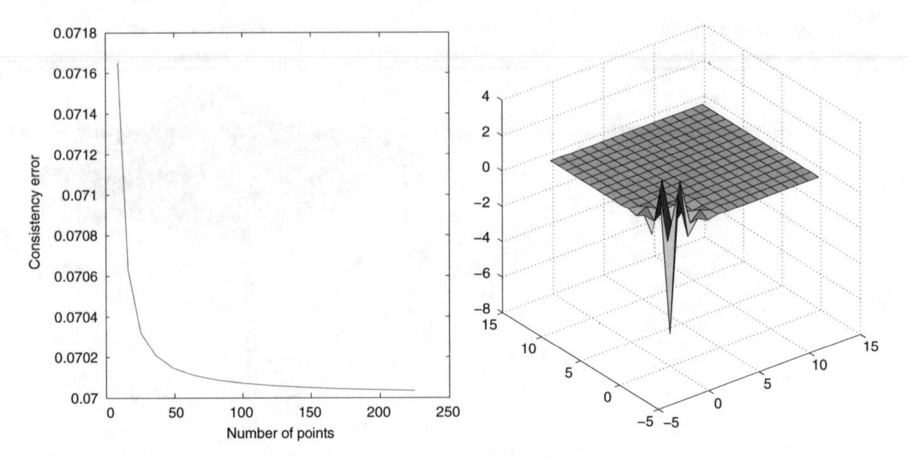

Fig. 3 Consistency error as a function of points offered, and stencil of optimal approximation for 225 nodes, as a function on the nodes

of [20] concerning exponential decay of Lagrangians of polyharmonic kernels, as used successfully in [14] to derive local inverse estimates. See [25] for an early reference on polyharmonic near-Lagrange functions.

We now show how the technique of this paper can be used to compare very different discretizations, while a smoothness order m is fixed, in the sense that the true solution lies in Sobolev space $W_2^m(\Omega)$. Because we have p-methods in mind, we take $m = 6$ for the standard Dirichlet problem for the Laplacian in 2D and can expect an optimal consistency order $m - d/2 - 2 = 3$ for a strong discretization. Weak discretizations will be at least one order better, but we omit such examples. The required order of polynomial exactness when using the polyharmonic kernel is $1 + m - d/2 = 6$, which means that one should use at least 21 nodes for local approximations, if nodes are in general position, without symmetries. The bandwidth of the generalized stiffness matrix must therefore be at least 21. For convenience, we go to the unit square and a regular grid of meshwidth h first, to define the nodes. But then we add uniformly distributed noise of $\pm h/4$ to each interior node, keeping the boundary nodes. Then we approximate the Laplacian at each interior node locally by taking $n \geq 25$ nearest neighbor nodes, including boundary nodes, and set up the reduced generalized square stiffness matrix $\mathbf{B}$ using the optimal polyharmonic approximation based on these neighboring nodes. On the boundary, we keep the given Dirichlet boundary values, following Sect. 4.

Table 3 shows results for local optimal approximations based on the polyharmonic kernel $H_{6,2}(r) = r^{10} \log r$ and $n = 30$ nearest neighbors. The stability constant was estimated via (23), for convenience and efficiency. One cannot expect to see an exact h^3 behavior in the penultimate column, since the nodes are randomly perturbed, but the overall behavior of the error is quite satisfactory. The computational complexity is roughly $O(Nn^3)$, and note that the linear system is not solved at all, because we used MATLAB's `condest`. Comparing with Table 4,

Table 3 Optimal polyharmonic approximations using 30 neighbors

$N = M$	$N_I = M_I$	h	$\tilde{C}_S(\mathbf{B})$	$\|\mathbf{c}_I\|_\infty$	$C_S(\mathbf{B})\|\mathbf{c}_I\|_\infty$
81	49	0.2500	2.3244	0.00075580	0.00175682
289	225	0.1250	0.3199	0.00005224	0.00001671
1089	961	0.0625	0.2964	0.00000872	0.00000259
4225	3969	0.0313	0.2961	0.00000147	0.00000044

Table 4 Optimal polyharmonic approximations using 25 neighbors

$N = M$	$N_I = M_I$	h	$\tilde{C}_S(\mathbf{B})$	$\|\mathbf{c}_I\|_\infty$	$C_S(\mathbf{B})\|\mathbf{c}_I\|_\infty$
81	49	0.2500	8.0180	0.00318328	0.02552351
289	225	0.1250	66.7176	0.00039055	0.02605641
1089	961	0.0625	417.8094	0.00003877	0.01620053
4225	3969	0.0313	75.5050	0.00000663	0.00050082

Table 5 Backslash approximation on 25 neighbors

$N = M$	$N_I = M_I$	h	$\tilde{C}_S(\mathbf{B})$	$\|\mathbf{c}_I\|_\infty$	$C_S(\mathbf{B})\|\mathbf{c}_I\|_\infty$
81	49	0.2500	9.0177	0.00354151	0.03193624
289	225	0.1250	25.6153	0.00058952	0.01510082
1089	961	0.0625	73.9273	0.00005482	0.00405249
4225	3969	0.0313	19.6458	0.00001186	0.00023305

it pays off to use a few more neighbors, and this also avoids instabilities. Users unaware of instabilities might think they can expect a similar behavior as in Table 3 when taking only 25 neighbors, but the third row of Table 4 should teach them otherwise. By resetting the random number generator, all tables were made to work on the same total set of points, but the local approximations still yield rather different results.

The computationally cheapest way to calculate approximations with the required polynomial exactness of order 6 on 25 neighbors is to solve the linear 20×25 system describing polynomial exactness via the MATLAB backslash operator. It will return a solution based on 21 points only, i.e. with minimal bandwidth, but the overall behavior in Table 5 may not be worth the computational savings, if compared to the optimal approximations on 30 neighbors.

A more sophisticated kernel-based *greedy* technique [27] uses between 21 and 30 points and works its way through the offered 30 neighbors to find a compromise between consistency error and support size. Table 6 shows the results, with an average of 23.55 neighbors actually used.

For these examples, one can plot the consistency error as a function of the nodes, and there usually is a factor of 5–10 between the error in the interior and on the boundary. Therefore it should be better to let the node density increase towards the boundary, though this may lead to instabilities that may call for overtesting, i.e. to use $N >> M$. For the same M and N as before, but with Chebyshev point

Table 6 Greedy polyharmonic approximations using at most 30 neighbors

$N = M$	$N_I = M_I$	h	$\tilde{C}_S(\mathbf{B})$	$\|\mathbf{c}_I\|_\infty$	$C_S(\mathbf{B})\|\mathbf{c}_I\|_\infty$
81	49	0.2500	3.6188	0.00104016	0.00376411
289	225	0.1250	0.6128	0.00006821	0.00004180
1089	961	0.0625	0.3061	0.00000961	0.00000294
4225	3969	0.0313	0.2980	0.00000123	0.00000037

Table 7 Greedy polyharmonic approximations using at most 30 neighbors, but in Chebyshev node arrangement

$N = M$	$N_I = M_I$	h	$\tilde{C}_S(\mathbf{B})$	$\|\mathbf{c}_I\|_\infty$	$C_S(\mathbf{B})\|\mathbf{c}_I\|_\infty$
81	49	0.2500	111.1016	0.00433490	0.48161488
289	225	0.1250	0.4252	0.00006541	0.00002781
1089	961	0.0625	1.2133	0.00000677	0.00000821
4225	3969	0.0313	0.4353	0.00000120	0.00000052

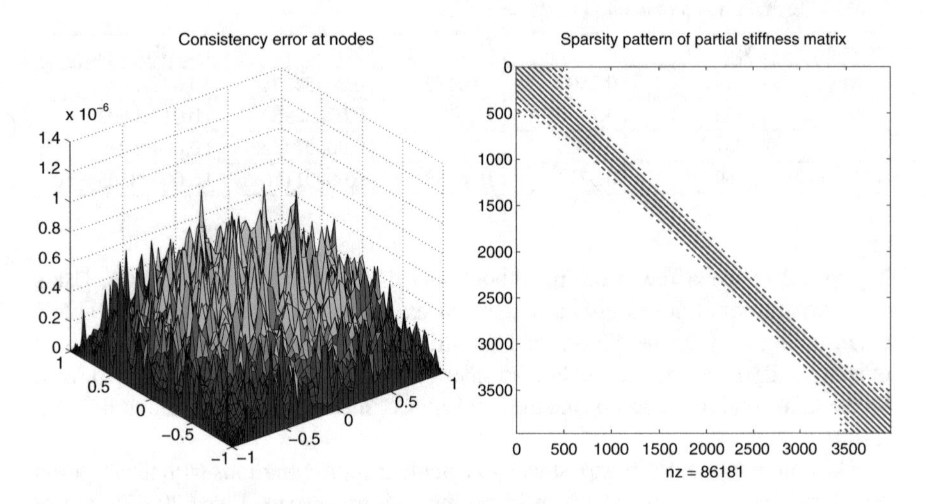

Fig. 4 Consistency plot and stiffness matrix **B** for Chebyshev situation

distribution, see Table 7. The additive noise on the interior points was 0.01, and we used the greedy method for up to 30 neighbors. This leads to a larger bandwidth near the corners, and to a consistency error that is now small at the boundary, see Fig. 4. The average number of neighbors used was 23.3. Unfortunately, the scaling laws of stencils go down the drain here, together with the proven consistency order, but the results are still unexpectedly good.

For reasons of space and readability, we provide no examples for local approximations to weak functionals, and no comparisons with local approximations obtained via Moving Least Squares or the Direct Meshless Petrov Galerkin Method.

9 Conclusion and Outlook

The tables of the preceding section show that the numerical calculation of relative error bounds for PDE solving in spaces of fixed Sobolev smoothness can be done efficiently and with good results. This provides a general tool to evaluate discretizations in a worst-case scenario, without referring to single examples and complicated theorems. Further examples should compare a large variety of competing techniques, the comparison being fair here as long as the smoothness m is fixed.

Users are strongly advised to use the cheap stability estimate (23) **anytime** to assess the stability of their discretization, if they have a square stiffness matrix. And, if they are not satisfied with the final accuracy, they should evaluate and plot the consistency error like in Fig. 4 to see where the discretization should be refined. For all of this, polyharmonic kernels are an adequate tool.

It is left to future research to investigate and improve the stability estimation technique via (23), and, if the effort is worth while, to prove general theorems on sufficient criteria for stability. These will include assumptions on the placement of the trial nodes, as well as on the selection of sufficiently many and well-placed test functionals. In particular, stabilization by overtesting should work in general, but the examples in this paper show that overtesting may not be necessary at all. However, this paper serves as a practical workaround, as long as there are no theoretical cutting-edge results available.

Acknowledgements This work was strongly influenced by helpful discussions and e-mails with Oleg Davydov and Davoud Mirzaei.

References

1. T. Aboiyar, E.H. Georgoulis, A. Iske, Adaptive ADER methods using kernel-based polyharmonic spline WENO reconstruction. SIAM J. Sci. Comput. **32**, 3251–3277 (2010)
2. M.G. Armentano, Error estimates in Sobolev spaces for moving least square approximations. SIAM J. Numer. Anal. **39**(1), 38–51 (2001)
3. M.G. Armentano, R.G. Durán, Error estimates for moving least square approximations. Appl. Numer. Math. **37**, 397–416 (2001)
4. S.N. Atluri, T.-L. Zhu, A new meshless local Petrov-Galerkin (MLPG) approach in computational mechanics. Comput. Mech. **22**, 117–127 (1998)
5. T. Belytschko, Y. Krongauz, D.J. Organ, M. Fleming, P. Krysl, Meshless methods: an overview and recent developments. Comput. Methods Appl. Mech. Eng. Spec. Issue **139**, 3–47 (1996)
6. D. Braess, *Finite Elements. Theory, Fast Solvers and Applications in Solid Mechanics*, 2nd edn. (Cambridge University Press, Cambridge, 2001)
7. O. Davydov, R. Schaback, Error bounds for kernel-based numerical differentiation. Numer. Math. **132**, 243–269 (2016)
8. O. Davydov, R. Schaback, Optimal stencils in Sobolev spaces, preprint 2016

9. J. Duchon, Splines minimizing rotation-invariate semi-norms in Sobolev spaces, in *Constructive Theory of Functions of Several Variables*, ed. by W. Schempp, K. Zeller (Springer, Berlin/Heidelberg, 1979), pp. 85–100
10. G. Fasshauer, Solving partial differential equations by collocation with radial basis functions, in *Surface Fitting and Multiresolution Methods*, ed. by A. LeMéhauté, C. Rabut, L.L. Schumaker (Vanderbilt University Press, Nashville, 1997), pp. 131–138
11. N. Flyer, E. Lehto, S. Blaise, G.B. Wright, A. St.-Cyr, A guide to RBF-generated finite differences for nonlinear transport: shallow water simulations on a sphere, preprint 2015
12. C. Franke, R. Schaback, Convergence order estimates of meshless collocation methods using radial basis functions. Adv. Comput. Math. **8**, 381–399 (1998)
13. C. Franke, R. Schaback, Solving partial differential equations by collocation using radial basis functions. Appl. Math. Comput. **93**, 73–82 (1998)
14. T. Hangelbroek, F.J. Narcowich, C. Rieger, J.D. Ward, An inverse theorem for compact Lipschitz regions using localized kernel bases, arXiv preprint arXiv:1508.02952v2 (2015)
15. Y.C. Hon, R. Schaback, On unsymmetric collocation by radial basis functions. Appl. Math. Comput. **119**, 177–186 (2001); MR MR1823674
16. A. Iske, On the approximation order and numerical stability of local Lagrange interpolation by polyharmonic splines, in *Modern Developments in Multivariate Approximation* (Birkhäuser, Basel, 2003), pp. 153–165
17. E.J. Kansa, Application of Hardy's multiquadric interpolation to hydrodynamics, in *Proceedings of the 1986 Summer Computer Simulation Conference*, vol. 4, pp. 111–117 (1986)
18. D.W. Kim, Y. Kim, Point collocation methods using the fast moving least-square reproducing kernel approximation. Int. J. Numer. Methods Eng. **56**, 1445–1464 (2003)
19. E. Larsson, E. Lehto, A. Heryodono, B. Fornberg, Stable computation of differentiation matrices and scattered node stencils based on Gaussian radial basis functions. SIAM J. Sci. Comput. **35**, A2096–A2119 (2013)
20. O.V. Matveev, Spline interpolation of functions of several variables and bases in Sobolev spaces. Trudy Mat. Inst. Steklov **198**, 125–152 (1992)
21. D. Mirzaei, R. Schaback, Direct Meshless Local Petrov-Galerkin (DMLPG) method: a generalized MLS approximation. Appl. Numer. Math. **68**, 73–82 (2013)
22. D. Mirzaei, R. Schaback, M. Dehghan, On generalized moving least squares and diffuse derivatives. IMA J. Numer. Anal. **32**(3), 983–1000 (2012). doi:10.1093/imanum/drr030
23. A.R. Mitchell, D.F. Griffiths, *The Finite Difference Method in Partial Differential Equations* (Wiley, Chichester, 1980), p. 233
24. B. Nayroles, G. Touzot, P. Villon, Generalizing the finite element method: diffuse approximation and diffuse elements. Comput. Mech. **10**, 307–318 (1992)
25. C. Rabut, Elementary M-harmonic cardinal B-splines. Numer. Algorithms **2**, 39–62 (1992)
26. B. Šarler, From global to local radial basis function collocation method for transport phenomena, in *Advances in Meshfree Techniques*. Computational Methods in Applied Sciences, vol. 5 (Springer, Dordrecht, 2007), pp. 257–282; MR 2433133 (2009i:65232)
27. R. Schaback, Greedy sparse linear approximations of functionals from nodal data. Numer. Algorithms **67**, 531–547 (2014)
28. R. Schaback, A computational tool for comparing all linear PDE solvers. Adv. Comput. Math. **41**, 333–355 (2015)
29. R. Schaback, All well–posed problems have uniformly stable and convergent discretizations. Numer. Math. **132**, 243–269 (2015)
30. R. Schaback, H. Wendland, Using compactly supported radial basis functions to solve partial differential equations, in *Boundary Element Technology XIII*, ed. by C.S. Chen, C.A. Brebbia, D.W. Pepper (WIT Press, Southampton/Boston, 1999), pp. 311–324
31. A.I. Tolstykh, On using radial basis functions in a "finite difference mode" with applications to elasticity problems. Comput. Mech. **33**, 68–79 (2003)
32. H. Wendland, Local polynomial reproduction and moving least squares approximation. IMA J. Numer. Anal. **21**, 285–300 (2001)
33. H. Wendland, *Scattered Data Approximation* (Cambridge University Press, Cambridge, 2005)

34. G.B. Wright, B. Fornberg, Scattered node compact finite difference-type formulas generated from radial basis functions. J. Comput. Phys. **212**(1), 99–123 (2006); MR MR2183606 (2006j:65320)
35. G.M. Yao, S. ul Islam, B. Šarler, A comparative study of global and local meshless methods for diffusion-reaction equation. CMES Comput. Model. Eng. Sci. **59**(2), 127–154 (2010); MR 2680809
36. G.M. Yao, B. Šarler, C.S. Chen, A comparison of three explicit local meshless methods using radial basis functions. Eng. Anal. Bound. Elem. **35**(3), 600–609 (2011); MR 2753822 (2012b:65150)

Generalizations of Simple Kriging Methods in Spatial Data Analysis

Qi Ye

Abstract In this article, we use the theory of meshfree approximation to generalize the simple kriging methods by kernel-based probabilities. The main idea is that the new kriging estimations are modeled by the Gaussian fields indexed by bounded linear functionals defined on Sobolev spaces. Moreover, the covariances of the Gaussian fields at the observed functionals can be computed by the given covariance kernels with respect to the related functionals, for example, Gaussian kernels evaluated at points and gradients. This guarantees that the generalized kriging estimations can be obtained by the same techniques of the simple kriging methods and the generalized kriging estimations can cover many kinds of the complex observed information. By the generalized kriging methods, we can model the geostatistics with the additional observations of gradients at the uncertain locations.

1 Introduction

The kriging method is a modern statistical tool to recover values at unknown locations by observed data, for example, application in geostatistics in [5, 14, 15]. In this article, we will generalize the simple kriging methods combining with the knowledge and techniques of approximation theory and numerical analysis in [4, 6, 16].

In the simple kriging methods, we will model the kriging estimations by the Gaussian field S with the mean 0 and the given covariance kernel K, for example, a Gaussian kernel

$$K(\boldsymbol{x},\boldsymbol{y}) := e^{-\theta^2\|\boldsymbol{x}-\boldsymbol{y}\|_2^2}, \quad \text{for } \boldsymbol{x},\boldsymbol{y} \in \mathbb{R}^d,$$

Q. Ye (✉)
School of Mathematical Sciences, South China Normal University, Guangzhou, Guangdong 510631, China
e-mail: yeqi@m.scnu.edu.cn

M. Griebel, M.A. Schweitzer (eds.), *Meshfree Methods for Partial Differential Equations VIII*, Lecture Notes in Computational Science and Engineering 115,
DOI 10.1007/978-3-319-51954-8_8

with a shape parameter $\theta > 0$. Generally speaking, we observe the locations

$$\boldsymbol{x}_1, \ldots, \boldsymbol{x}_n,$$

to obtain the observed data values

$$f_1, \ldots, f_n,$$

which can be viewed as the realizations of the multivariate normal random variables

$$S_{\boldsymbol{x}_1}, \ldots, S_{\boldsymbol{x}_n}, \tag{1}$$

that is, the discretization of the Gaussian field S at the data points $\boldsymbol{x}_1, \ldots, \boldsymbol{x}_n$. Thus, we can compute the prediction at the unknown location $\boldsymbol{x}_0$ by the covariance kernel K in Eqs. (15)–(16). We know that the kriging estimation can be also seen as a spline function for the interpolation. In approximation theory, the interpolation is related to the point evaluation functions

$$\delta_{\boldsymbol{x}_1}, \ldots, \delta_{\boldsymbol{x}_n}.$$

Here, the point evaluation function $\delta_{\boldsymbol{x}}$ is defined by $\delta_{\boldsymbol{x}} f := f(\boldsymbol{x})$. Thus, we can rewrite the multivariate normal random variables in Eq. (1) as

$$\delta_{\boldsymbol{x}_1} S, \ldots, \delta_{\boldsymbol{x}_n} S, \tag{2}$$

In this article, we have a new idea to generalize the simple kriging models such as the generalization of classical interpolation to Hermite-Birkhoff interpolation. Since the Sobolev imbedding theorem guarantees that the point evaluation function $\delta_{\boldsymbol{x}}$ is a bounded linear functional L defined on the Sobolev spaces, we will extend the multivariate normal random variables in Eq. (2) into another multivariate normal random variables which are introduced by the general bounded linear functionals

$$L_1, \ldots, L_n$$

defined on the Sobolev spaces, that is,

$$L_1 S, \ldots, L_n S. \tag{3}$$

Here, we call a real-scalar linear operator a linear functional, for example, a derivative and a integral in Remark 3. In Theorem 1, we have the Gaussian field LS indexed by the bounded linear functional L of which covariances can be computed by the given positive definite kernel K with the related functional L. We can observe that the covariance matrix $\mathsf{A}_{K,\mathcal{L}}$ of $L_1 S, \ldots, L_n S$ in Eq. (8) is the generalization of the classical covariance matrix $\mathsf{A}_{K,X}$ of $S_{\boldsymbol{x}_1}, \ldots, S_{\boldsymbol{x}_n}$ in Eq. (5). This provides that the covariances of the multivariate normal random variables in Eq. (3) are known to construct the generalized kriging models by the same techniques of simple kriging method and Bayesian estimation; hence we can obtain the new kernel-based estimators given in Eq. (11) and Eqs. (12)–(13) to model the generalized kriging estimation.

Moreover, we can also apply the generalized kriging methods into the geostatistics with more kinds of observed information at uncertain data points in Sect. 4. To be more precise, the uncertain data points mean that the observed data values are missed at these locations. Sometimes we may still have the increasing or decreasing rates at some uncertain data points. For the typical example, we can not observe the heights at some locations of the mountains while we know whether the heights increase or decrease along some directions at these locations, that is, the mathematical model is that the value $f(\boldsymbol{x})$ is unknown at the observed point $\boldsymbol{x}$ while the range of the gradient $\nabla f(\boldsymbol{x})$ along the direction $\boldsymbol{e}$ is given such as $\boldsymbol{e}^T\nabla f(\boldsymbol{x}) \geq 0$ or $\boldsymbol{e}^T\nabla f(\boldsymbol{x}) \leq 0$, where $\nabla := \left(\frac{\partial}{\partial x_1}, \cdots, \frac{\partial}{\partial x_d}\right)^T$. This shows that the new geostatistical model will be set up by the Gaussian fields indexed by the bounded linear functionals composed of $L := \delta_{\boldsymbol{x}}$ and $L := \delta_{\boldsymbol{x}} \circ \boldsymbol{e}^T\nabla$ same as the constructions of the generalized kriging estimations, that is, Eqs. (32)–(33). Different from the classical geostatistical models, the covariance matrix in Eq. (20) is also set up by the gradient and the preconditioned observed data values in Eq. (31) are the averages of possible observed data values by kernel-based probabilities. By the numerical example in Sect. 5, we find that the generalized kriging estimation is better than the simple kriging estimation. In this article, we mainly show the big picture of the new idea of the kriging methods. So, we only discuss the comparisons of the simple kriging methods and the classical geostatistics. In our next research proposals, we will investigate another kriging methods by the same ideas and methods shown here.

2 Initial Ideas

Let us look at a simple example of the one-dimensional interpolation in Fig. 1 such as we have the observed data values $f_1, \ldots, f_7 \in \mathbb{R}$ at the data points $x_1, \ldots, x_7 \in [0, 1] \subseteq \mathbb{R}$. Our target is to predict the value at the unknown location $x_0 \in [0, 1]$ based on the interpolation conditions by the statistical and numerical techniques.

In the spatial statistics, we can predict the value at the unknown location x_0 by the simple kriging method in [14], that is, the simple kriging model set up by the Gaussian field S with the mean 0 and the covariance kernel K, where K is a Gaussian kernel defined on $[0, 1] \times [0, 1]$.

Remark 1 A stochastic field $S : \mathcal{D} \times \Omega \to \mathbb{R}$ defined on a probability space $(\Omega, \mathcal{F}, \mathbb{P})$ is called a Gaussian field with a mean 0 and a covariance kernel $K : \mathcal{D} \times \mathcal{D} \to \mathbb{R}$ if, for any $\boldsymbol{x} \in \mathcal{D} \subseteq \mathbb{R}^d$, the random variable $S_{\boldsymbol{x}}$ is a normal random variable with the mean 0 and the variance $K(\boldsymbol{x}, \boldsymbol{x})$ in [3, Definition 3.28].

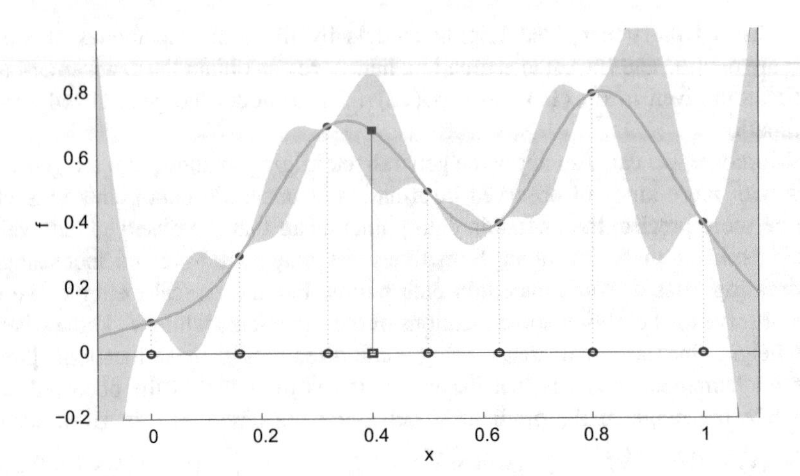

Fig. 1 The 1D example: The *red circles* represent the observed data values $f_1, \ldots, f_7$ at the data points $x_1, \ldots, x_7$ shown in the *blue circles*, and the *blue square* represents the unknown location x_0. The covariance kernel of the Gaussian field S is the Gaussian kernel K with the shape parameter $\theta = 6$. The best prediction $s(x_0)$ of S_{x_0} conditioned on the observed data is shown in the *red square*. The kriging estimations shown in the *green curve* run along the means of the normally distributed confidence intervals of 99% shown in *gray*

For example in Fig. 1, the simple kriging method provides the best linear unbiased prediction $s(x_0)$, that is,

$$s(x_0) := \mathrm{E}\left(S_{x_0} | S_{x_1} = f_1, \ldots, S_{x_7} = f_7\right) = \sum_{k=1}^{7} w_k(x_0) f_k,$$

where the elements $w_1(x_0), \ldots, w_7(x_0)$ are uniquely solved by the linear system

$$\begin{pmatrix} K(x_1, x_1) & \cdots & K(x_1, x_7) \\ & \ddots & \\ K(x_7, x_1) & \cdots & K(x_7, x_7) \end{pmatrix} \begin{pmatrix} w_1(x_0) \\ \vdots \\ w_7(x_0) \end{pmatrix} = \begin{pmatrix} K(x_0, x_1) \\ \vdots \\ K(x_0, x_7) \end{pmatrix}.$$

Recently, the meshfree methods (radial basis functions) give a numerical tool to construct the kernel-based interpolant u by the non-polynomial basis in [6, 16], for example, the kernel-based interpolant u is a linear combination of the Gaussian-kernel basis, that is,

$$u(x) := \sum_{k=1}^{7} c_k K(x, x_k), \quad \text{for } x \in [0, 1],$$

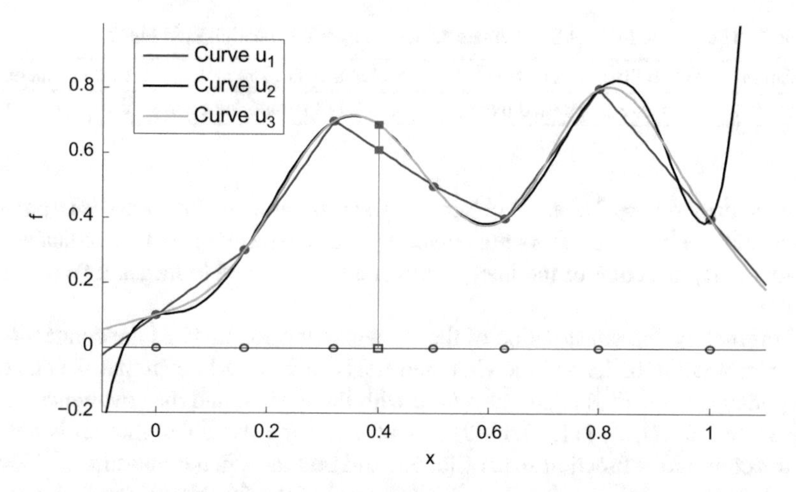

Fig. 2 The example of the initial ideas: The observed data $f_1, \ldots, f_7$ and $x_1, \ldots, x_7$ (*red* and *blue circles*) are the same as in Fig. 1. The *blue*, *green*, and *black curves* represent the piecewise linear spline u_1, the polynomial interpolant u_2, and the kernel-based interpolant u_3, respectively. The *red* and *pink squares* represent the different estimate values v_1, v_2 at the location x_0

where the coefficients $c_1, \ldots, c_7$ are uniquely solved by the linear system

$$\begin{pmatrix} K(x_1, x_1) & \cdots & K(x_1, x_7) \\ & \ddots & \\ K(x_7, x_1) & \cdots & K(x_7, x_7) \end{pmatrix} \begin{pmatrix} c_1 \\ \vdots \\ c_7 \end{pmatrix} = \begin{pmatrix} f_1 \\ \vdots \\ f_7 \end{pmatrix}.$$

We also find that $u(x_0)$ is equal to $s(x_0)$ for any $x_0 \in [0, 1]$ (see the green curve in Fig. 1).

Moreover, the paper [13] and the book [7] show that the formulas of the simple kriging method and the meshfree approximation are the same for the general positive definite kernels. Based on the new discoveries of Scheuerer et al. [13], we renew the kernel-based methods combing with the knowledge of statistics and probability, stochastic analysis, approximation theory, and numerical analysis in the recent papers [17, 18].

Now let us look at the initial idea of this article in Fig. 2. In numerical analysis, many choices of the continuous functions satisfy the interpolations such as the piecewise linear spline u_1, the polynomial interpolant u_2, and the kernel-based interpolant u_3 in Fig. 2; hence we will have the different estimate values, for example, the values v_1, v_2 at the unknown location x_0 in Fig. 2. In stochastic analysis, we will view the interpolating paths u_1, u_2, u_3 as the sample events. Thus, the estimate values v_1, v_2 are supported by the interpolating paths u_1, u_2, u_3 such that the probabilities p_1, p_2 at v_1, v_2 are endowed with $2/3, 1/3$, respectively in Table 1. Based on the similar techniques of the Bayesian estimation in [2], the best estimator $\hat{v}$ is the average of v_1, v_2 weighted by the probabilities p_1, p_2, that

Table 1 The probabilities of the estimate values v_1, v_2 at the location x_0 in Fig. 2

Location	Probability at v_1 in red	Probability at v_2 in pink	Best estimator
$x_0 := 0.4$	$p_1 := 2/3$ (supported by u_2, u_3)	$p_2 := 1/3$ (supported by u_1)	$\hat{v} := v_1 p_1 + v_2 p_2$

is, $\hat{v} := v_1 p_1 + v_2 p_2$. Here, we obtain the best estimators by the global interpolating paths and the best estimators are solved by the local averages. This indicates that a probability structure of the interpolating paths is needed to measure the estimate values.

Fortunately, the construction of the Brownian motion inspires the connection of the interpolating paths and the Gaussian fields. It is well-known that the standard Brownian motion W is a Gaussian field with the mean 0 and the covariance kernel $K(t,s) := \min\{t,s\}$. [11, Chap. 2] provides that the Brownian motion is defined on the continuous function space $\mathrm{C}[0,\infty)$ and that the Wiener measure $\mathbb{P}_*$ is well-posed on the sample space $(\Omega_*, \mathcal{F}_*)$ composed of the function space $\mathrm{C}[0,\infty)$ and the Borel σ-algebra $\mathcal{B}(\mathrm{C}[0,\infty))$. By [11, Theorem 4.20], the coordinate mapping process $W_t(\omega) := \omega(t)$ for $t \in [0,\infty)$ and $\omega \in \Omega_*$ is a standard Brownian motion on the probability space $(\Omega_*, \mathcal{F}_*, \mathbb{P}_*)$. Moreover, we find that the initial condition $Y_0 = y_0$ of the simple stochastic ordinary differential equation $\mathrm{d}Y_t = \mathrm{d}W_t$ is equivalent to the interpolation at the origin. This indicates that we can extend the interpolating paths u_1, u_2, u_3 in Fig. 2 to all interpolating paths in $\mathrm{C}[0,\infty)$ such that

$$\mathcal{A}(f) := \{\omega \in \mathrm{C}[0,\infty) : \omega(x_1) = f_1, \ldots, \omega(x_7) = f_7\},$$

can be equivalently measured by the multivariate normal random variables

$$W_{x_1}, \ldots, W_{x_7}.$$

Then we can make a connection of the interpolations and the Gaussian fields, that is,

$$\mathcal{A}(f) = \{\omega \in \Omega_* : W_{x_1}(\omega) = f_1, \ldots, W_{x_7}(\omega) = f_7\}. \tag{4}$$

In numerical analysis, we can extend the interpolations at the points to the Hermite-Birkhoff interpolations at the derivatives, for example, meshfree approximation for partial differential equations in [6]. The idea of Eq. (4) let us generalize the kriging methods for the Hermite-Birkhoff data with the differential operators similar as the equivalent formulas of the kriging methods and the meshfree approximation in the previous example. In the papers [17, 18], we extend the initial idea in Fig. 2 to all interpolating paths in the Sobolev spaces such that the generalized Hermite-Birkhoff interpolations are connected to the Gaussian fields indexed by the bounded linear functionals in Theorem 1. Therefore, we can obtain the best estimators by Gaussian fields at the observed linear functionals in the following sections.

3 Kernel-Based Approximations via Kernel-Based Probabilities

In this section, we will study the kernel-based approximation by the kernel-based probabilities. These kernel-based probabilities are introduced by the positive definite kernels such that we can construct the normal random variables on the Sobolev spaces by the bounded linear functionals.

Remark 2 The positive definite kernel $K : \mathcal{D} \times \mathcal{D} \to \mathbb{R}$ is defined same as in [16, Definition 6.24], that is, the quadratic form $\sum_{j,k=1}^{n} c_j c_k K(\boldsymbol{x}_j, \boldsymbol{x}_k) > 0$, for any $n \in \mathbb{N}$, any distinct points $X := \{\boldsymbol{x}_1, \ldots, \boldsymbol{x}_n\} \subseteq \mathcal{D}$, and any $\boldsymbol{c} := (c_1, \cdots, c_n)^T \in \mathbb{R}^n \setminus \{\boldsymbol{0}\}$. This is equivalent that all matrixes

$$\mathsf{A}_{K,X} := \begin{pmatrix} K(\boldsymbol{x}_1, \boldsymbol{x}_1) & \cdots & K(\boldsymbol{x}_1, \boldsymbol{x}_n) \\ & \ddots & \\ K(\boldsymbol{x}_n, \boldsymbol{x}_1) & \cdots & K(\boldsymbol{x}_n, \boldsymbol{x}_n) \end{pmatrix}, \tag{5}$$

are strictly positive definite.

Firstly, we review the theorems of the constructions of normal random variables over kernel-based probabilities in the recent paper [18].

Theorem 1 ([18, Theorem 2.1]) *Suppose that $\mathcal{D} \subseteq \mathbb{R}^d$ is a regular and compact domain and the positive definite kernel $K \in \mathrm{C}^{2m,1}(\mathcal{D} \times \mathcal{D})$ for $m > d/2$. Let L be a bounded linear functional on the L_2-based Sobolev space $\mathcal{H}^m(\mathcal{D})$. Then there exists a probability measure $\mathbb{P}_K$ on the measurable space*

$$(\Omega_m, \mathcal{F}_m) := (\mathcal{H}^m(\mathcal{D}), \mathcal{B}(\mathcal{H}^m(\mathcal{D}))),$$

such that the normal random variable

$$LS(\omega) := L\omega, \quad \text{for } \omega \in \Omega_m,$$

is well-defined on the probability space $(\Omega_m, \mathcal{F}_m, \mathbb{P}_K)$ and that this random variable LS has the mean 0 and the variance $L_{\boldsymbol{x}} L_{\boldsymbol{y}} K(\boldsymbol{x}, \boldsymbol{y})$. Moreover, the probability measure $\mathbb{P}_K$ is independent of the bounded linear functional L.

Remark 3 In Theorem 1, the Sobolev space $\mathcal{H}^m(\mathcal{D})$ is endowed with the inner product $(\omega_1, \omega_2)_{\mathcal{H}^m(\mathcal{D})} := \sum_{|\boldsymbol{\alpha}| \leq m} \int_{\mathcal{D}} D^{\boldsymbol{\alpha}} \omega_1(\boldsymbol{x}) D^{\boldsymbol{\alpha}} \omega_2(\boldsymbol{x}) \mathrm{d}\boldsymbol{x}$, the collection $\mathcal{B}(\mathcal{H}^m(\mathcal{D}))$ represents the Borel σ-algebra in the Sobolev space $\mathcal{H}^m(\mathcal{D})$, the element $\omega \in \Omega_m$ represents the sample path (trajectory), and the space $\mathrm{C}^{2m,1}(\mathcal{D} \times \mathcal{D}) \subseteq \mathrm{C}^{2m}(\mathcal{D} \times \mathcal{D})$ consists of all functions which have the continuous derivatives up to the order $2m$ and of which the $2m$th partial derivatives satisfy the Lipschitz condition. A linear functional $L : \mathcal{H}^m(\mathcal{D}) \to \mathbb{R}$ is called bounded if there exists a constant $C > 0$ such that $|L\omega| \leq C \|\omega\|_{\mathcal{H}^m(\mathcal{D})}$ for all $\omega \in \mathcal{H}^m(\mathcal{D})$. Its equivalent concept is that a bounded linear functional L is continuous on $\mathcal{H}^m(\mathcal{D})$. Moreover, the notations L_x and L_y denote the linear functional L associated to the first and second arguments of $\boldsymbol{x}$ and $\boldsymbol{y}$, respectively, that is, $L_xK(\boldsymbol{x},\boldsymbol{y}) = L(K(\cdot,\boldsymbol{y}))$ and $L_yK(\boldsymbol{x},\boldsymbol{y}) = L(K(\boldsymbol{x},\cdot))$.

In Theorem 1, the Sobolev space $\mathcal{H}^m(\mathcal{D})$ and the Borel σ-algebra $\mathcal{B}(\mathcal{H}^m(\mathcal{D}))$ are thought as the sample space $(\Omega_m, \mathcal{F}_m)$ of the probability space, and this $(\Omega_m, \mathcal{F}_m)$ is endowed with the probability measure $\mathbb{P}_K$. Since the Sobolev space $\mathcal{H}^m(\mathcal{D})$ and its dual space $\mathcal{H}^m(\mathcal{D})'$ are isometrically isomorphic, there exists a unique element $g_L \in \mathcal{H}^m(\mathcal{D})$, which is equivalent to the given $L \in \mathcal{H}^m(\mathcal{D})'$, such that

$$LS(\omega) = L\omega = (\omega, g_L)_{\mathcal{H}^m(\mathcal{D})}, \quad \text{for } \omega \in \Omega_m = \mathcal{H}^m(\mathcal{D});$$

hence the probability measure $\mathbb{P}_K$ is also a Gaussian measure in [3, Definition 3.29]. Moreover, the probability measure $\mathbb{P}_K$ is only dependent of the positive definite kernel K. When the kernel K is fixed, then the probability space $(\Omega_m, \mathcal{F}_m, \mathbb{P}_K)$ will not be changed by any $L \in \mathcal{H}^m(\mathcal{D})'$. In another hand, the probability distributions of the normal random variable LS are affected by the linear functional L and the kernel K, for example, the variance of LS. So, we can call $\mathbb{P}_K$ the *kernel-based probability measure* of K.

In this article, we will not discuss the choices of the best kernels and the positive definite kernel $K \in \mathrm{C}^{2m,1}(\mathcal{D} \times \mathcal{D})$ is always given and fixed such that the probability measure $\mathbb{P}_K$ is uniquely defined on the Sobolev space $\mathcal{H}^m(\mathcal{D})$ by Theorem 1. Here, the degree m is always larger than $d/2$ for the Sobolev imbedding theorem [1, Theorem 4.12].

Let the collection $\mathcal{G}$ be composed of all normal random variables LS given in Theorem 1, that is, $\mathcal{G} := \{LS : L \in \mathcal{H}^m(\mathcal{D})'\}$. Clearly, the dual space $\mathcal{H}^m(\mathcal{D})'$ is a Hilbert spaces; hence $\mathcal{G}$ is a Gaussian Hilbert space and the linear isometry $L \mapsto LS$ is a Gaussian field indexed by the Hilbert space $\mathcal{H}^m(\mathcal{D})'$ in [10, Definition 1.18 and 1.19]. By the Sobolev imbedding theorem, the point evaluation function $\delta_{\boldsymbol{x}}$ at any $\boldsymbol{x} \in \mathcal{D}$ is a bounded linear functional on $\mathcal{H}^m(\mathcal{D})$, where $\delta_{\boldsymbol{x}}\omega = \omega(\boldsymbol{x})$ for all $\omega \in \mathcal{H}^m(\mathcal{D})$. Thus, the normal random variable $\delta_{\boldsymbol{x}}S$ is well-defined for any $\boldsymbol{x} \in \mathcal{D}$. We can also observe that $\{\delta_{\boldsymbol{x}}S : \boldsymbol{x} \in \mathcal{D}\}$ is equivalent to a classical Gaussian field with the mean 0 and the covariance kernel K. This indicates that the Gaussian fields indexed by the bounded linear functionals give a new tool to construct the generalized kriging models.

Kernel-Based Approximation: Now let us look at the generalized Hermite-Birkhoff interpolation. Suppose that we have the observed data values

$$f_1, \ldots, f_n \in \mathbb{R},$$

evaluated by some unknown function $f \in \mathcal{H}^m(\mathcal{D})$ at the bounded linear functionals

$$L_1, \ldots, L_n \in \mathcal{H}^m(\mathcal{D})',$$

on the Sobolev space $\mathcal{H}^m(\mathcal{D})$, that is,

$$f_1 := L_1 f, \ldots, f_n := L_n f.$$

Denote that the vector data value and the vector bounded linear operator

$$\boldsymbol{f} := (f_1, \cdots, f_n)^T \in \mathbb{R}^n, \quad \mathcal{L} := (L_1, \cdots, L_n)^T \in \bigotimes_{k=1}^{n} \mathcal{H}^m(\mathcal{D})'.$$

Then $\mathcal{L}f = \boldsymbol{f}$. By the observed data $\mathcal{L}$ and $\boldsymbol{f}$, we will predict the unknown value

$$f_0 := L_0 f \in \mathbb{R},$$

evaluated at another bounded linear functional $L_0 \in \mathcal{H}^m(\mathcal{D})'$. For example, if the bounded linear functionals are the evaluation functions at the unknown data point $\boldsymbol{x}_0 \in \mathcal{D}$ and the observed data points $X := \{\boldsymbol{x}_1, \ldots, \boldsymbol{x}_n\} \subseteq \mathcal{D}$, that is,

$$L_0 := \delta_{\boldsymbol{x}_0}, \; L_1 := \delta_{\boldsymbol{x}_1}, \ldots, L_n := \delta_{\boldsymbol{x}_n},$$

then this typical interpolation is the same as the classical geostatistical problem.

Since a lot of functions in $\mathcal{H}^m(\mathcal{D})$ satisfy the interpolation conditions, we will construct the estimators based on the collection of all interpolating paths in $\mathcal{H}^m(\mathcal{D})$, that is,

$$\mathcal{A}_{\mathcal{L}}(\boldsymbol{f}) := \{\omega \in \mathcal{H}^m(\mathcal{D}) : \mathcal{L}\omega = \boldsymbol{f}\}.$$

Obviously, the unknown function f always belongs to the subset $\mathcal{A}_{\mathcal{L}}(\boldsymbol{f})$. Actually, there are many choices of the prediction $v \in \mathbb{R}$ of the unknown value f_0 such that we need to investigate the different subset

$$\mathcal{A}_{L_0}(v) := \{\omega \in \mathcal{H}^m(\mathcal{D}) : L_0\omega = v\}.$$

If we can confirm that f belongs to $\mathcal{A}_{L_0}(\hat{v}) \cap \mathcal{A}_{\mathcal{L}}(\boldsymbol{f})$ for some prediction $\hat{v}$, then $\hat{v} = f_0$. But, it is so difficult to check which prediction $\hat{v}$ satisfies the sufficient condition of $f \in \mathcal{A}_{L_0}(\hat{v}) \cap \mathcal{A}_{\mathcal{L}}(\boldsymbol{f})$.

The kernel-based probability measure $\mathbb{P}_K$ given in Theorem 1 provides a new numerical tool to measure the probability of $f \in \mathcal{A}_{L_0}(v) \cap \mathcal{A}_{\mathcal{L}}(\boldsymbol{f})$ for the different prediction v, that is, $\mathbb{P}_K(\mathcal{A}_{L_0}(v) \cap \mathcal{A}_{\mathcal{L}}(\boldsymbol{f}))$. We also find that

$$\mathbb{P}_K(\mathcal{A}_{L_0}(v) \cap \mathcal{A}_{\mathcal{L}}(\boldsymbol{f})) = \mathbb{P}_K(\mathcal{A}_{L_0}(v) \,|\mathcal{A}_{\mathcal{L}}(\boldsymbol{f}))\, \mathbb{P}_K(\mathcal{A}_{\mathcal{L}}(\boldsymbol{f})).$$

Since the observed data $\mathcal{L}$ and $\boldsymbol{f}$ are already given, the interpolation $\mathcal{A}_{\mathcal{L}}(\boldsymbol{f})$ is thought to happen; hence the prediction v can be equivalently weighted by the conditional probability $\mathbb{P}_K(\mathcal{A}_{L_0}(v) \,|\mathcal{A}_{\mathcal{L}}(\boldsymbol{f}))$. By the same techniques of the

Bayesian estimation, the best estimator $\hat{v}_{K,L_0|\mathcal{L}f}$ is the average of the prediction v weighted by the conditional probability $\mathbb{P}_K\left(\mathcal{A}_{L_0}(v)\,|\mathcal{A}_{\mathcal{L}}(f)\right)$, that is,

$$\hat{v}_{K,L_0|\mathcal{L}f} := \sum_{v\in\mathbb{R}} v\mathbb{P}_K\left(\mathcal{A}_{L_0}(v)\,|\mathcal{A}_{\mathcal{L}}(f)\right). \tag{6}$$

Theorem 1 guarantees that the normal random variables

$$L_0S,\ L_1S,\ldots,L_nS,$$

are well-defined on the probability space $(\Omega_m, \mathcal{F}_m, \mathbb{P}_K)$. We denote the multivariate normal random vector

$$\mathcal{L}S := (L_1S,\cdots,L_nS)^T.$$

According to the constructions of L_0S and $\mathcal{L}S$, we have

$$\mathcal{A}_{L_0}(v) = \{\omega\in\Omega_m : L_0S(\omega) = v\}, \quad \mathcal{A}_{\mathcal{L}}(f) = \{\omega\in\Omega_m : \mathcal{L}S(\omega) = f\}.$$

Thus, the conditional probability $\mathbb{P}_K\left(\mathcal{A}_{L_0}(v)\,|\mathcal{A}_{\mathcal{L}}(f)\right)$ can be computed by the normal random variables L_0S and $\mathcal{L}S$, that is,

$$\mathbb{P}_K\left(\mathcal{A}_{L_0}(v)\,|\mathcal{A}_{\mathcal{L}}(f)\right) = \mathbb{P}_K\left(L_0S = v|\mathcal{L}S = f\right);$$

hence

$$\sum_{v\in\mathbb{R}} \mathbb{P}_K\left(\mathcal{A}_{L_0}(v)\,|\mathcal{A}_{\mathcal{L}}(f)\right) = \sum_{v\in\mathbb{R}} \mathbb{P}_K\left(L_0S = v|\mathcal{L}S = f\right) = 1,$$

such that the best estimator $\hat{v}_{K,L_0|\mathcal{L}f}$ in Eq. (6) can be rewritten as the conditional mean

$$\hat{v}_{K,L_0|\mathcal{L}f} = \sum_{v\in\mathbb{R}} v\mathbb{P}_K\left(L_0S = v|\mathcal{L}S = f\right) = \mathrm{E}\left(L_0S|\mathcal{L}S = f\right). \tag{7}$$

Since L_0S has the mean 0 and the covariance $L_{0,x}L_{0,y}K(\boldsymbol{x},\boldsymbol{y})$ and $\mathcal{L}S$ has the vector mean $\mathbf{0}$ and the covariance matrix

$$\mathsf{A}_{K,\mathcal{L}} := \mathcal{L}_x\mathcal{L}_y^T K(\boldsymbol{x},\boldsymbol{y}) = \begin{pmatrix} L_{1,x}L_{1,y}K(\boldsymbol{x},\boldsymbol{y}) & \cdots & L_{1,x}L_{n,y}K(\boldsymbol{x},\boldsymbol{y}) \\ & \ddots & \\ L_{n,x}L_{1,y}K(\boldsymbol{x},\boldsymbol{y}) & \cdots & L_{n,x}L_{n,y}K(\boldsymbol{x},\boldsymbol{y}) \end{pmatrix}, \tag{8}$$

the conditional probability density function $p_{K,L_0|\mathcal{L}}$ of L_0S given $\mathcal{L}S$ be represented as

$$p_{K,L_0|\mathcal{L}}(v|\boldsymbol{f}) := \frac{1}{\sigma_{K,L_0|\mathcal{L}}\sqrt{2\pi}} \exp\left(-\frac{\left(v - m_{K,L_0|\mathcal{L}\boldsymbol{f}}\right)^2}{2\sigma^2_{K,L_0|\mathcal{L}}}\right),$$

where the mean

$$m_{K,L_0|\mathcal{L}\boldsymbol{f}} := L_0\boldsymbol{k}^T_{K,\mathcal{L}}\mathsf{A}^\dagger_{K,\mathcal{L}}\boldsymbol{f},$$

and the standard deviation

$$\sigma_{K,L_0|\mathcal{L}} := \sqrt{L_{0,\boldsymbol{x}}L_{0,\boldsymbol{y}}K(\boldsymbol{x},\boldsymbol{y}) - L_0\boldsymbol{k}^T_{K,\mathcal{L}}\mathsf{A}^\dagger_{K,\mathcal{L}}L_0\boldsymbol{k}_{K,\mathcal{L}}}, \tag{9}$$

such as the proofs of Ye [18, Corollaries 2.7 and 2.8]. Here, the vector

$$L_0\boldsymbol{k}_{K,\mathcal{L}} = \left(L_{0,\boldsymbol{x}}L_{1,\boldsymbol{y}}K(\boldsymbol{x},\boldsymbol{y}), \cdots, L_{0,\boldsymbol{x}}L_{n,\boldsymbol{y}}K(\boldsymbol{x},\boldsymbol{y})\right)^T,$$

is computed by the kernel basis

$$\boldsymbol{k}_{K,\mathcal{L}}(\boldsymbol{x}) := \mathcal{L}_{\boldsymbol{y}}K(\boldsymbol{x},\boldsymbol{y}) = \left(L_{1,\boldsymbol{y}}K(\boldsymbol{x},\boldsymbol{y}), \cdots, L_{n,\boldsymbol{y}}K(\boldsymbol{x},\boldsymbol{y})\right)^T, \quad \text{for } \boldsymbol{x} \in \mathcal{D}.$$

Therefore, the conditional mean $\hat{v}_{K,L|\mathcal{L}\boldsymbol{f}}$ in Eq. (7) can be computed by the conditional probability density function $p_{K,L_0|\mathcal{L}}$, that is,

$$\hat{v}_{K,L_0|\mathcal{L}\boldsymbol{f}} = \int_{\mathbb{R}} p_{K,L_0|\mathcal{L}}(v|\boldsymbol{f})\mathrm{d}v = m_{K,L_0|\mathcal{L}\boldsymbol{f}} = L_0\boldsymbol{k}^T_{K,\mathcal{L}}\mathsf{A}^\dagger_{K,\mathcal{L}}\left(\boldsymbol{f} - \mathcal{L}\mu\right). \tag{10}$$

Remark 4 Since the covariance matrix $\mathsf{A}_{K,\mathcal{L}}$ is always positive definite, the pseudo inverse $\mathsf{A}^\dagger_{K,\mathcal{L}}$ of $\mathsf{A}_{K,\mathcal{L}}$ is well-defined by the eigen-decomposition of $\mathsf{A}_{K,\mathcal{L}}$.

We know that the kriging estimations can be modeled by a kernel-based function. Now we show that the best estimator $\hat{v}_{K,L_0|\mathcal{L}\boldsymbol{f}}$ in Eq. (10) can be computed by a function $s_{K,\mathcal{L}\boldsymbol{f}}$ such as

$$\hat{v}_{K,L_0|\mathcal{L}\boldsymbol{f}} = L_0 s_{K,\mathcal{L}\boldsymbol{f}}. \tag{11}$$

Observing Eq. (10), the function $s_{K,\mathcal{L}\boldsymbol{f}}$ can be represented as

$$s_{K,\mathcal{L}\boldsymbol{f}}(\boldsymbol{x}) := \boldsymbol{k}_{K,\mathcal{L}}(\boldsymbol{x})^T\boldsymbol{c}, \quad \text{for } \boldsymbol{x} \in \mathcal{D}, \tag{12}$$

where the coefficient $\boldsymbol{c}$ is uniquely solved by the linear system

$$\mathsf{A}_{K,\mathcal{L}}\boldsymbol{c} = \boldsymbol{f}. \tag{13}$$

Here, when the covariance matrix $\mathsf{A}_{K,\mathcal{L}}$ is singular, then $\boldsymbol{c}$ is the least-square solution of linear system (13). Thus, we call this estimation $L_0 s_{K,\mathcal{L}f}$ the *kernel-based estimator* for the observed data $\mathcal{L}$ and $\boldsymbol{f}$. The kernel-based estimators can be viewed as the general estimation in [12].

In particular, when $L_0 := \delta_{x_0}$ and $\mathcal{L} := (\delta_{x_1}, \cdots, \delta_{x_n})^T$, then

$$\boldsymbol{k}_{K,\delta_X}(\boldsymbol{x}) = \boldsymbol{k}_{K,X}(\boldsymbol{x}) := \begin{pmatrix} K(\boldsymbol{x}, \boldsymbol{x}_1) \\ \vdots \\ K(\boldsymbol{x}, \boldsymbol{x}_n) \end{pmatrix}, \quad \text{for } \boldsymbol{x} \in \mathcal{D}, \tag{14}$$

and

$$\mathsf{A}_{K,\delta_X} = \mathsf{A}_{K,X};$$

hence the kernel-based estimator $L_0 s_{K,\mathcal{L}f}$ can be rewritten as

$$L_0 s_{K,\mathcal{L}f} = s_{K,Xf}(\boldsymbol{x}_0) = \boldsymbol{w}_{K,X}(\boldsymbol{x}_0)^T \boldsymbol{f}, \tag{15}$$

where the Lagrange basis $\boldsymbol{w}_{K,X}(\boldsymbol{x})$ is solved by the linear system

$$\mathsf{A}_{K,X} \boldsymbol{w}_{K,X}(\boldsymbol{x}) = \boldsymbol{k}_{K,X}(\boldsymbol{x}), \quad \text{for } \boldsymbol{x} \in \mathcal{D}. \tag{16}$$

It is clear that the kernel-based estimator $L_0 s_{K,\mathcal{L}f}$ in Eqs. (15)–(16) is equivalent to the simple kriging estimation. So, we call the general kernel-based estimator $L_0 s_{K,\mathcal{L}_n f_n}$ in Eq. (11) and Eqs. (12)–(13) the *generalized kriging estimation*.

Convergence: Finally, we look at the convergence of the kernel-based estimators by the same techniques in [18]. Suppose that the vector operator $\mathcal{L}_\infty := (L_1, \cdots, L_n, \cdots)^T$ composes of the countable bounded linear functionals $L_1, \ldots, L_n, \ldots$ on the Sobolev space $\mathcal{H}^m(\mathcal{D})$ and the observed data values $\boldsymbol{f}_\infty := (f_1, \cdots, f_n, \cdots)^T$ composes of the countable values $f_1, \ldots, f_n, \ldots$ such that there exists a unique solution $f \in \mathcal{H}^m(\mathcal{D})$ satisfying the interpolation conditions, that is, $\mathcal{L}_\infty f = \boldsymbol{f}_\infty$. For example, the vector operator $\mathcal{L}_\infty$ composes of the countable point evaluation functions $\delta_{x_1}, \ldots, \delta_{x_n}, \ldots$ at the data points $X_\infty := \{\boldsymbol{x}_n\}_{n=1}^\infty$ which is dense in the domain $\mathcal{D}$. Let $\mathcal{L}_n := (L_1, \cdots, L_n)^T$ and $\boldsymbol{f}_n := (f_1, \cdots, f_n)^T$ for all $n \in \mathbb{N}$. Then we have

$$\mathcal{A}_{\mathcal{L}_1}(\boldsymbol{f}_1) \supseteq \cdots \supseteq \mathcal{A}_{\mathcal{L}_n}(\boldsymbol{f}_n) \supseteq \cdots \supseteq \bigcap_{n=1}^{\infty} \mathcal{A}_{\mathcal{L}_n}(\boldsymbol{f}_n) = \mathcal{A}_{\mathcal{L}_\infty}(\boldsymbol{f}_\infty) = \{f\};$$

hence we can obtain the convergence of the kernel-based estimators such as

$$\begin{aligned} \lim_{n\to\infty} L_0 s_{K,\mathcal{L}_n f_n} &= \lim_{n\to\infty} \mathrm{E}(L_0 S | \mathcal{L}_n S = \boldsymbol{f}_n) = \lim_{n\to\infty} \mathrm{E}(L_0 S | \mathcal{A}_{\mathcal{L}_n}(\boldsymbol{f}_n)) \\ &= \mathrm{E}(L_0 S | \mathcal{A}_{X_\infty}(\boldsymbol{f}_\infty)) = L_0 f, \end{aligned}$$

for any bounded linear functional L_0 on $\mathcal{H}^m(\mathcal{D})$. It is obvious that the convergence is independent of the bounded linear functional L_0. Therefore, we have:

Proposition 1 *Let L_0 be any bounded linear functional on $\mathcal{H}^m(\mathcal{D})$. If there exists a unique solution $f \in \mathcal{H}^m(\mathcal{D})$ satisfying the interpolation conditions for all observed data $\mathcal{L}_n$ and $\boldsymbol{f}_n$ discussed above such as $\mathcal{L}_\infty f = \boldsymbol{f}_\infty$, then the kernel-based estimator $L_0 s_{K,\mathcal{L}_n \boldsymbol{f}_n}$ in Eq. (11) and Eqs. (12)–(13) converges to the unknown value $L_0 f$ when $n \to \infty$.*

In particular, when L_0 is the point evaluation function $\delta_{\boldsymbol{x}_0}$ at any $\boldsymbol{x}_0 \in \mathcal{D}$, then

$$\lim_{n\to\infty} s_{K,\mathcal{L}_n \boldsymbol{f}_n}(\boldsymbol{x}_0) = \lim_{n\to\infty} \delta_{\boldsymbol{x}_0} s_{K,\mathcal{L}_n \boldsymbol{f}_n} = \delta_{\boldsymbol{x}_0} f = f(\boldsymbol{x}_0) .$$

Remark 5 We compute the conditional mean square

$$\mathrm{E}\left(\left| L_0 S - L_0 s_{K,\mathcal{L}_n \boldsymbol{f}_n} \right|^2 \middle| \mathcal{L}_n S = \boldsymbol{f}_n \right) = \sigma^2_{K,L_0|\mathcal{L}_n},$$

where the conditional standard deviation $\sigma_{K,L_0|\mathcal{L}_n}$ is given in Eq. (9). We can also check that $\sigma_{K,L_0|\mathcal{L}_n}$ is equivalent to the (generalized) power function in meshfree approximation in [16, Sects. 11.1 and 16.1]. The details of the error bounds of the kernel-based estimators will be discussed in our next paper.

4 Applications in Geostatistics

In this section, we apply the kernel-based approximation to model the geostatistics. In the classical geostatistical problems, we have the observed data values

$$f_1, \ldots, f_n \in \mathbb{R},$$

at the known locations

$$\boldsymbol{x}_1, \ldots, \boldsymbol{x}_n \in \mathcal{D} \subseteq \mathbb{R}^d,$$

that is,

$$f_1 := f(\boldsymbol{x}_1), \ldots, f_n := f(\boldsymbol{x}_n),$$

for some function $f \in \mathcal{H}^m(\mathcal{D})$. Let

$$\boldsymbol{\delta}_X := \begin{pmatrix} \delta_{\boldsymbol{x}_1} \\ \vdots \\ \delta_{\boldsymbol{x}_n} \end{pmatrix}, \quad \boldsymbol{f} := \begin{pmatrix} f_1 \\ \vdots \\ f_n \end{pmatrix}, \quad X := \{\boldsymbol{x}_1, \ldots, \boldsymbol{x}_n\} .$$

Same as in Sect. 3, the positive definite kernel $K \in \mathrm{C}^{2m,1}(\mathcal{D} \times \mathcal{D})$ is fixed here to construct the kernel-based estimators (kriging estimations), for example, the Gaussian kernel $K(\boldsymbol{x},\boldsymbol{y}) := e^{-\theta^2\|x-y\|_2^2}$ belongs to $\mathrm{C}^{\infty}(\mathcal{D} \times \mathcal{D})$.

By the simple kriging method, we can obtain the best prediction

$$s_{K,X,f}(\boldsymbol{x}_0) = \boldsymbol{w}_{K,X}(\boldsymbol{x}_0)^T \boldsymbol{f} = \boldsymbol{k}_{K,X}(\boldsymbol{x}_0)^T \boldsymbol{\alpha}, \tag{17}$$

in Eqs. (15)–(16) to measure the value at the unknown location $\boldsymbol{x}_0 \in \mathcal{D}$, where the coefficients $\boldsymbol{\alpha}$ are uniquely solved by the linear system

$$\mathsf{A}_{K,X}\boldsymbol{\alpha} = \boldsymbol{f}. \tag{18}$$

Usually, we may not have the observed data values at another locations

$$z_1, \ldots, z_t \in \mathcal{D} \subseteq \mathbb{R}^d.$$

But, we can still observe the rates of the changes along the directions

$$\boldsymbol{e}_1, \ldots, \boldsymbol{e}_t \in \mathbb{R}^d \text{ subjected to } \|\boldsymbol{e}_1\|_2 = \cdots = \|\boldsymbol{e}_t\|_2 = 1,$$

at these locations $z_1, \ldots, z_t$. To be more precisely, we have the additional rates of the changes

$$g_1, \ldots, g_t \in \mathbb{R},$$

along the directions $\boldsymbol{e}_1, \ldots, \boldsymbol{e}_t$ at $z_1, \ldots, z_t$, that is,

$$g_1 := \boldsymbol{e}_1^T \nabla f(z_1), \ldots, g_t := \boldsymbol{e}_t^T \nabla f(z_t),$$

where ∇ is the gradient composing of the partial derivatives $\frac{\partial}{\partial x_1}, \ldots, \frac{\partial}{\partial x_d}$. Let

$$\boldsymbol{\delta}_Z \circ \mathsf{E}\nabla := \begin{pmatrix} \delta_{z_1} \circ \boldsymbol{e}_1^T \nabla \\ \vdots \\ \delta_{z_t} \circ \boldsymbol{e}_t^T \nabla \end{pmatrix}, \quad \boldsymbol{g} := \begin{pmatrix} g_1 \\ \vdots \\ g_t \end{pmatrix}, \quad Z := \{z_1, \ldots, z_t\}, \quad \mathsf{E} := (\boldsymbol{e}_1 \cdots \boldsymbol{e}_t)^T.$$

Here $\delta_z \circ \boldsymbol{e}^T \nabla$ is the gradient along the direction $\boldsymbol{e}$ at any $z \in \mathcal{D}$, that is, $\delta_z \circ \boldsymbol{e}^T \nabla f = \boldsymbol{e}^T \nabla f(z)$. Usually, we can not obtain the exact values of the gradients. But, we can easily obtain the positive or negative (increasing or decreasing) rates of the gradients along the observed directions, for example,

$$\boldsymbol{e}^T \nabla f(z) \geq 0 \text{ or } \boldsymbol{e}^T \nabla f(z) \leq 0.$$

For convenience, we give a *new* notation of the equality "$\triangleq$" in this article for 0^+ and 0^-. We define the equalities of $c \in \mathbb{R}$ and $0^+, 0^-$ in the following way:

$$u \triangleq c, \quad u \triangleq 0^+, \quad u \triangleq 0^-,$$

are equivalent to

$$u = c, \quad u \geq 0, \quad u \leq 0.$$

So, we further suppose that $g_1, \ldots, g_t$ are endowed with 0^+ or 0^-, that is,

$$g_1 := 0^+ \text{ or } 0^-, \ldots, g_t := 0^+ \text{ or } 0^-.$$

In this section, we assume the degree

$$m > 1 + d/2,$$

such that the point evaluation function $\delta_{\boldsymbol{x}}$ and the point gradient $\delta_z \circ \boldsymbol{e}^T \nabla$ are bounded on the Sobolev space $\mathcal{H}^m(\mathcal{D})$ for any $\boldsymbol{x}, z \in \mathcal{D}$ by the Sobolev imbedding theorem.

Thus, the vector bounded linear operator and the vector observed value are well-defined by

$$\mathcal{L} := \begin{pmatrix} \boldsymbol{\delta}_X \\ \boldsymbol{\delta}_Z \circ \mathsf{E}\nabla \end{pmatrix}, \quad \boldsymbol{\xi} := \begin{pmatrix} \boldsymbol{f} \\ \boldsymbol{g} \end{pmatrix}.$$

According to Theorem 1, the normal random variable $\delta_{\boldsymbol{x}_0} S$ and the multivariate normal random vector $\mathcal{L}S$ are well-defined on the probability space $(\Omega_m, \mathcal{F}_m, \mathbb{P}_K)$. In this section, we will construct the kernel-based estimator $s_{K,\mathcal{L},\boldsymbol{\xi}}(\boldsymbol{x}_0) = \delta_{\boldsymbol{x}_0} s_{K,\mathcal{L},\boldsymbol{\xi}}$ same as in Eq. (7) to measure the prediction at the unknown location $\boldsymbol{x}_0$ based on the observed data $\mathcal{L}$ and $\boldsymbol{\xi}$.

Since some elements of the data values $\boldsymbol{\xi}$ include 0^+ or 0^- such as $\boldsymbol{g}$, we need to recompute the conditional mean

$$s_{K,\mathcal{L},\boldsymbol{\xi}}(\boldsymbol{x}_0) = \delta_{\boldsymbol{x}_0} s_{K,\mathcal{L},\boldsymbol{\xi}} = \mathrm{E}\left(\delta_{\boldsymbol{x}_0} S \middle| \mathcal{L}S \triangleq \boldsymbol{\xi}\right), \tag{19}$$

by the same techniques of Eq. (10). Here, when $k = 1, \ldots, n$ then $L_k S \triangleq \xi_k$ is equivalent to $\delta_{\boldsymbol{x}_k} S = f_k$, and when $k = n+1, \ldots, n+t$ then $L_k S \triangleq \xi_k$ is equivalent to $\delta_{z_{k-n}} \circ \boldsymbol{e}_{k-n}^T \nabla S \geq 0$ or $\delta_{z_{k-n}} \circ \boldsymbol{e}_{k-n}^T \nabla S \leq 0$ dependent of $g_{k-n} = 0^+$ or 0^-. For computing the conditional mean, we need the probability density function $p_{K,\mathcal{L}}$ of $\mathcal{L}S$, that is,

$$p_{K,\mathcal{L}}(\boldsymbol{u}) := \frac{1}{\sqrt{\det_{\dagger}(2\pi \mathsf{A}_{K,\mathcal{L}})}} \exp\left(-\frac{1}{2}\boldsymbol{u}^T \mathsf{A}_{K,\mathcal{L}}^{\dagger} \boldsymbol{u}\right),$$

where $\det_\dagger$ is the *pseudo determinant*, that is, the product of all nonzero eigenvalues of a positive definite matrix. Here, the covariance matrix $\mathsf{A}_{K,\mathcal{L}}$ can be precisely written as

$$\mathsf{A}_{K,\mathcal{L}} = \begin{pmatrix} \mathsf{A}_{K,X} & \mathsf{B}'^{T}_{K,XZ} \\ \mathsf{B}'_{K,XZ} & \mathsf{A}''_{K,Z} \end{pmatrix}, \tag{20}$$

where

$$\mathsf{A}''_{K,Z} := \mathsf{A}_{K,\delta_Z \circ \mathsf{E}\nabla} = \begin{pmatrix} e_1^T \nabla_x \nabla_z^T e_1 K(z_1, z_1) & \cdots & e_1^T \nabla_x \nabla_z^T e_t K(z_1, z_t) \\ & \ddots & \\ e_t^T \nabla_x \nabla_z^T e_1 K(z_t, z_1) & \cdots & e_t^T \nabla_x \nabla_z^T e_t K(z_t, z_t) \end{pmatrix}, \tag{21}$$

and

$$\mathsf{B}'_{K,XZ} := \begin{pmatrix} e_1^T \nabla_x K(z_1, x_1) & \cdots & e_1^T \nabla_x K(z_1, x_n) \\ & \ddots & \\ e_t^T \nabla_x K(z_t, x_1) & \cdots & e_t^T \nabla_x K(z_t, x_n) \end{pmatrix} = \begin{pmatrix} e_1^T \nabla_z^T K(x_1, z_1) & \cdots & e_t^T \nabla_z^T K(x_1, z_t) \\ & \ddots & \\ e_1^T \nabla_z^T K(x_n, z_1) & \cdots & e_t^T \nabla_z^T K(x_n, z_t) \end{pmatrix}^T. \tag{22}$$

For simplifying the complexities of the notations, we redefine some notations of the integrals as follows:

$$\int_{u \triangleq c} \phi(u) \mathrm{d}u = \phi(c), \quad \int_{u \triangleq 0+} \phi(u) \mathrm{d}u = \int_0^{+\infty} \phi(u) \mathrm{d}u,$$
$$\int_{u \triangleq 0-} \phi(u) \mathrm{d}u = \int_{-\infty}^{0} \phi(u) \mathrm{d}u.$$

Then the conditional mean $s_{K,\mathcal{L},\xi}(x_0)$ in Eq. (19) can be rewritten as

$$s_{K,\mathcal{L},\xi}(x_0) = \frac{1}{\hat{q}} \int_{\mathbb{R}} \int_{u \triangleq \xi} v p_{K,\delta_{x_0},\mathcal{L}}(v, u) \mathrm{d}u \mathrm{d}v, \tag{23}$$

where $p_{K,\delta_{x_0},\mathcal{L}}$ is the joint probability density function of $\delta_{x_0} S$ and $\mathcal{L}S$, and

$$\hat{q} := \int_{u \triangleq \xi} p_{K,\mathcal{L}}(u) \mathrm{d}u.$$

(Here $\hat{q}$ is not the probability.) Since

$$p_{K,\delta_{x_0},\mathcal{L}}(v, u) = p_{K,\delta_{x_0}|\mathcal{L}}(v|u) p_{K,\mathcal{L}}(u),$$

we have

$$\int_{\mathbb{R}} v p_{K,\delta_{x_0},\mathcal{L}}(v,\boldsymbol{u})\mathrm{d}v = p_{K,\mathcal{L}}(\boldsymbol{u}) \int_{\mathbb{R}} v p_{K,\delta_{x_0}|\mathcal{L}}(v|\boldsymbol{u})\mathrm{d}v = \boldsymbol{k}_{K,\mathcal{L}}(\boldsymbol{x}_0)^T \mathsf{A}_{K,\mathcal{L}}^{\dagger}\boldsymbol{u}, \quad (24)$$

where the kernel basis $\boldsymbol{k}_{K,\mathcal{L}}(\boldsymbol{x})$ can be represented as

$$\boldsymbol{k}_{K,\mathcal{L}}(\boldsymbol{x}) := \begin{pmatrix} \boldsymbol{k}_{K,X}(\boldsymbol{x}) \\ \boldsymbol{k}'_{K,Z}(\boldsymbol{x}) \end{pmatrix}, \quad \text{for } \boldsymbol{x} \in \mathcal{D},$$

and

$$\boldsymbol{k}'_{K,Z}(\boldsymbol{x}) := \boldsymbol{k}_{K,\delta_Z \circ \mathrm{E}\nabla}(\boldsymbol{x}) = \begin{pmatrix} \boldsymbol{e}_1^T \nabla_z K(\boldsymbol{x}, z_1) \\ \vdots \\ \boldsymbol{e}_t^T \nabla_z K(\boldsymbol{x}, z_t) \end{pmatrix}, \quad \text{for } \boldsymbol{x} \in \mathcal{D}. \quad (25)$$

Putting Eq. (24) into Eq. (23), we have that

$$s_{K,\mathcal{L},\xi}(\boldsymbol{x}_0) = \boldsymbol{k}_{K,\mathcal{L}}(\boldsymbol{x}_0)^T \mathsf{A}_{K,\mathcal{L}}^{\dagger}\hat{\boldsymbol{\xi}}, \quad (26)$$

where

$$\hat{\boldsymbol{\xi}} := \frac{1}{\hat{q}} \int_{\boldsymbol{u} \triangleq \boldsymbol{\xi}} \boldsymbol{u} p_{K,\mathcal{L}}(\boldsymbol{u})\mathrm{d}z = \mathrm{E}\left(\mathcal{L}S|\mathcal{L}S \triangleq \boldsymbol{\xi}\right). \quad (27)$$

Clearly, the kernel-based estimator $s_{K,\mathcal{L},\xi}(\boldsymbol{x}_0)$ is a linear combination of the kernel basis $\boldsymbol{k}_{K,\mathcal{L}}(\boldsymbol{x}_0)$, that is,

$$s_{K,\mathcal{L},\xi}(\boldsymbol{x}_0) = \boldsymbol{k}_{K,\mathcal{L}}(\boldsymbol{x}_0)^T \boldsymbol{c}, \quad (28)$$

where the coefficients $\boldsymbol{c}$ are uniquely solved by the linear system

$$\mathsf{A}_{K,\mathcal{L}}\boldsymbol{c} = \hat{\boldsymbol{\xi}}. \quad (29)$$

Moreover, we simply the right-hand side $\hat{\boldsymbol{\xi}}$ of linear system (29). When $k = 1, \ldots, n$, then $\xi_k = f_k$; hence

$$\hat{\xi}_k = \frac{\int_{u_1 \triangleq \xi_1} \cdots \int_{u_k = f_k} \cdots \int_{u_{n+t} \triangleq \xi_{n+t}} u_k p_{K,\mathcal{L}}(\boldsymbol{u})\mathrm{d}\boldsymbol{u}}{\int_{u_1 \triangleq \xi_1} \cdots \int_{u_k = f_k} \cdots \int_{u_{n+t} \triangleq \xi_{n+t}} p_{K,\mathcal{L}}(\boldsymbol{u})\mathrm{d}\boldsymbol{u}} = f_k.$$

When $k = n + 1, \ldots, n + t$, then $\xi_k = g_k = 0^+$ or 0^-; hence

$$\hat{\xi}_k = \frac{\int_{u_1 \triangleq \xi_1} \cdots \int_{u_k \geq 0 \text{ or } u_k \leq 0} \cdots \int_{u_{n+t} \triangleq \xi_{n+t}} u_k p_{K,\mathcal{L}}(\boldsymbol{u})\mathrm{d}\boldsymbol{u}}{\int_{u_1 \triangleq \xi_1} \cdots \int_{u_k \geq 0 \text{ or } u_k \leq 0} \cdots \int_{u_{n+t} \triangleq \xi_{n+t}} p_{K,\mathcal{L}}(\boldsymbol{u})\mathrm{d}\boldsymbol{u}} =: \hat{g}_{k-n}.$$

Thus,

$$\hat{\xi} = \begin{pmatrix} f \\ \hat{g} \end{pmatrix}, \quad \text{where } \hat{g} := \begin{pmatrix} \hat{g}_1 \\ \vdots \\ \hat{g}_t \end{pmatrix}. \tag{30}$$

We will also simplify the computation of $\hat{g}$ by the conditional probability density function $p_{K,\delta_Z \circ \mathsf{E}\nabla|\delta_X}$ of $\delta_Z \circ \mathsf{E}\nabla S$ given $\delta_X S$, that is,

$$p_{K,\delta_Z \circ \mathsf{E}\nabla|\delta_X}(v|f) := \frac{1}{\sqrt{\det_\dagger\left(2\pi \Sigma_{K,\delta_Z \circ \mathsf{E}\nabla|\delta_X}\right)}} \exp\left(-\frac{1}{2}\left(v - m_{K,\delta_Z \circ \mathsf{E}\nabla|\delta_X f}\right)^T \times \Sigma^\dagger_{K,\delta_Z \circ \mathsf{E}\nabla|\delta_X}\left(v - m_{K,\delta_Z \circ \mathsf{E}\nabla|\delta_X f}\right)\right),$$

where

$$m_{K,\delta_Z \circ \mathsf{E}\nabla|\delta_X f} := \mathsf{B}'_{K,XZ} \mathsf{A}^{-1}_{K,X} f,$$

and

$$\Sigma_{K,\delta_Z \circ \mathsf{E}\nabla|\delta_X} := \mathsf{A}''_{K,Z} - \mathsf{B}'_{K,XZ} \mathsf{A}^{-1}_{K,X} \mathsf{B}'^T_{K,XZ}.$$

Since

$$p_{K,\mathcal{L}}(v, f) = p_{K,\delta_Z \circ \mathsf{E}\nabla|\delta_X}(v|f) p_{K,\delta_X}(f),$$

we have

$$\hat{g} = \frac{\int_{v \triangleq g} v p_{K,\delta_Z \circ \mathsf{E}\nabla|\delta_X}(v|f) \mathrm{d}v}{\int_{v \triangleq g} p_{K,\delta_Z \circ \mathsf{E}\nabla|\delta_X}(v|f) \mathrm{d}v}. \tag{31}$$

Approximation of the preconditioned $\hat{g}$: Finally, we will study the efficient algorithm to approximate $\hat{g}$. We generalize the multivariate normal random vectors

$$v_1, \ldots, v_s \sim \text{i.i.d.} \mathcal{N}\left(m_{K,\delta_Z \circ \mathsf{E}\nabla|\delta_X f}, \Sigma_{K,\delta_Z \circ \mathsf{E}\nabla|\delta_X}\right),$$

and we choose all random vectors $v_{s_1}, \ldots, v_{s_r} \in \left\{v_k : v_k \triangleq g,\ k = 1, \ldots, s\right\}$; hence we have

$$\hat{g} \approx \frac{1}{r} \sum_{j=1}^{r} v_{s_j}.$$

Remark 6 We compare the observed data values $\boldsymbol{g}$ and the preconditioned data values $\hat{\boldsymbol{g}}$. The data values $\boldsymbol{g}$ and $\hat{\boldsymbol{g}}$ have the same sizes. The elements of $\boldsymbol{g}$ are chosen from $\{0^+, 0^-\}$ while the elements of $\hat{\boldsymbol{g}}$ are chosen from $\mathbb{R}$. Thus, the equality of $\boldsymbol{v} \triangleq \boldsymbol{g}$ represents the range, that is, $v_k \geq 0$ if $g_k = 0^+$ or $v_k \leq 0$ if $g_k = 0^-$ for $k = 1, \ldots, t$. In the other hand, the equality of $\boldsymbol{v} \triangleq \hat{\boldsymbol{g}}$ represents the point. Moreover $\hat{\boldsymbol{g}}$ can be viewed as the average of $\boldsymbol{g}$ over the area of $\boldsymbol{v} \triangleq \boldsymbol{g}$ by the weight $p_{K,\delta_Z \circ E\nabla|\delta_X}(\boldsymbol{v}|\boldsymbol{f})\mathrm{d}\boldsymbol{v}$.

In conclusions, we can combine Eqs. (28)–(30) to obtain the generalized kriging method to measure the prediction at the unknown location $\boldsymbol{x}_0 \in \mathcal{D}$ by the kernel-based model

$$s_{K,X,Z,f,g}(\boldsymbol{x}_0) := s_{K,\mathcal{L},\xi}(\boldsymbol{x}_0) = \boldsymbol{k}_{K,X}(\boldsymbol{x}_0)^T \boldsymbol{\alpha} + \boldsymbol{k}'_{K,Z}(\boldsymbol{x}_0)^T \boldsymbol{\beta}, \tag{32}$$

where the coefficients $\boldsymbol{\alpha}$ and $\boldsymbol{\beta}$ are uniquely solved by the linear system

$$\begin{pmatrix} \mathsf{A}_{K,X} & \mathsf{B}'^T_{K,XZ} \\ \mathsf{B}'_{K,XZ} & \mathsf{A}''_{K,Z} \end{pmatrix} \begin{pmatrix} \boldsymbol{\alpha} \\ \boldsymbol{\beta} \end{pmatrix} = \begin{pmatrix} \boldsymbol{f} \\ \hat{\boldsymbol{g}} \end{pmatrix}. \tag{33}$$

Here, the kernel bases $\boldsymbol{k}_{K,X}(\boldsymbol{x}), \boldsymbol{k}'_{K,Z}(\boldsymbol{x})$, the matrixes $\mathsf{A}_{K,X}, \mathsf{A}''_{K,Z}, \mathsf{B}'_{K,XZ}$, and the vector $\hat{\boldsymbol{g}}$ are defined in Eqs. (14), (25), (5), (21), (22), and (31), respectively.

5 Numerical Examples

In this section, we will study the numerical example of the generalized kriging method in geostatistics given in Eqs. (32)–(33). For convenience, we look at an example given by the Franke test function $f \in \mathrm{C}^\infty(\mathbb{R}^2)$ given in [6]. Then we can obtain the observed data $\boldsymbol{y}_1, \ldots, \boldsymbol{y}_r \in [0,1]^2$ and $f(\boldsymbol{y}_1), \ldots, f(\boldsymbol{y}_r) \in \mathbb{R}$ in the left side of Fig. 3.

It is well known that the Gaussian kernel $K \in \mathrm{C}^\infty(\mathbb{R}^2 \times \mathbb{R}^2)$ is a positive definite kernel. Therefore, the geostatistical models given in Sect. 4 are well-defined for the Gaussian kernels. In this section, all numerical tests of the (generalized) kriging methods are set up by the Gaussian kernel

$$K(\boldsymbol{x}, \boldsymbol{y}) := e^{-\theta^2 \|\boldsymbol{x}-\boldsymbol{y}\|_2^2}, \quad \text{for } \boldsymbol{x}, \boldsymbol{y} \in [0,1]^2,$$

with the shape parameter $\theta := 3$. Many other positive definite kernels can be found in the papers [8, 9] such as Sobolev-spline kernels, cubic-spline kernels, and (generalized) min kernels.

By the simple kriging method, we can obtain the kriging estimations at the unknown location $\boldsymbol{x}_0 \in [0,1]^2$ in Eqs. (17)–(18) dependent of the observed data $\boldsymbol{y}_1, \ldots, \boldsymbol{y}_r$ and $f(\boldsymbol{y}_1), \ldots, f(\boldsymbol{y}_r)$, for example in the right side of Fig. 3.

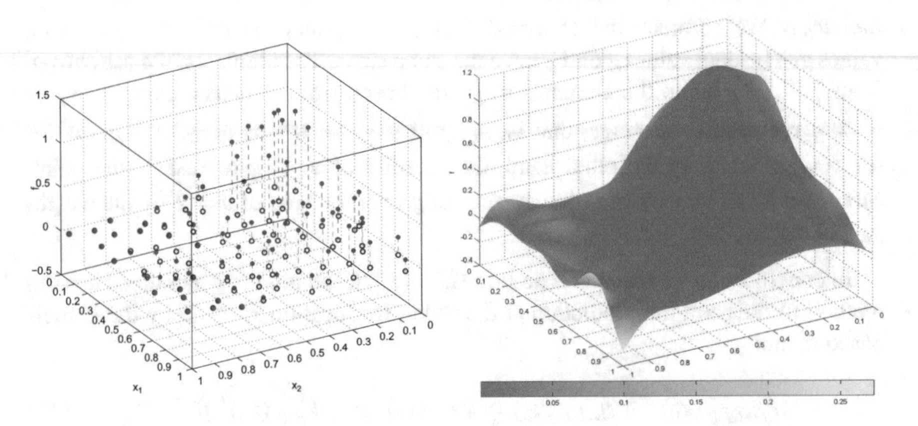

Fig. 3 In the *left panel*, the *blue circles* represent the Halton points $\boldsymbol{y}_1, \ldots, \boldsymbol{y}_r$ and the *red circles* represent the data values $f(\boldsymbol{y}_1), \ldots, f(\boldsymbol{y}_r)$ given by the Franke function f, where $r = 81$. The *right panel* is the 2D example of the simple kriging method. The kriging estimations given in Eqs. (17)–(18) are calculated by the observed data $\boldsymbol{y}_k$ and $f(\boldsymbol{y}_k)$ for $k = 1, \ldots, r$. The *bottom color bar* represents the values of the absolute errors

Now we study the numerical example for the missing observed data values at the data points $z_1, \ldots, z_t \in [0,1]^2$. But, we still have the positive or negative rates $g_1, \ldots, g_t \in \{0^+, 0^-\}$ of the gradient ∇ along the directions $\boldsymbol{e}_1, \ldots, \boldsymbol{e}_t$ at $z_1, \ldots, z_t$, where $\boldsymbol{e}_1 = \cdots = \boldsymbol{e}_t := (1,0)^T$ and $\nabla := \left(\frac{\partial}{\partial x_1}, \frac{\partial}{\partial x_2}\right)^T$. Here $g_k := 0^+$ if $\boldsymbol{e}_k^T \nabla f(z_k) \geq 0$ or $g_k := 0^-$ if $\boldsymbol{e}_k^T \nabla f(z_k) \leq 0$ for $k = 1, \ldots, t$. In the following numerical tests, we only have the observed data values $f(\boldsymbol{x}_1), \ldots, f(\boldsymbol{x}_n) \in \mathbb{R}$ at the data points $\boldsymbol{x}_1, \ldots, \boldsymbol{x}_n \in [0,1]^2$ such that $n + t = r$ and $\{\boldsymbol{x}_1, \ldots, \boldsymbol{x}_n\} \cup \{z_1, \ldots, z_t\} = \{\boldsymbol{y}_1, \ldots, \boldsymbol{y}_r\}$, for example in the left side of Fig. 4.

Therefore, we can obtain the generalized kriging models given in Eqs. (32)–(33) for the data $\boldsymbol{x}_1, \ldots, \boldsymbol{x}_n, z_1, \ldots, z_t$ and $f(\boldsymbol{x}_1), \ldots, f(\boldsymbol{x}_n), g_1, \ldots, g_t$, for example in the right side of Fig. 4. Comparing the right sides of Figs. 4 and 3, we find that the generalized kriging estimations are closed to the simple kriging estimations even though more than 30% observed data values are missed at $z_1, \ldots, z_t$.

Finally, we look at the root-mean-square errors of the simple and generalized kriging methods for the different observed data points in Table 2. By the computational results, the generalized kriging method is better than the simple kriging method. The reason is that the generalized kriging method can cover the observed gradients at the uncertain data points which the simple kriging method can not process. More numerical tests will appear in the personal website of the author.

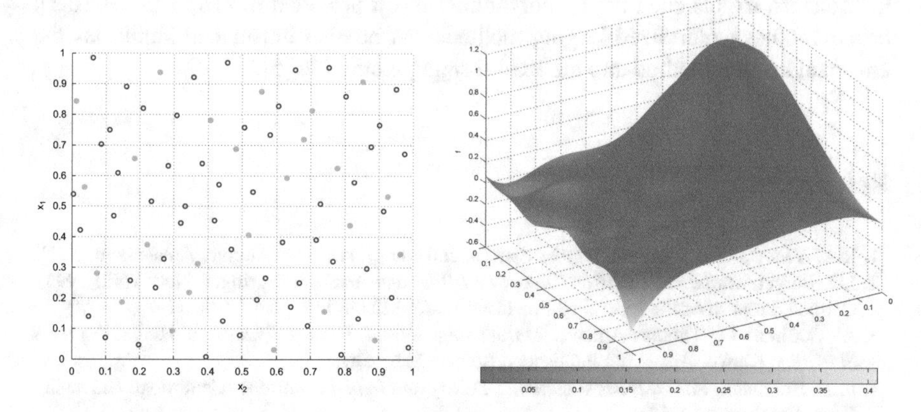

Fig. 4 The *left panel* shows the observation locations. The *blue circles* represent the data point $\boldsymbol{x}_1, \ldots, \boldsymbol{x}_n$ with the observed data values. The *green circles* represent the data points $z_1, \ldots, z_t$ without the observed data values but with the gradients observation. Here $n = 56$ and $t = 25$. All data points $\boldsymbol{x}_1, \ldots, \boldsymbol{x}_n$ and $z_1, \ldots, z_t$ are the same as the Halton points $\boldsymbol{y}_1, \ldots, \boldsymbol{y}_r$ in Fig. 3. The *right panel* is the 2D example of the generalized kriging method. The generalized kriging estimations given in Eqs. (32)–(33) are calculated by the discrete observation of the Franke function f at the data points $\boldsymbol{x}_1, \ldots, \boldsymbol{x}_n$ and $z_1, \ldots, z_t$. The *bottom color bar* represents the values of the absolute errors

Table 2 The root-mean-square errors of the simple and generalized kriging methods for the different observed data evaluated by the Franke function f

	Simple kriging	Generalized kriging	Simple kriging	Generalized kriging
Data points	$n = 56$	$n = 56$ and $t = 25$	$n = 45$	$n = 45$ and $t = 36$
Errors	0.0196	0.0184	0.0294	0.0288

6 Final Remarks

In this article, we generalize the kriging methods to estimate the predictions at the unknown locations by the mixtures of the observed data values and observed gradients at the chosen data points. Same as the ideas of the papers [17, 18], we combine the knowledge of regression analysis, stochastic analysis, approximation theory, and numerical analysis to obtain the kernel-based estimators induced by the Gaussian fields indexed by bounded linear functionals defined on Sobolev spaces. Therefore, we can obtain the new models in geostatistics.

To reduce the complexities of the discussions, we only consider the generalization of the simple kriging methods for the non-noised data in this article. So, we can always obtain the exact information of the positive or negative rates of the changes at the uncertain data points. Actually, the generalized kriging methods can also cover the noised data. In our next paper, we will show that the kernel-based estimators in Eq. (11) and Eqs. (12)–(13) can be updated to solve the inverse problems for the noised data. Moreover, the Gaussian fields given in Theorem 1 can be endowed with the nonzero means such as the polynomial mean provided in [18, Theorem 6.1]; hence another generalized kriging methods can be also introduced similar as the universal kriging method and the IRFk-kriging method in [5].

References

1. R.A. Adams, J.J.F. Fournier, *Sobolev Spaces*, 2nd edn. (Elsevier/Academic, Amsterdam, 2003)
2. J.O. Berger, *Statistical Decision Theory and Bayesian Analysis* (Springer, New York, 1993). Corrected reprint of the second edition (1985). MR 1234489 (94j:62002)
3. A. Berlinet, C. Thomas-Agnan, *Reproducing Kernel Hilbert Spaces in Probability and Statistics* (Kluwer Academic Publishers, Boston, MA, 2004)
4. M.D. Buhmann, *Radial Basis Functions: Theory and Implementations* (Cambridge University Press, Cambridge, 2003)
5. J.-P. Chilés, P. Delfiner, *Geostatistics: Modeling Spatial Uncertainty*, 2nd edn. (Wiley, Hoboken, 1999)
6. G.E. Fasshauer, *Meshfree Approximation Methods with* MATLAB (World Scientific Publishing, Hackensack, NJ, 2007)
7. G.E. Fasshauer, M.J. McCourt, *Kernel-Based Approximation Methods Using MATLAB* (World Scientific, Hackensack, NJ, 2015)
8. G.E. Fasshauer, Q. Ye, Reproducing kernels of generalized Sobolev spaces via a Green function approach with distributional operators. Numer. Math. **119**(3), 585–611 (2011)
9. G.E. Fasshauer, Q. Ye, Reproducing kernels of Sobolev spaces via a green kernel approach with differential operators and boundary operators. Adv. Comput. Math. **38**(4), 891–921 (2013)
10. S. Janson, *Gaussian Hilbert Spaces* (Cambridge University Press, Cambridge, 1997)
11. I. Karatzas, S.E. Shreve, *Brownian Motion and Stochastic Calculus*, 2nd edn. (Springer, New York, 1991)
12. B.A. Lockwood, M. Anitescu, Gradient-enhanced universal kriging for uncertainty propagation. Nucl. Sci. Eng. **170**(2), 168–195 (2012)
13. M. Scheuerer, R. Schaback, M. Schlather, Interpolation of spatial data—a stochastic or a deterministic problem? Eur. J. Appl. Math. **24**, 601–629 (2013)
14. M.L. Stein, *Interpolation of Spatial Data* (Springer, New York, 1999)
15. G. Wahba, *Spline Models for Observational Data* (Society for Industrial and Applied Mathematics, Philadelphia, PA, 1990)
16. H. Wendland, *Scattered Data Approximation* (Cambridge University Press, Cambridge, 2005)
17. Q. Ye, Kernel-based methods for stochastic partial differential equations (2015). arXiv:1303.5381v8
18. Q. Ye, Optimal designs of positive definite kernels for scattered data approximation. Appl. Comput. Harmon. Anal. **41**(1), 214–236 (2016)

Evaluation of Local Multiscale Approximation Spaces for Partition of Unity Methods

Marc Alexander Schweitzer and Sa Wu

Abstract The simulation of the behavior of heterogeneous and composite materials poses a number of challenges to numerical methods e.g. due to the presence of discontinuous material coefficients. Moreover, the material properties of fibers and inclusions are significantly different from those of the surrounding matrix. Thus, the gradients of the solution feature a substantial discontinuity at the material interface between inclusions and matrix. Hence, materials with many fine scale inclusions need a very high resolution mesh in the context of traditional finite element (FE) analysis. However, many approaches within the context of numerical homogenization have been proposed to tackle and overcome this need for a large number of degrees of freedom. To this end, either discontinuous coefficients are replaced by smooth effective coefficients or, standard FE shape functions are replaced by more complex, numerically computed shape functions while the overall quality of the approximation is retained. In this paper we study two-dimensional examples of heat transfer and (linear) elasticity in composite materials using a number of different homogenization approaches with the overall goal of evaluating and comparing their performance when used for the construction of multiscale enrichment functions for a partition of unity method (PUM).

M.A. Schweitzer
Institut für Numerische Simulation, Rheinische Friedrich-Wilhelms-Universität Bonn, Wegelerstr. 6, 53115 Bonn, Germany

Fraunhofer Institute for Algorithms and Scientific Computing SCAI Sankt Augustin, Germany
e-mail: schweitzer@ins.uni-bonn.de; marc.alexander.schweitzer@scai.fraunhofer.de

S. Wu (✉)
Institut für Numerische Simulation, Rheinische Friedrich-Wilhelms-Universität Bonn, Bonn, Germany
e-mail: wu@ins.uni-bonn.de

M. Griebel, M.A. Schweitzer (eds.), *Meshfree Methods for Partial Differential Equations VIII*, Lecture Notes in Computational Science and Engineering 115,
DOI 10.1007/978-3-319-51954-8_9

1 Introduction

In a classical FE analysis approach to the simulation of heterogeneous and composite materials, a major challenge is the need for a very high resolution mesh i.e., a large number of degrees of freedom, to resolve the discontinuities of the material parameters associated with the heterogeneity of the material. Thus, the resulting linear systems are very large and often numerically challenging to solve due to a large variation of the material parameters. A number of approaches [2, 7–9, 11, 16] have been proposed over the years to overcome these issues. Typically, in these so-called numerical homogenization techniques, the global heterogeneous problem is tackled by a two-step approach: First, we either construct simplified effective material parameters (an homogenized model) or we construct more effective multiscale basis functions that respect the material's heterogeneity implicitly. Then, the global material response is approximated either via the homogenized model using standard FE basis functions or the heterogeneous model is approximated directly with the help of the constructed multiscale basis functions. Note that the second step, the so-called online phase, requires a much smaller number of degrees of freedom to attain the required accuracy than the much more involved offline phase in the first step.

Even though all numerical homogenization approaches share this general procedure their specific goals may be quite different. For instance, some techniques [2, 7, 11] aim at the construction of multiscale basis functions that provide spectral convergence properties while most other approaches settle for linear convergence with respect to the energy norm. Note, however, that for most practical purposes the asymptotic behavior of the considered construction are of limited value since we seek to employ a very small number of degrees of freedom in the online phase only. Thus, we usually stay in the pre-asymptotic regime in the online phase in practice and there it is not clear which homogenization approach provides the most efficient approximation overall to the global problem.

The goal of our work is the evaluation of various numerical homogenization techniques for the construction of multiscale basis functions with respect to their effectiveness in the context of so-called generalized finite element methods (GFEM), i.e., partition of unity methods (PUM). To this end, we compare the approximation properties of the considered numerical homogenization approaches using tools from approximation theory (Kolmogorov n-widths and $\sup - \inf$s), the computational costs associated with the construction of the respective multiscale basis functions as well as their performance within a PUM in this paper. In particular, we consider a bi-material composite with circular/spherical inclusions and study the robustness of the different multiscale basis functions with respect to the contrast of the material parameters and the distribution of the inclusions.

The remainder of the paper is structured as follows. First, we describe the considered model problem in Sect. 1.1 before we shortly review the PUM framework for the discretization of the global model problem in Sect. 2. Then, we discuss the approximation theoretic tools employed for the comparison of different multiscale basis systems and present the respective general construction principles in Sect. 4 before we introduce the concrete approximation spaces from numerical homogenization considered in our numerical experiments. Here, we focus on the numerical computation of two quantities, the n-widths, which characterizes the accuracy that can be obtained using an optimal choice of n basis functions, and the sup − infs, which characterize the accuracy that is obtained with a specific choice of n particular basis functions. We summarize the results of our numerical experiments in Sect. 6. Finally, we close with some remarks in Sect. 7.

1.1 Model Problem

To introduce our model problem, we consider a heterogeneous or composite bi-material defined via two disjoint open sets Ω_1, Ω_2 such that

$$\overline{\Omega} = \overline{\Omega_1} \cup \overline{\Omega_2} \subset \mathbb{R}^2, \quad \kappa : \Omega \to \mathbb{R}, \quad \kappa(x) = \begin{cases} 1 & x \in \Omega_1, \\ C & x \in \Omega_2, \end{cases} \tag{1}$$

where Ω_1 denotes the matrix phase and Ω_2 the union of all inclusions, compare Fig. 1. When we study the elastic response of the material, we assume a Poisson ratio of $\nu = 0.3$ in both phases and choose the Young modulus by $E(x) := \kappa(x)$ given in (1). Thus, the Lamé coefficients are given by

$$\lambda(x) = \frac{\nu\kappa(x)}{(1+\nu)(1-2\nu)}, \quad \mu(x) = \frac{\kappa(x)}{2(1+\nu)}.$$

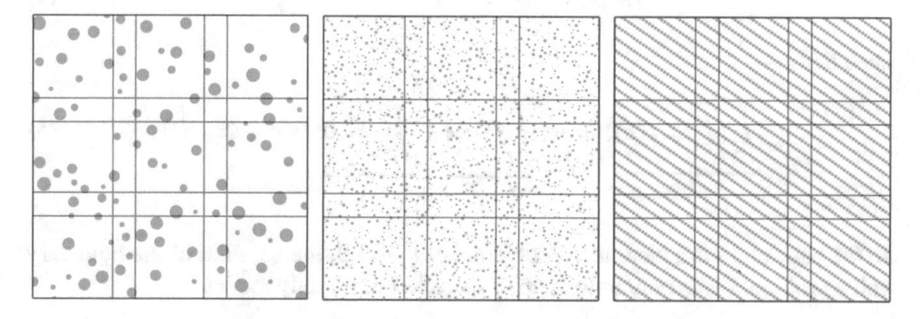

Fig. 1 Square domain Ω with Dirichlet boundary Γ_D (*red*), Neumann boundary Γ_N (*magenta*) and 121 randomly distributed circular inclusion of random size, 2500 randomly distributed and 2597 regularly distributed circular inclusions of the same size (*light green circles*). The *gray lines* depict the internal patch boundaries $\partial\omega_l \setminus \partial\Omega$ of a 3×3 PUM cover. We will refer to these material configurations as cases $\mathbb{A}$, $\mathbb{B}$, $\mathbb{C}$ (*left to right*)

The equations for both heat transfer and linear elasticity can be written within the same notation

$$\begin{aligned} -\nabla \cdot \sigma(u) &= f \text{ in } \Omega, \\ u &= g \text{ on } \Gamma_D \subset \partial\Omega, \\ \sigma(u) \cdot \mathfrak{n} &= h \text{ on } \Gamma_N = \partial\Omega \setminus \Gamma_D \end{aligned} \tag{2}$$

with outer normal $\mathfrak{n}$. For ease of notation we do not differentiate explicitly between the scalar solution $u : \Omega \to \mathbb{R}$ to the heat equation, where

$$\epsilon(u) := \nabla u, \qquad \sigma(u) := \kappa\epsilon(u), \qquad \langle \sigma(u), \epsilon(u) \rangle := \sigma(u) \cdot \epsilon(u),$$

and the vector-valued linear elastic response $u : \Omega \to \mathbb{R}^2$ with

$$\epsilon(u) := \frac{1}{2}\left(\nabla u + \nabla u^T\right), \; \sigma(u) := 2\mu\epsilon(u) + \lambda \operatorname{tr} \epsilon(u)\mathbb{I}, \; \langle \sigma(u), \epsilon(u) \rangle := \sigma(u) : \epsilon(u).$$

But, we rather employ a more abstract notation that encompasses both cases by defining

$$\begin{aligned} &\mathcal{H}^1(\Omega) := H^1(\Omega) \text{ for the heat equation and} \\ &\mathcal{H}^1(\Omega) := \left(H^1(\Omega)\right)^2 \text{ for the linear elasticity equation.} \end{aligned} \tag{3}$$

Then, with the standard approach of handling the boundary and load data by particular solutions, see Sect. 4.1, we arrive at the weak form of (2): Find a solution $u \in \mathcal{H}^1_D(\Omega)$ with

$$\mathcal{H}^1_D(\Omega) := \left\{ v \in \mathcal{H}^1(\Omega) : v|_{\Gamma_D} = 0 \right\}, \tag{4}$$

such that

$$\underbrace{\int_\Omega \langle \sigma(u), \epsilon(v) \rangle \, \mathrm{d}x}_{=:a(u,v)} = \underbrace{\int_\Omega f v \, \mathrm{d}x + \int_{\Gamma_N} h v \, \mathrm{d}s}_{=:b(v)} \; \forall v \in \mathcal{H}^1_D(\Omega). \tag{5}$$

In the case of a pure Neumann problem, i.e. $\Gamma_D = \emptyset$, the solution of the boundary value problem (2) is only unique up to so-called rigid body modes

$$\mathcal{R} := \left\{ v \in \mathcal{H}^1(\Omega) : a(v, v) = 0 \right\} = \operatorname{span}\left\{ \rho_r : r < R \right\} \tag{6}$$

with $r, R \in \mathbb{N}_0$. A basis of $\mathcal{R}$ is for instance given by

$$\rho_0 = (1, 0), \quad \rho_1 = (0, 1), \quad \rho_2 = (-y, x)$$

for the two-dimensional linear elasticity equation.

In the remainder of this work we focus on rectangular domains with circular inclusions

$$\Omega = [-A, A] \times [-B, B], \qquad \Omega_2 = \cup_{i<I} B_{\delta_i}(x_i)$$

with $i, I \in \mathbb{N}_0$. In our experiments we consider e.g. $I = 121, I = 2500$ (randomly distributed inclusions of different size), $I = 2567$ (identical inclusions on grid), see also Fig. 1. Note that we assume that all inclusions are separated by some minimum distance $\delta > 0$, i.e.

$$\text{dist}(B_{\delta_i}(x_i), B_{\delta_j}(x_j)) > \delta \quad \forall i \neq j \,.$$

Moreover, we assign Dirichlet and Neumann boundary conditions on

$$\Gamma_D := \{-A, A\} \times [-B, B], \qquad \Gamma_N := [-A, A] \times \{-B, B\}.$$

respectively, see also Fig. 1.

For the global discretization of these model problems in the online step we employ the PUM scheme discussed in the following section.

2 Partition of Unity Method

The notion of a partition of unity method (PUM) was coined in [3, 4] and is based on the special finite element methods developed in [5]. The abstract ingredients of a PUM are a partition of unity (PU) $\{\varphi_l : l = 0, \ldots, N\}$ (compare Definition 1) and a collection of local approximation spaces $\mathcal{V}_l(\omega_l) := \text{span}\langle \vartheta_l^m \rangle_m$ defined on the patches $\omega_l := \text{supp}(\varphi_l)$ for $l = 0, \ldots, N$ (Fig. 2). With these two ingredients we define the PUM space

$$V^{\text{PU}} := \sum_{l=0}^{N} \varphi_l \mathcal{V}_l = \text{span}\langle \varphi_l \vartheta_l^m \rangle; \tag{7}$$

i.e., the shape functions of a PUM space are simply defined as the products of the PU functions φ_l and the local approximation functions ϑ_l^m. For PUM spaces (7) which employ a PU $\{\varphi_l\}$ satisfying Definition 1 there hold the error estimates of Theorem 1 due to [4, 20].

Definition 1 (Partition of Unity) Let $\Omega \subset \mathbb{R}^d$ be an open set and let $\{\varphi_l : l = 0, \ldots, N\}$ be a collection of N Lipschitz functions with

$$\sum_{l=0}^{N} \varphi_l \equiv 1 \text{ on } \overline{\Omega}, \quad \|\varphi_l\|_{L^\infty(\mathbb{R}^d)} \leq C_\infty, \quad \|\nabla \varphi_l\|_{L^\infty(\mathbb{R}^d)} \leq \frac{C_\nabla}{\text{diam}(\omega_l)},$$

where C_∞ and C_∇ are two positive constants and $\omega_l := \text{supp}(\varphi_l)^\circ$ is a Lipschitz domain. The collection of functions $\{\varphi_l : l = 0, \dots, N\}$ is referred to as a partition of unity (PU). The sets ω_l are called patches and their collection is referred to as a cover $C_\Omega := \{\omega_l : l = 0, \dots, N\}$ of the domain Ω.

Theorem 1 *Let $\Omega \subset \mathbb{R}^d$ be a Lipschitz domain. Let $\{\varphi_l : l = 0, \dots, N\}$ be a partition of unity according to Definition 1 which is moreover non-negative, i.e.*

$$0 \le \varphi_l(x) \le 1 \quad \text{for all } x \in \Omega, l = 0, \dots, N.$$

Let us further introduce the covering index $\lambda_{C_\Omega} : \Omega \to \mathbb{N}$ such that

$$\lambda_{C_\Omega}(x) := \text{card}(\{\omega_l \in C_\Omega : x \in \omega_l\})$$

and let us assume that

$$\lambda_{C_\Omega}(x) \le \Lambda \in \mathbb{N} \quad \text{for all } x \in \Omega$$

with Λ independent of the number of patches N. Let a collection of local approximation spaces $\mathcal{V}_l = \text{span}\langle \vartheta_l^n \rangle \subset H^1(\omega_l)$ be given. Let $u \in H^1(\Omega)$ be the function to be approximated. Assume that the local approximation spaces $\mathcal{V}_l$ have the following approximation properties: On each patch $\Omega \cap \omega_l$, the function u can be approximated by a function $v_l \in \mathcal{V}_l$ such that

$$\|u - v_l\|_{L^2(\Omega\cap\omega_l)} \le \hat{\epsilon}_l \quad \text{and} \quad \|\nabla(u - v_l)\|_{L^2(\Omega\cap\omega_l)} \le \tilde{\epsilon}_l \tag{8}$$

hold for all $l = 0, \dots, N$. Then, the function

$$v^{\text{PU}} := \sum_{l=0}^{N} \varphi_l v_l \in V^{\text{PU}} \subset H^1(\Omega)$$

satisfies the global estimates

$$\|u - v^{\text{PU}}\|_{L^2(\Omega)} \le \sqrt{C_\infty}\Big(\sum_{l=0}^{N} \hat{\epsilon}_l^2\Big)^{1/2},$$

$$\|\nabla(u - v^{\text{PU}})\|_{L^2(\Omega)} \le \sqrt{2}\Big(\sum_{l=0}^{N} \Lambda\Big(\frac{C_\nabla}{\text{diam}(\omega_l)}\Big)^2 \hat{\epsilon}_l^2 + C_\infty \tilde{\epsilon}_l^2\Big)^{1/2}.$$

Thus, the PU functions provide the locality and global regularity of the product functions whereas the local spaces $\mathcal{V}_l$, which are all independent of each other, equip V^{PU} with its approximation power. Thus, the standard choice of $\mathcal{V}_l$ for the approximation of a smooth function u with PUM are classical polynomial approximation spaces $\mathcal{P}^{p_l} = \mathcal{V}_l$ of degree p_l on the patches ω_l. If the function

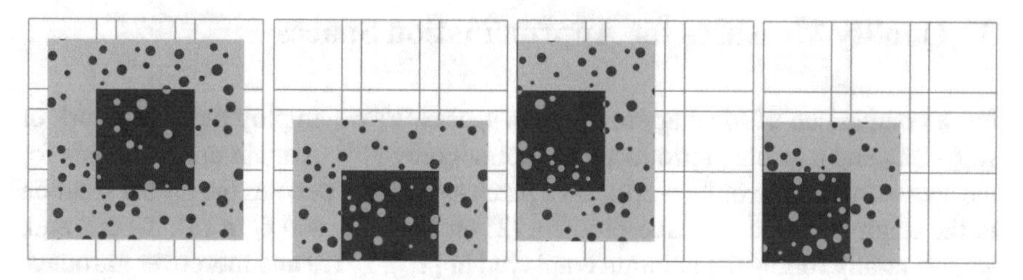

Fig. 2 Overlapping patches $\omega_l \subset \omega_l^+$ from 3×3 grid for material configuration $\mathbb{A}$, depicted are $\Omega_2 \cap \omega_l$ (*light green circles*) and $\Omega_2 \cap (\omega_l^+ \setminus \omega_l)$ (*red circles*), matrix $\Omega_1 \cap \omega_l$ (*blue*) and $\Omega_1 \cap (\omega_l^+ \setminus \omega_l)$ (*green*), Dirichlet boundaries $\Gamma_{l,D}$ (*red*), Neumann boundaries $\Gamma_{l,N}$ (*magenta*). We have ω_4 (*first*, no bc), ω_1 (*second*, Neumann bc), ω_3 (*third*, Dirichlet bc), ω_0 (*last*, mixed bc)

u, however, has some irregular features such as singularities or discontinuities locally these polynomial spaces are often enriched by appropriate non-smooth basis functions, i.e. $\mathcal{V}_l = \mathcal{P}^{p_l} + \mathcal{E}_l$ for some patches ω_l, see e.g. [19]. In our setting, where we are concerned with the approximation of solutions of (2) with highly oscillatory and discontinuous material coefficients, such irregular features however arise everywhere in the domain and thus we will employ enrichment on every patch ω_l.[1] Unfortunately, appropriate enrichment spaces $\mathcal{E}_l$ for heterogeneous materials are in general not known analytically and depend e.g. on the distribution of the inclusions and the material properties. Thus, the major challenge here is to pre-compute enrichment spaces with good approximation properties, i.e. allowing for small local errors $\hat{\epsilon}_l$ and $\tilde{\epsilon}_l$ in (8), for a particular material configuration efficiently and to identify the most cost effective approach overall.

For the construction of such local enrichment spaces $\mathcal{E}_i$ we employ a number of different numerical homogenization techniques and compare their approximation properties and the computational cost involved in the construction. Moreover, we are concerned with the robustness of the resulting enrichments with respect to varying loading conditions, distribution of inclusions and the contrast of the material parameters. Note that the local enrichment spaces $\mathcal{E}_i$ on different patches ω_i are completely independent of each other and thus can be computed in parallel using an arbitrary homogenization approach. Throughout this paper, however, we will use the same homogenization technique for all patches and we consider in particular the approaches from [2, 7, 11]. To evaluate and compare the approximation qualities of the enrichment spaces arising from these homogenization techniques we employ tools from approximation theory which we shortly introduce in the following section.

[1]Note that we consider so-called flat-top PU functions only since they allow to control the stability of an enriched basis withe arbitrary enrichments [18].

3 Quality Measures for Approximation Spaces

For a comparison of local approximation spaces $\mathcal{V}_l$ we employ the framework of sup – infs and n-widths introduced by Kolmogorov [14] to attain upper bounds for the approximation error $\|u - v_l\|_{\omega_l}$ on a patch ω_l where u denotes the global solution to the original boundary value problem (2) on Ω, i.e. $\hat{\epsilon}_l$ and $\tilde{\epsilon}_l$ in (8). To this end, we essentially follow the notation employed in [2, 6, 7, 12] and moreover introduce $u|_{\omega_l}$ to denote the local restriction of the global solution u to a particular patch ω_l and assume that these local restrictions satisfy $u|_{\omega_l} \in \mathcal{W}_l(\omega_l)$ for some local Hilbert space $(\mathcal{W}_l, (\cdot,\cdot)_{\mathcal{W}_l})$ given on ω_l.

Then, we can characterize the quality of a local approximation space $\mathcal{V}_l$ for its use in our PUM, suitable for the approximation of (2), also by studying the best-approximation $v_l \in \mathcal{V}_l$ to any element of the unit ball

$$B_{\mathcal{W}_l} := \left\{u \in \mathcal{W}_l : (u,u)_{\mathcal{W}_l} = 1\right\}.$$

in $\mathcal{W}_l$ with respect to any desired norm $\|\cdot\|$. We fix this arbitrary norm for the measurement of the error $\|u|_{\omega_l} - v_l\|$ via the assumption that $\mathcal{W}_l \subset \mathcal{H}_l$ is contained in a larger Hilbert space $(\mathcal{H}_l, (\cdot,\cdot)_{\mathcal{H}_l})$ and consider its induced norm for the error $\|u|_{\omega_l} - v_l\|_{\mathcal{H}_l}$ in the following, i.e. we assume $\mathcal{V}_l \subset \mathcal{H}_l$. Then, there holds the obvious upper bound for the best-approximation from $\mathcal{V}_l$ to data in $\mathcal{W}_l$ measured in $\mathcal{H}_l$

$$\begin{aligned}
\inf_{v\in\mathcal{V}_l} \|v - u|_{\omega_l}\|_{\mathcal{H}_l} &= \|u|_{\omega_l}\|_{\mathcal{W}_l} \inf_{v\in\mathcal{V}_l} \left\| \frac{v}{\|u|_{\omega_l}\|_{\mathcal{W}_l}} - \frac{u|_{\omega_l}}{\|u|_{\omega_l}\|_{\mathcal{W}_l}} \right\|_{\mathcal{H}_l} \\
&= \|u|_{\omega_l}\|_{\mathcal{W}_l} \inf_{v\in\mathcal{V}_l} \left\| v - \frac{u|_{\omega_l}}{\|u|_{\omega_l}\|_{\mathcal{W}_l}} \right\|_{\mathcal{H}_l} \\
&\le \|u|_{\omega_l}\|_{\mathcal{W}_l} \sup_{\|w\|_{\mathcal{W}_l}=1} \inf_{v\in\mathcal{V}_l} \|v - w\|_{\mathcal{H}_l} \\
&= \|u|_{\omega_l}\|_{\mathcal{W}_l} \Psi(\mathcal{V}_l)
\end{aligned}$$

where

$$\Psi(\mathcal{V}_l) = \Psi\left(\mathcal{V}_l, \mathcal{W}_l, (\cdot,\cdot)_{\mathcal{W}_l}, \mathcal{H}_l, (\cdot,\cdot)_{\mathcal{H}_l}\right) := \sup_{\|w\|_{\mathcal{W}_l}=1} \inf_{v\in\mathcal{V}_l} \|w - v\|_{\mathcal{H}_l}. \tag{9}$$

With the help of (9) we can moreover identify the best local approximation space for a fixed dimension n via

$$\mathcal{V}^*_{l,n} := \operatorname*{arg\,inf}_{\mathcal{V}_l\subset\mathcal{H}_l \dim\mathcal{V}_l=n} \Psi(\mathcal{V}_l, \mathcal{W}_l, (\cdot,\cdot)_{\mathcal{W}_l}, \mathcal{H}_l, (\cdot,\cdot)_{\mathcal{H}_l}) \tag{10}$$

and introduce the so-called Kolmogorov n-width[2]

$$d_n(\mathcal{W}_l, (\cdot,\cdot)_{\mathcal{W}_l}, \mathcal{H}_l, (\cdot,\cdot)_{\mathcal{H}_l}) := \inf_{\mathcal{V}_l \subset \mathcal{H}_l \dim \mathcal{V}_l = n} \Psi(\mathcal{V}_l, \mathcal{W}_l, (\cdot,\cdot)_{\mathcal{W}_l}, \mathcal{H}_l, (\cdot,\cdot)_{\mathcal{H}_l}). \tag{11}$$

These abstract quantities can in fact be computed via the solution of generalized eigenvalue problems [14, 17] if $\mathcal{W}_l \stackrel{C}{\hookrightarrow} \mathcal{H}_l$ is dense and compactly embedded. The eigenpairs $(\eta_{l,k}, \chi_{l,k}) \in \mathcal{W}_l \times \mathbb{R}$ of

$$(\eta_{l,k}, v)_{\mathcal{H}_l} = \chi_{l,k} (u, v)_{\mathcal{W}_l} \quad \forall v \in \mathcal{W}_l \tag{12}$$

are comprised of orthogonal eigenvectors $\eta_{l,k}$ [12] with respective eigenvalues $\chi_{l,k}$ such that

$$\chi_{l,0} \geq \chi_{l,1} \geq \ldots \searrow 0$$

holds if $\mathcal{W}_l \stackrel{C}{\hookrightarrow} \mathcal{H}_l$. Thus, we can directly compute the n-width as

$$d_n(\mathcal{W}_l, (\cdot,\cdot)_{\mathcal{W}_l}, \mathcal{H}_l, (\cdot,\cdot)_{\mathcal{H}_l}) := \sqrt{\chi_{l,n}} \tag{13}$$

and identify the associated optimal approximation space by

$$\mathcal{V}^*_{l,n} := \operatorname{span}\{\eta_{l,k} : k < n\}.$$

Note, however, that the optimal subspace $\mathcal{V}^*_{l,n}$ which realizes the Kolmogorov n-width is not unique [14]. Thus, there might be another subspace $\mathcal{V}_l$ with comparable approximation properties which may be determined with less computational work. In particular, its construction might not require the solution of a generalized eigenvalue problem such as (12).

We can employ a similar technique to compute the sup – infs (9). To this end, let I denote the identity operator and $\pi_{\mathcal{V}_l,(\cdot,\cdot)_{\mathcal{H}_l}}$ the $(\cdot,\cdot)_{\mathcal{H}_l}$-orthogonal projection onto $\mathcal{V}_l$. Then, for any $w \in \mathcal{W}_l$ the best approximation from $\mathcal{V}_l$ is given by $\pi_{\mathcal{V}_l,(\cdot,\cdot)_{\mathcal{H}_l}} w$. Thus, the non-vanishing eigenvalues ψ of

$$\left((I - \pi_{\mathcal{V}_l,(\cdot,\cdot)_{\mathcal{H}_l}})w, (I - \pi_{\mathcal{V}_l,(\cdot,\cdot)_{\mathcal{H}_l}})v\right)_{\mathcal{H}_l} = \psi\,(w, v)_{\mathcal{W}_l} \quad \forall v \in \mathcal{W}_l \tag{14}$$

[2]In the literature the inner products usually not part of the definitions of the sup – inf and n-width but assumed to be fixed for $\mathcal{H}_l$, $\mathcal{W}_l$. We want to point out this dependence and later on characterize the approximation in energy norm, H^1- and L^2-norms and will, thus, employ different definitions for the inner product $(\cdot,\cdot)_{\mathcal{H}_l}$ on $\mathcal{H}_l$.

satisfy [12]

$$\psi_0 \geq \psi_1 \geq \ldots \searrow 0$$

and it holds

$$\psi_0 = \sup_{w \in \mathcal{W}_l} \frac{\left\| w - \pi_{\mathcal{V}_l,(\cdot,\cdot)_{\mathcal{H}_l}} w \right\|^2_{\mathcal{H}_l}}{\|w\|^2_{\mathcal{W}_l}}$$

such that

$$\Psi(\mathcal{V}_l, \mathcal{W}_l, (\cdot,\cdot)_{\mathcal{W}_l}, \mathcal{H}_l, (\cdot,\cdot)_{\mathcal{H}_l}) = \sqrt{\psi_0}. \tag{15}$$

Note that for an actual computation of (13) and (15) by solving (12) and (14) we need access to the basis functions of $\mathcal{V}_l$ and $\mathcal{W}_l$ as well as the inner product $(\cdot,\cdot)_{\mathcal{H}_l}$ and its induced norm $\|\cdot\|_{\mathcal{H}_l}$. Yet, there is no need for an explicit access to basis of $\mathcal{H}_l$.

4 Construction of Spaces $\mathcal{W}_l$ and $\mathcal{H}_l$

Thus, let us now consider the construction of appropriate spaces $\mathcal{W}_l$ and $\mathcal{H}_l$ on a particular patch ω_l, compare [2, 6, 7, 12, 14, 17]. To this end, we first restrict our global problem (2) on Ω to the patch ω_l. Moreover, we reduce this localized problem to a homogeneous problem to allow for a direct comparison of different constructions of local approximation spaces $\mathcal{V}_l \subset \mathcal{H}_l$ irrespective of the employed treatment of the data f, g, h, compare [2, 8, 9].

4.1 $\mathcal{H}_l$ via Reduction to Homogeneous Problem

Therefore, we employ the following reduction to a homogeneous problem on a patch ω_l. Let us assume the existence of a solution u on Ω to (2) such that

$$u|_{\omega_l} = u_{l,f} + u_{l,g} + u_{l,h} + u_l$$

holds, where $u_{l,f}, u_{l,g}, u_{l,h}$ denote particular solutions for the load f and the boundary data g, h. Thus, the particular solutions $u_{l,f}$ satisfy

$$\begin{aligned} \nabla \cdot \sigma(u_{l,f}) &= f \text{ in } \omega_l, \\ u_{l,f} &= 0 \text{ on } \Gamma_{l,D} := \partial\omega_l \cap \Gamma_D, \\ \sigma(u_{l,f}) \cdot \mathfrak{n} &= 0 \text{ on } \Gamma_{l,N} := \partial\omega_l \cap \Gamma_N. \end{aligned}$$

Similarly, the equations

$$\begin{aligned} \nabla \cdot \sigma(u_{l,g}) &= 0 & \nabla \cdot \sigma(u_{l,h}) &= 0 & &\text{in } \omega_l, \\ u_{l,g} &= g & u_{l,h} &= 0 & &\text{on } \Gamma_{l,D} \\ \sigma(u_{l,g}) \cdot \mathfrak{n} &= 0 & \sigma(u_{l,h}) \cdot \mathfrak{n} &= h & &\text{on } \Gamma_{l,N}, \end{aligned}$$

hold for $u_{l,g}$ and $u_{l,h}$. Assume that these particular solutions are known, then it remains to solve either the Neumann problem

$$\begin{aligned} \nabla \cdot \sigma(u_l) &= 0 \text{ in } \omega_l, & \sigma(u_l) \cdot \mathfrak{n} &= 0 \text{ on } \Gamma_{l,N}, \\ u_l &= 0 \text{ on } \Gamma_{l,D}, & \sigma(u_l) \cdot \mathfrak{n} &= h_l \text{ on } \Gamma_{l,0} \end{aligned} \tag{16}$$

with an unknown traction $h_l := \sigma(u - u_{l,f} - u_{l,g} - u_{l,h}) \cdot \mathfrak{n}$ on $\Gamma_{l,0} := \partial\omega_l \setminus (\Gamma_D \cup \Gamma_N)$, or the Dirichlet problem

$$\begin{aligned} \nabla \cdot \sigma(u_l) &= 0 \text{ in } \omega_l, & \sigma(u_l) \cdot \mathfrak{n} &= 0 \text{ on } \Gamma_{l,N}, \\ u_l &= 0 \text{ on } \Gamma_{l,D}, & u_l &= g_l \text{ on } \Gamma_{l,0} \end{aligned} \tag{17}$$

with unknown trace $g_l := u - u_{l,f} - u_{l,g} - u_{l,h}$ on $\Gamma_{l,0}$. Thus, with known particular solutions $u_{l,f}$, $u_{l,g}$ and $u_{l,h}$, it is sufficient to consider the construction of a suitable approximation space $\mathcal{V}_l$ for the homogeneous problems (16) or (17). Throughout this paper, we will employ this approach and will be concerned with the construction of spaces $\mathcal{V}_l$ for the efficient approximation of solutions u_l to (16) or (17).

In analogy to (3) and (4), we then have $u_l \in \mathcal{H}^1_{l,D}$ where

$$\mathcal{H}^1_{l,D} := \left\{ v \in \mathcal{H}^1(\omega_l) : v|_{\Gamma_{l,D}} = 0 \right\}. \tag{18}$$

Given a suitable traction h_l on $\Gamma_{l,0}$ the weak form of (16) is to find $u_l \in \mathcal{H}^1_{l,D}$ such that

$$\underbrace{\int_{\omega_l} \langle \sigma(u_l), \epsilon(v) \rangle \, dx}_{=:a_l(u_l,v)} = \underbrace{\int_{\Gamma_{l,0}} h_l v \, ds}_{=:b_l(v)} \quad \forall v \in \mathcal{H}^1_{l,D}. \tag{19}$$

Note that this problem, however, does not admit a unique solution in the case of $\Gamma_{l,D} = \emptyset$, i.e. $\mathcal{H}^1_{l,D} = \mathcal{H}^1(\omega_l)$. We therefore restrict ourselves to $\mathcal{H}^1(\omega_l)/\mathcal{R}$ in this case.

The space of all solutions to (16) up to rigid body modes (6), i.e., the space of all so-called $\mathcal{L}$-harmonic functions, obviously is a suitable space $\mathcal{H}_l$, i.e.,

$$\mathcal{H}_l := \left\{ v \in \mathcal{H}^1_{l,D} : \mathcal{L}(v) = -\nabla \cdot \sigma(v) = 0 \right\},$$

see (19). With rigid modes removed a_l defines an inner product on $\mathcal{H}_l$ and induces a norm

$$\|v\|_{\mathcal{H}_l} := \|v\|_{a_l} := \sqrt{a_l(v,v)} \tag{20}$$

so that the closure of $(\mathcal{H}_l, \|\cdot\|_{\mathcal{H}_l})$ gives our desired Hilbert space. In the remainder of the paper, we will be concerned with the following three choices of $\mathcal{H}_l$

$$(u,v)_{\mathcal{H}_l} = (u,v)_{l,E} := a_l(u,v), \qquad \|u\|_{\mathcal{H}_l} = \|u\|_{l,E} := \sqrt{a_l(u,v)}, \tag{21}$$

$$(u,v)_{\mathcal{H}_l} = (u,v)_{l,1} := \int_{\omega_l} \langle \nabla u, \nabla v \rangle \,\mathrm{d}x, \quad \|u\|_{\mathcal{H}_l} = \|u\|_{l,1} := \sqrt{(u,u)_{l,1}} \tag{22}$$

and

$$(u,v)_{\mathcal{H}_l} = (u,v)_{l,0} := \int_{\omega_l} uv \,\mathrm{d}x, \qquad \|u\|_{\mathcal{H}_l} = \|u\|_{l,0} := \sqrt{(u,u)_{l,0}} \tag{23}$$

for a comparison of spaces $\mathcal{V}_l$ with respect to the energy norm $\|\cdot\|_{l,E}$, H^1-norm $\|\cdot\|_{l,1}$ and the L^2-norm $\|\cdot\|_{l,0}$ on a particular patch ω_l.

4.2 $\mathcal{W}_l$ via Oversampling

We now define the space $\mathcal{W}_l$ of target functions (cmp. Sect. 3) for which we then want to construct appropriate approximation spaces $\mathcal{V}_l \subset \mathcal{H}_l$. To this end, we follow the approach of Babuška and Lipton [2], Babuška et al. [7] which is based on the observation that the global solution u of the underlying boundary value problem (2) is in general more regular than the solutions to the local homogeneous problems (16) or (17), compare also Theorem 1 and (8). Hence, we only need to be concerned with the construction of a good approximation space for (16) or (17), i.e. for all local solutions u_l which admit an $\mathcal{L}$-harmonic extension to the whole domain Ω. This approach, however, requires global operations on Ω and thus is computationally too expensive. Therefore, we employ a so-called oversampling approach, i.e., we localize this construction from the global domain Ω to an enlarged patch $\omega_l^+ \supset \omega_l$ (Fig. 2) and consider the $\mathcal{L}$-harmonic extension to $\omega_l^+ \subset \Omega$ only. To this end, let us introduce the following notation

$$\begin{aligned} \omega_l^+ &:= \beta\omega_l, & \Gamma_{l,D}^+ &:= \Gamma_D \cap \partial\omega_l^+, \\ \Gamma_{l,N}^+ &:= \Gamma_N \cap \partial\omega_l^+, & \Gamma_{l,0}^+ &:= \partial\omega_l^+ \setminus (\Gamma_{l,D} \cup \Gamma_{l,N}), \end{aligned} \tag{24}$$

where we make the assumption that $\Gamma_{l,D} = \emptyset \Leftrightarrow \Gamma_{l,D}^{+} = \emptyset$ to avoid any artificial influence of essential boundary conditions on $\Gamma_D \cap (\omega_l^+ \setminus \omega_l)$ when $\Gamma_{l,D} = \Gamma_D \cap \omega_l = \emptyset$. Moreover, we employ the following construction to define the overlapping patches ω_l and their extension ω_l^+. We define ω_l by overlaying the cells $C_H(x_l)$ of a regular grid of meshwidth H of our domain Ω centered at x_l with extended concentric cells $B_{\alpha H}(x_l) \supset C_H(x_l)$ with $\alpha > 1$. Then, the enlarged patches ω_l^+ are obtained by scaling these patches $\omega_l := B_{\alpha H}(x_l)$ by $\beta > 1$, i.e., $\omega_l^+ := B_{\beta\alpha H}(x_l)$. In our numerical experiments we employ $\alpha = 1.25$ and $\beta = 2$.

Altogether, we consider (16), which is defined on ω_l, and its extension to ω_l^+ in the following. Let us introduce the convention, that a super-script $+$ indicates that we refer to the extended problem on ω_l^+, e.g., we denote the respective trial and test space on ω_l^+ as

$$\mathcal{H}_{l,D}^{1,+} := \left\{ v \in \mathcal{H}^1(\omega_l^+) : v|_{\Gamma_{l,D}^+} = 0 \right\}. \tag{25}$$

Moreover, we introduce the restriction operator

$$T_l : V(\omega_l^+) \to W(\omega_l) \tag{26}$$

which operates on any function space $V(\omega_l^+)$ defined on the extended patch ω_l^+ and restricts its elements $v \in V(\omega_l^+)$ to the patch ω_l, i.e. $T_l(v) = v|_{\omega_l} \in W(\omega_l)$.

From [2, 7] we have the compactness and denseness of the image of T_l that is required for the generalized eigenvalue characterization of the sup $-$ inf and n-width with respect to the energy norm. Thus, for the analysis with respect to the H^1- and L^2-norm, we need the inequalities $\|u\|_{l,1} \leq C_1 \|u\|_{l,E}$ and $\|u\|_{l,0} \leq C_2 \|u\|_{l,E}$ to hold for some constants C_1 and C_2.

For the scalar heat equation C_1 is readily obtained from the boundedness of a_l and a_l^+ with respect to $\|\cdot\|_{l,1}$. For C_2 we additionally need the Poincaré- or Friedrichs-inequality for $W_0^{1,p}$, see [15, Theorem 12.17] or [10, Theorem 6.1-2], for $\Gamma_{l,D} \neq 0$ and the classical Poincaré- or Poincaré-Wirtinger inequality for $W^{1,p}$, see [15, Theorem 12.23] or [10, Theorem 6.1-8], for $\Gamma_{l,D} = \emptyset$. In the case of linear elasticity, C_1 is directly obtained from Korn's inequality, see [13] or [10, Theorem 6.3-3] and C_2 is provided by the Poincaré-type inequalities [10, 15] as in the scalar case. Thus, we may consider the best-approximation with respect to the energy-, the H^1- and the L^2-norm via the restriction operator T_l of (26) with $V(\omega_l^+) = (\mathcal{H}_l^+, (\cdot,\cdot)_{l,E}^+)$ and $W(\omega_l)$ one of the spaces $(\mathcal{H}_l, (\cdot,\cdot)_{l,E})$, $(\mathcal{H}_l, (\cdot,\cdot)_{l,1})$, $(\mathcal{H}_l, (\cdot,\cdot)_{l,0})$.

Hence, $\mathcal{H}_l^+ \overset{C}{\hookrightarrow} \mathcal{H}_l$ holds in all cases and we can set

$$\mathcal{W}_l := \mathcal{H}_l^+, \quad \text{and} \quad (\cdot,\cdot)_{\mathcal{W}_l} := a_l^+(\cdot,\cdot),$$

so that the resulting generalized eigenvalue problem (12) for the n-widths turns into

$$a_l\left(T_l\eta_{l,k}^{E,+}, T_l v\right) = \chi_{l,k}^{E,+} a_l^+(\eta_{l,k}^{E,+}, v), \qquad \forall v \in \mathcal{H}_l^+ \tag{27}$$

for the characterization with respect to $\|\cdot\|_{l,E}$,

$$\left(T_l\eta_{l,k}^{1,+}, T_l v\right)_{l,1} = \chi_{l,k}^{1,+} a_l^+(\eta_{l,k}^{1,+}, v), \qquad \forall v \in \mathcal{H}_l^+ \tag{28}$$

for the characterization with respect to $\|\cdot\|_{l,1}$ and

$$\left(T_l\eta_{l,k}^{0,+}, T_l v\right)_{l,0} = \chi_{l,k}^{0,+} a_l^+(\eta_{l,k}^{0,+}, v), \qquad \forall v \in \mathcal{H}_l^+ \tag{29}$$

for the characterization with respect to $\|\cdot\|_{l,0}$. Similarly, for the computation of the sup – inf we get the generalized eigenvalue problems

$$a_l\left((I - \pi_{\mathcal{V}_l,a_l})T_l u, (I - \pi_{\mathcal{V}_l,a_l})T_l v\right) = \psi a_l^+(u, v) \qquad \forall v \in \mathcal{H}_l^+ \tag{30}$$

for the characterization with respect to $\|\cdot\|_{l,E}$,

$$\left((I - \pi_{\mathcal{V}_l,(\cdot,\cdot)_{l,1}})T_l u, (I - \pi_{\mathcal{V}_l,(\cdot,\cdot)_{l,1}})T_l v\right)_{l,1} = \psi a_l^+(u, v) \qquad \forall v \in \mathcal{H}_l^+ \tag{31}$$

for the characterization with respect to $\|\cdot\|_{l,1}$ and

$$\left((I - \pi_{\mathcal{V}_l,(\cdot,\cdot)_{l,0}})T_l u, (I - \pi_{\mathcal{V}_l,(\cdot,\cdot)_{,0}})T_l v\right)_{l,0} = \psi a_l^+(u, v) \qquad \forall v \in \mathcal{H}_l^+ \tag{32}$$

for the characterization with respect to $\|\cdot\|_{l,0}$.

Having defined $\mathcal{W}_l, \mathcal{H}_l$ and the inner products necessary for our analysis, we further define, for fixed $\mathbb{N} \ni n := \dim \mathcal{V}_l$

$$\begin{aligned} \Psi_l^E(\mathcal{V}_l) &:= \Psi(\mathcal{V}_l, \mathcal{H}_l^+, a_l^+, \mathcal{H}_l, a_l), & \mathcal{V}_l^E &:= \mathcal{V}_{l,n}^*(\mathcal{H}_l^+, a_l^+, \mathcal{H}_l, a_l), \\ d_{l,n}^E &:= d_n(\mathcal{H}_l^+, a_l^+, \mathcal{H}_l, a_l), & \Lambda_l^E(\mathcal{V}_l) &:= \frac{\Psi_l^E(\mathcal{V}_l)}{d_{l,n}^E} \end{aligned} \tag{33}$$

and

$$\begin{aligned} \Psi_l^1(\mathcal{V}_l) &:= \Psi(\mathcal{V}_l, \mathcal{H}_l^+, a_l^+, \mathcal{H}_l, (\cdot,\cdot)_{l,1}), & \mathcal{V}_l^1 &:= \mathcal{V}_{l,n}^*(\mathcal{H}_l^+, a_l^+, \mathcal{H}_l, (\cdot,\cdot)_{l,1}), \\ d_{l,n}^1 &:= d_n(\mathcal{H}_l^+, a_l^+, \mathcal{H}_l, (\cdot,\cdot)_{l,1}), & \Lambda_l^1(\mathcal{V}_l) &:= \frac{\Psi_l^1(\mathcal{V}_l)}{d_{l,n}^1} \end{aligned} \tag{34}$$

and

$$\Psi_l^0(\mathcal{V}_l) := \Psi(\mathcal{V}_l, \mathcal{H}_l^+, a_l^+, \mathcal{H}_l, (\cdot,\cdot)_{l,0}), \qquad \mathcal{V}_l^0 := \mathcal{V}_{l,n}^*(\mathcal{H}_l^+, a_l^+, \mathcal{H}_l, (\cdot,\cdot)_{l,0}),$$

$$d_{l,n}^0 := d_n(\mathcal{H}_l^+, a_l^+, \mathcal{H}_l, (\cdot,\cdot)_{l,0}), \qquad \Lambda_0^l(\mathcal{V}_l) := \frac{\Psi_l^0(\mathcal{V}_l)}{d_{l,n}^0} \tag{35}$$

where we refer to $\Lambda_l^E, \Lambda_l^1, \Lambda_l^0 \geq 1$ as optimality ratios.

4.3 Practical Construction of $\mathcal{W}_l$

Note that, our definitions of $\mathcal{H}_l, \mathcal{W}_l$ and their respective norms do not enable us to perform computations in practical applications since we only assume to have $\mathcal{V}_l = \text{span}\{\eta_{l,k} : k < n\} \subset \mathcal{H}_l$ from some homogenization approach given via some basis $\eta_{l,k}, k < n$. Yet, for the computation of the sup – infs $\Psi_l^E, \Psi_l^1, \Psi_l^0$ and the n-widths $d_{l,n}^E, d_{l,n}^1, d_{l,n}^0$ we in fact need to have a basis for $\mathcal{W}_l = \mathcal{H}_l^+$ as well.

Throughout this paper, we employ classical FE discretizations with linear elements on simplicial meshes to approximate the respective function spaces. Here, we construct the respective meshes on the enlarged patches ω_l^+ independently of each other, i.e., we do not assume the overlapping meshes to be aligned. Moreover, we assume that the meshwidth of these meshes is small enough to resolve all inclusions and desired features of the solution with sufficient accuracy. We denote the respective FE spaces defined on these meshes by $\mathcal{V}_l^{\text{FE},+}$ and refer to the standard nodal basis functions of $\mathcal{V}_l^{\text{FE},+}$ as $\phi_{l,m}^+$. Moreover, we introduce the subspace $\mathcal{V}_{l,0}^{\text{FE},+} \subset \mathcal{V}_l^{\text{FE},+}$ defined by the nodal basis functions $\phi_{l,m}^+$ associated with the boundary nodes $x_l \in \Gamma_{l,0}$ on the free boundary. Finally, we make the assumption that the restrictions $T_l(\mathcal{V}_l^{\text{FE},+})$ are given by a subspace $\mathcal{V}_l^{\text{FE}}$ of $\mathcal{V}_l^{\text{FE},+}$, i.e., that the mesh on ω_l^+ resolves the patch ω_l and its boundary $\partial\omega_l$. With the help of these discrete spaces we can now construct approximate spaces $\mathcal{W}_l = \mathcal{H}_l^+$ and respective basis functions.

Let us first consider a rather expensive brute force approach. Here, we simply follow the definition of $\mathcal{H}_l^+$ and either compute all solutions $u_{l,N}^+$, see (16), to the boundary value problem

$$\begin{aligned} \nabla \cdot \sigma(u_{l,N}^+) = 0, \quad \text{in } \omega_l^+, \quad \sigma(u_{l,N}^+) \cdot \mathfrak{n} = 0, \quad \text{on} \Gamma_{l,N}^+, \\ u_{l,N}^+ = 0, \quad \text{on } \Gamma_{l,D}^+, \quad \sigma(u_{l,N}^+) \cdot \mathfrak{n} = h_l^+ \quad \text{on} \Gamma_{l,0}^+ \end{aligned} \tag{36}$$

with solution operator $h_l^+ \mapsto u_{l,N}^+ = H_{l,N}^+ h_l^+$ and an exhaustive set of tractions h_l^+ on $\Gamma_{l,0}^+$ or all solutions $u_{l,D}^+$, see (17) to

$$\begin{aligned} \nabla \cdot \sigma(u_{l,D}^+) = 0, \quad \text{in } \omega_l^+, \quad & \sigma(u_{l,D}^+) \cdot \mathfrak{n} = 0, \quad \text{on } \Gamma_{l,N}^+, \\ u_{l,D}^+ = 0, \quad \text{on } \Gamma_{l,D}^+, \quad & u_{l,D}^+ = g_l^+ \quad \text{on } \Gamma_{l,0}^+ \end{aligned} \tag{37}$$

with solution operator $g_l^+ \mapsto u_{l,D}^+ = H_{l,D}^+ g_l^+$ and an exhaustive set of traces g_l^+ on $\Gamma_{l,0}^+$. Thus, we can formally define

$$\mathcal{H}_l^+ := \langle H_{l,N}^+ \phi_{l,m} : \phi_{l,n} \in \mathcal{V}_{l,0}^{\mathrm{FE},+} \rangle$$

containing all FE approximations of (37).

Note that we employ this rather expensive construction for the computation of all sup – infs and n-widths throughout this paper. Moreover, we employ this definition of $\mathcal{H}_l^+$ for the construction of the optimal approximation spaces $\mathcal{V}_l^E, \mathcal{V}_l^1, \mathcal{V}_l^0$.

To reduce the computational effort involved in the definition of approximate spaces $\mathcal{H}_l^+$, we also consider the much cheaper approach of Babuška and Lipton [2], Babuška et al. [7]. Here, the basic idea is to find a subspace $\mathcal{H}_l^{+,s} \subset \mathcal{H}_l^+$ with $\dim \mathcal{H}_l^+ \gg \dim \mathcal{H}_l^{+,s} \geq n_l \gg n$ that approximates the first n eigenvectors of the n-width problems (27)–(29) sufficiently well. We then use this subspace $\mathcal{H}_l^{+,s}$ instead of $\mathcal{H}_l^+$ in the construction of the respective local approximation spaces $\mathcal{V}_l^{E,s}, \mathcal{V}_l^{1,s}, \mathcal{V}_l^{0,s}$. In the remainder of this work, we choose

$$n_l = \gamma \max_{\mathcal{V}_l} n = \gamma \max_{\mathcal{V}_l} \dim \mathcal{V}_l \tag{38}$$

with $\gamma = 2$ to be twice the maximum dimension of employed local spaces $n = \dim \mathcal{V}_l$. Note, however, that the choice of γ is not obvious and may impact the results substantially.

Let us now consider the concrete choice of this subspace $\mathcal{H}_l^{+,s}$. To this end, we consider polynomials $\mathcal{P}_d$ of degree $\leq d$ which provide a spectral basis for problems with smooth coefficients. Moreover, for the scalar heat equation with constant coefficients, its polynomial solutions, the so-called harmonic polynomials, are explicitly known. They span the $2d + 1$ dimensional space

$$\mathcal{P}_{\Delta,d} := \{v \in \mathcal{P}_d : \Delta v = 0\} = \operatorname{span}\{\Re(x+iy)^e, \Im(x+iy)^e : e \leq d\}. \tag{39}$$

In the vectorial elasticity case, we simply take each component to be a harmonic polynomial and obtain a $4d + 2$ dimensional space $\mathcal{P}_{\Delta,d}$, respectively. We now consider $h_l^+ \in \mathcal{P}_{\Delta,d}$ with d large enough such that $\dim \mathcal{P}_{\Delta,d} \geq n_l$ to define

$$\mathcal{H}_l^{+,s} := \left\{H_{l,N}^+ h_l^+ : h_l^+ \in \mathcal{P}_{\Delta,d}\right\}. \tag{40}$$

5 Concrete Choices of $\mathcal{V}_l$

With all necessary definitions for the spaces $\mathcal{H}_l$ and $\mathcal{W}_l$ available, let us now shortly summarize the choices we consider for the local approximation spaces $\mathcal{V}_l$ in the following.

- Optimal spaces of eigenvectors $\eta_{l,k}^{E,+}, \eta_{l,k}^{1,+}, \eta_{l,k}^{0,+} \in \mathcal{H}_l^+$ associated with the n largest eigenvalues

$$\mathcal{V}_l^E := \{T_l \eta_{l,k}^{E,+}, k < n\}, \quad a_l(T_l \eta_{l,k}^{E,+}, T_l v^+) = \chi a_l^+(\eta_{l,k}^{E,+}, v^+) \quad \forall v^+ \in \mathcal{H}_l^+,$$
$$\mathcal{V}_l^1 := \{T_l \eta_{l,k}^{1,+}, k < n\}, \quad (T_l \eta_{l,k}^{1,+}, T_l v^+)_{l,1} = \chi a_l^+(\eta_{l,k}^{1,+}, v^+) \quad \forall v^+ \in \mathcal{H}_l^+,$$
$$\mathcal{V}_l^0 := \{T_l \eta_{l,k}^{0,+}, k < n\}, \quad (T_l \eta_{l,k}^{0,+}, T_l v^+)_{l,0} = \chi a_l^+(\eta_{l,k}^{0,+}, v^+) \quad \forall v^+ \in \mathcal{H}_l^+.$$

- Cheaper approximation of optimal spaces of eigenvectors $\eta_{l,k}^{E,+,s}, \eta_{l,k}^{1,+,s}$, $\eta l, k^{0,+,s} \in \mathcal{H}_l^{+,s}$ associated with the n largest eigenvalues

$$\mathcal{V}_l^{E,s} := \{T_l \eta_{l,k}^{E,+,s}, k < n\}, \quad a_l(T_l \eta_{l,k}^{E,+,s}, T_l v^+) = \chi a_l^+(\eta_{l,k}^{E,+,s}, v^+) \quad \forall v^+ \in \mathcal{H}_l^{+,s},$$
$$\mathcal{V}_l^{1,s} := \{T_l \eta_{l,k}^{1,+,s}, k < n\}, \quad (T_l \eta_{l,k}^{1,+,s}, T_l v^+)_{l,1} = \chi a_l^+(\eta_{l,k}^{1,+,s}, v^+) \quad \forall v^+ \in \mathcal{H}_l^{+,s},$$
$$\mathcal{V}_l^{0,s} := \{T_l \eta_{l,k}^{0,+,s}, k < n\}, \quad (T_l \eta_{l,k}^{0,+,s}, T_l v^+)_{l,0} = \chi a_l^+(\eta_{l,k}^{0,+,s}, v^+) \quad \forall v^+ \in \mathcal{H}_l^{+,s}.$$

- In analogy to [8, 9], the harmonic polynomials $\mathcal{H}_l^{+,s}$

$$\mathcal{V}_l^h := \{T_l H_{l,N}^+ h_l^+ : h_l^+ \in \mathcal{P}_{\Delta,d^-}, \dim \mathcal{P}_{\Delta,d^-} = n\} \subset \mathcal{H}_l^{+,s}.$$

- The multiscale spaces introduced in [11] can be interpreted as optimal spaces for $H^1(\omega_l) \overset{C}{\hookrightarrow} L^2(\omega_l)$ and thus fit into our framework. From a computational point of view, it is more convenient to construct these spaces via the eigenvectors $\eta_{l,k}^e \in \mathcal{V}_l^{\mathrm{FE}}$ of

$$\mathcal{V}_l^e := \{\eta_{l,k}^e, k < n\}, \quad a_l(\eta_{l,k}^e, v) = \chi\left(\kappa\, \eta_{l,k}^e, v\right)_{l,0} \forall v \in \mathcal{V}_l^{\mathrm{FE}}$$

associated with the smallest non-vanishing eigenvalues. Moreover, we study the effects of oversampling with $\omega_l^+ \supset \omega_l$ we also consider the eigenvectors $\eta_{l,k}^{e,+}$ associated with the smallest non-vanishing eigenvalues of

$$\mathcal{V}_l^{e,+} := \{\eta_{l,k}^{e,+}, k < n\}, \quad a_l^+(\eta_{l,k}^{e,+}, v^+) = \chi\left(\kappa\, \eta_{l,k}^{e,+}, v^+\right)_{l,0} \quad \forall v^+ \in \mathcal{V}_l^{\mathrm{FE},+}.$$

6 Numerical Results

Let us now summarize the results of our numerical experiments using the local approximation spaces $\mathcal{V}_l$ defined in the previous section. Here, we used the FE framework `FEniCS` [1] to these spaces, i.e. the multiscale basis functions, which are then used to enrich the PUM framework PUMA [21] developed by Fraunhofer SCAI. All eigenvalue problems . . . 10^{-7} ALEXALEXALEX.

We consider the three different configurations of inclusions (denoted as $\mathbb{A}$, $\mathbb{B}$, $\mathbb{C}$ in the following) depicted in Fig. 1 in our experiments to study the robustness of the considered homogenization approaches with respect to inclusion distribution and the material contrast C, see (1). Moreover, we study the computational costs associated with the different approaches and the overall performance of the constructed enrichments within a global PUM approximation of our model problem introduced in Sect. 1.1.

Besides these issues we also consider more fundamental topics such as: How many degrees of freedom do we need in the optimal case to achieve a certain local approximation error ϵ (according to the n-widths $d^E_{l,n}$ and $d^1_{l,n}$, $d^0_{l,n}$)? In particular, we are interested to see if this number is robust with respect to the contrast C and the distribution of the inclusions.

In all numerical experiments, we compute the n-widths and sup $-$ infs for space dimensions up to $n \leq 100$ in the scalar case and $n \leq 200$ for the linear elasticity model problem. Recall also that we use $\dim \mathcal{H}_l^{+,s} = \gamma\, n = 2n$, compare Sect. 4.3. The employed meshes defined on the extended patches ω_l^+ are sufficiently refined to resolve all features of interest. For instance, the center patch ω_4 for the most complex pattern of random inclusions of varying sizes was meshed using 514,912 cells and $M = 258{,}096$ nodes, yielding $\dim \mathcal{H}_l^+ = 1278$, compare Sect. 4.3.

6.1 General Observations

At first, we focus on the sup $-$ infs obtained for all considered homogenization approaches and the optimal n-width for material configuration A, see Fig. 3. Comparing the depicted plots for the scalar model problem and the two-dimensional elasticity problem, we find essentially an identical behavior with the only difference that the approximation of a two-dimensional vector field obviously requires twice as many degrees of freedom as the approximation of a scalar field.

Moreover, we see that for a qualitative comparison we can restrict ourselves to the plots with respect to the energy norm. This is in accordance with the equivalence of $\|\cdot\|_{l,E}$ and $\|\cdot\|_{l,1}$ and with $\|\cdot\|_{l,0}$ being bounded by $\|\cdot\|_{l,E}$.

Note also, that the optimal spaces $\mathcal{V}_l^E, \mathcal{V}_l^1, \mathcal{V}_l^0$ seem to perform equally well for all quality measures $\Psi_l^{\{E,1,0\}}$, i.e., with our particular choice of $\mathcal{W}_l = \mathcal{H}_l^+$ and $\mathcal{H}_l$ there appears to be no benefit of computing an optimal basis for, say, $\|\cdot\|_{l,1}$ instead of $\|\cdot\|_{l,E}$, as both bases will perform equally well for both norms. Thus, to remove

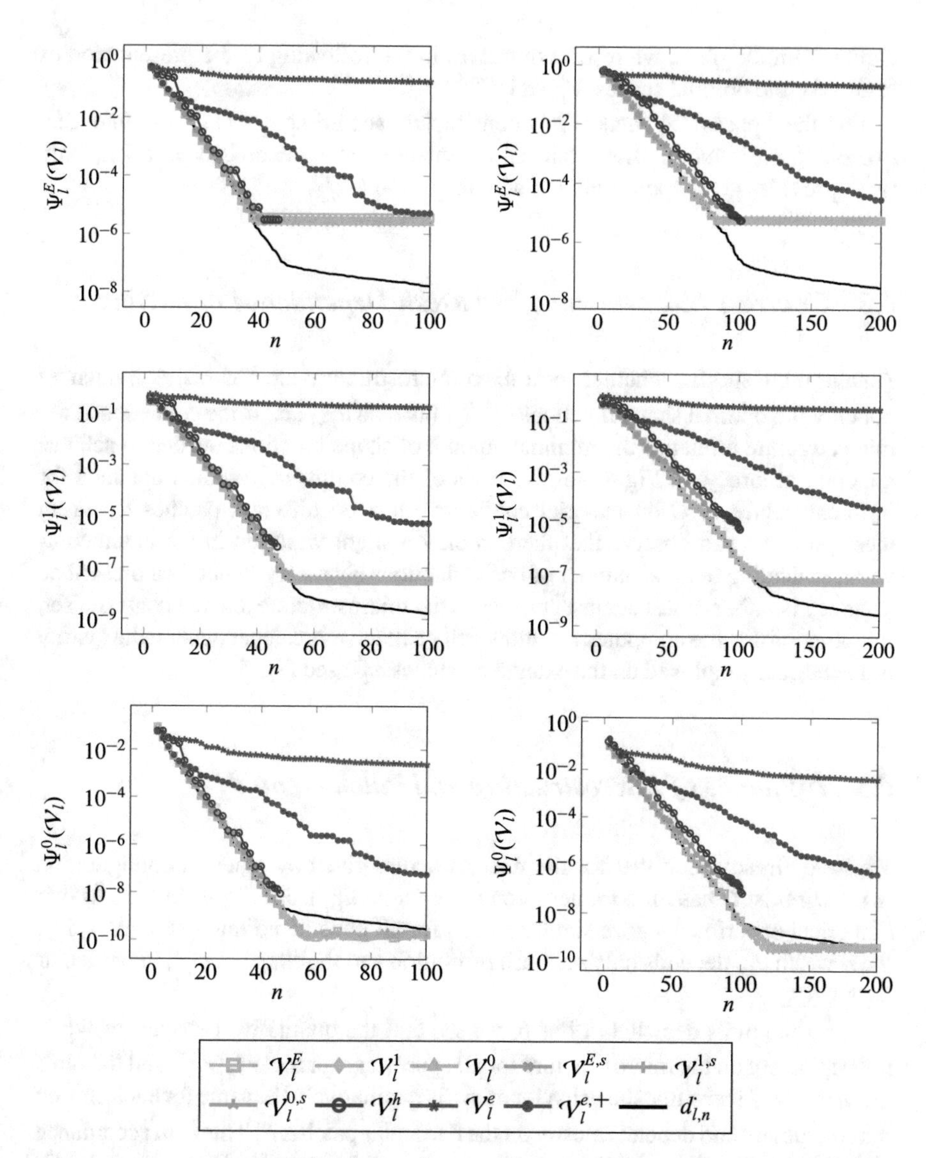

Fig. 3 Computed sup − infs Ψ_l^E, Ψ_l^1 and Ψ_l^0 (*top to bottom*) for the central patch ω_4 for our scalar model problem (*left*) and the linear elasticity equation (*right*) for material configuration $\mathbb{A}$ with 121 randomly distributed inclusions of random size. Note that in the Ψ_l^0 plots we find values of $\Psi_l^0(V_l^{\{E,1,0\}})$ below the computed n-width $d_{l,n}^0$. This is a result of using two different numerical generalized eigenvalue problems for the computation of $d_{l,n}^0$ and Ψ_l^0. In some sense this shows that the numerical computation of $d_{l,n}^0$ via the basis of eigenvectors V_l^0 is less affected by round-off than using $\mathcal{H}_l^+$

clutter from the plots, we restrict ourselves in the following to the presentation of results for the optimal spaces $\mathcal{V}_l^E$ and $\mathcal{V}_l^{E,s}$ only.

For the spaces $\mathcal{V}_l^e$ and $\mathcal{V}_l^{e,+}$ constructed according to [11] we find that oversampling is required to attain an acceptable convergence behavior, i.e., only $\mathcal{V}_l^{e,+}$ provides good approximation with respect to $\|\cdot\|_{l,E}^+$.

6.2 Contrast Independence and Mesh Dependence of n-Widths

A natural question is whether for a fixed microstructure the contrast C influences not only the optimal shape functions $\eta_{l,k}^{\{E,1,0\}}$ but also $d_{l,n}^E$, i.e., if the contrast and the microstructure influence the minimal amount of shape functions needed to achieve a certain accuracy. In Fig. 4 we give plots of the computed n-widths obtained for different values of C on material configuration A on different patches ω_l. From these plots we can observe that there is only a slight variation in the obtained n-width indicating that the number of basis functions necessary to obtain a prescribed accuracy is rather robust against changes in the microstructure and the contrast. Yet, it is obvious that the computed n-widths will be much more dependent on the quality of the meshes employed on the extended patches ω_l^+, see Fig. 5.

6.3 Influence of Microstructure and Patch ω_l on $d_{l,n}^E$

We have already seen that for the microstructure given by material configuration $\mathbb{A}$ the contrast C has little influence on the n-width $d_{l,n}^E$ from Fig. 4. In fact, further numerical experiments gave similar results for all considered microstructures, i.e., the n-width $d_{l,n}^E$ depends only on mesh resolution and the shape of ω_l^+, in particular $\partial\omega_l^+ \setminus \partial\Omega$.

From the plots depicted in Fig. 6, we see that the qualitative behavior of Ψ_l^E is not dependent on the microstructure for all spaces except for $\mathcal{V}_l^e, \mathcal{V}_l^{e,+}$ and the plots given in Fig. 7 show that the behavior of $d_{l,n}^E$ is qualitatively the same for each chosen microstructure and depends mostly on the particular patch ω_l^+. This is in accordance with the construction of $\mathcal{H}_l^+$ using the degrees of freedom on $\Gamma_{l,0}^+ = \partial\omega_l^+ \setminus \partial\Omega$. Moreover, the exact type of boundary condition on $\partial\omega_l^+ \cap \partial\Omega$ (Dirichlet, Neumann, or mixed) appears to play little role in the behavior of the n-widths.

The computed n-widths $d_{l,n}^E$ given in Fig. 8 clearly show that $d_{l,n}^E$ is actually quantitatively the same for each microstructure up to machine accuracy.

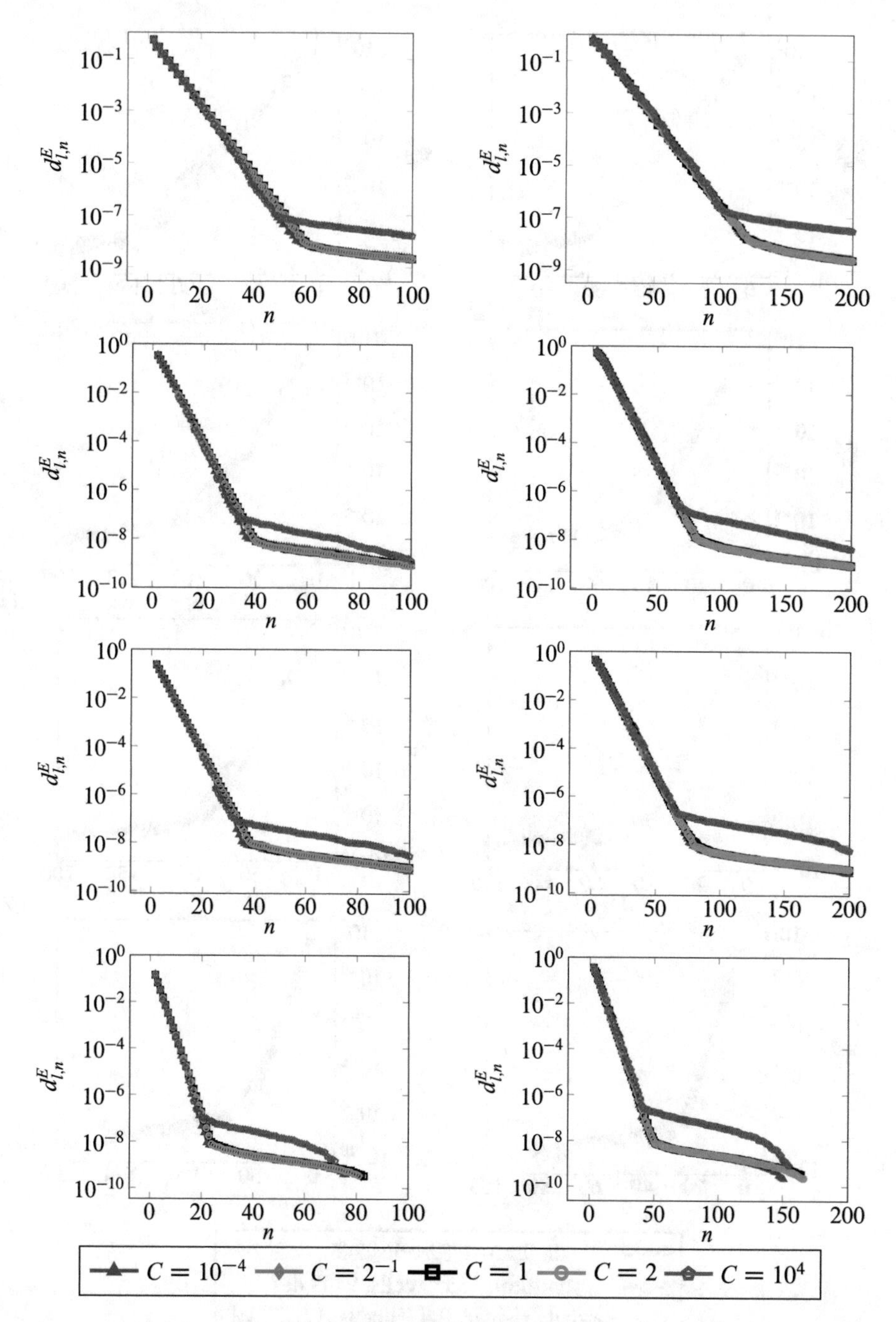

Fig. 4 Computed n-widths $d^E_{l,n}$ for patches ω_4, ω_1, ω_3, and ω_0 (*top to bottom*) for different contrasts C for our scalar model problem (*left*) and linear elasticity (*right*)

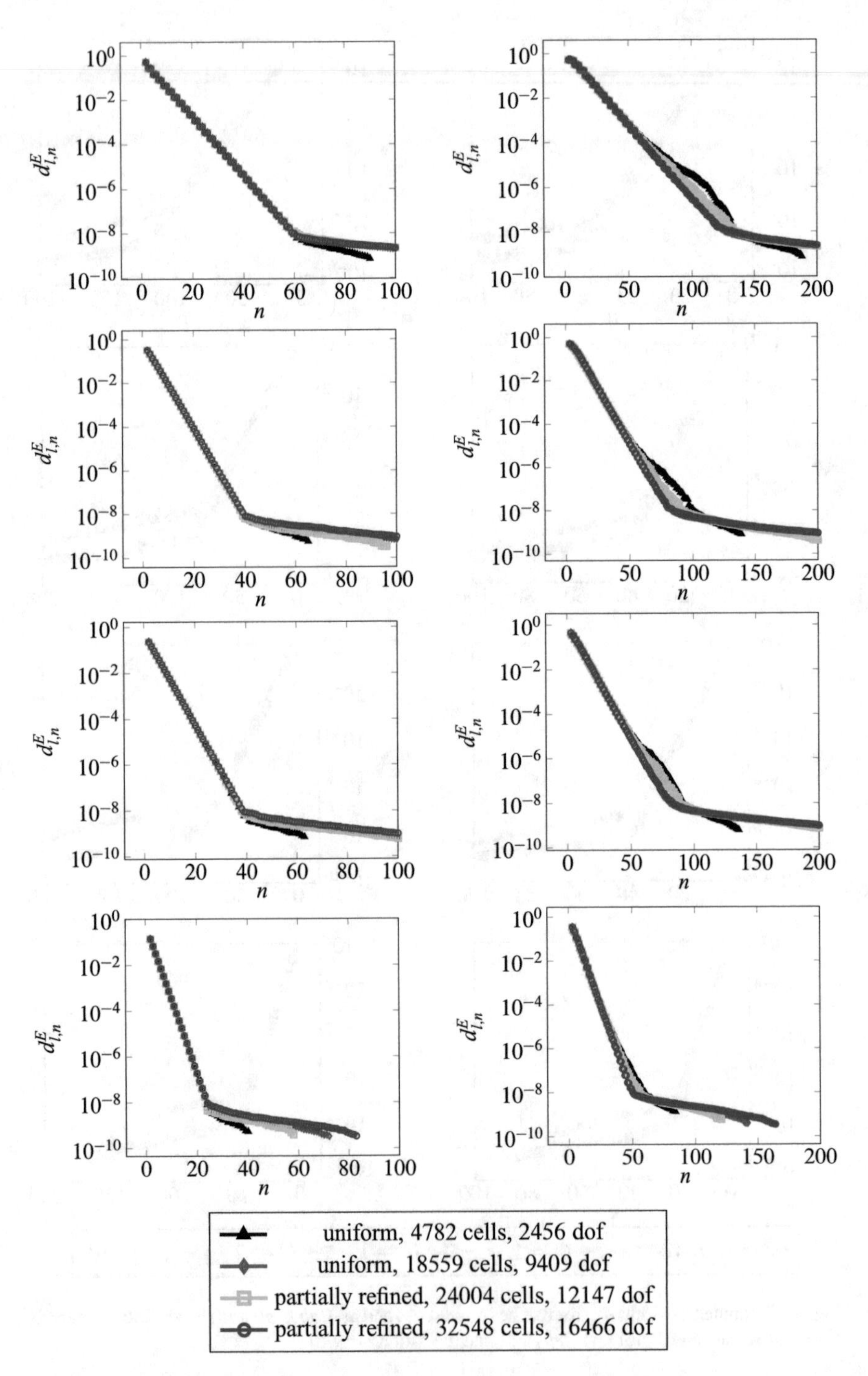

Fig. 5 Computed n-widths $d^E_{l,n}$ for the smooth case $C = 1$ (i.e. no inclusions) for patches ω_4, ω_1, ω_3, and ω_0 (*top to bottom*) on four different meshes for our scalar model problem (*left*) and linear elasticity (*right*)

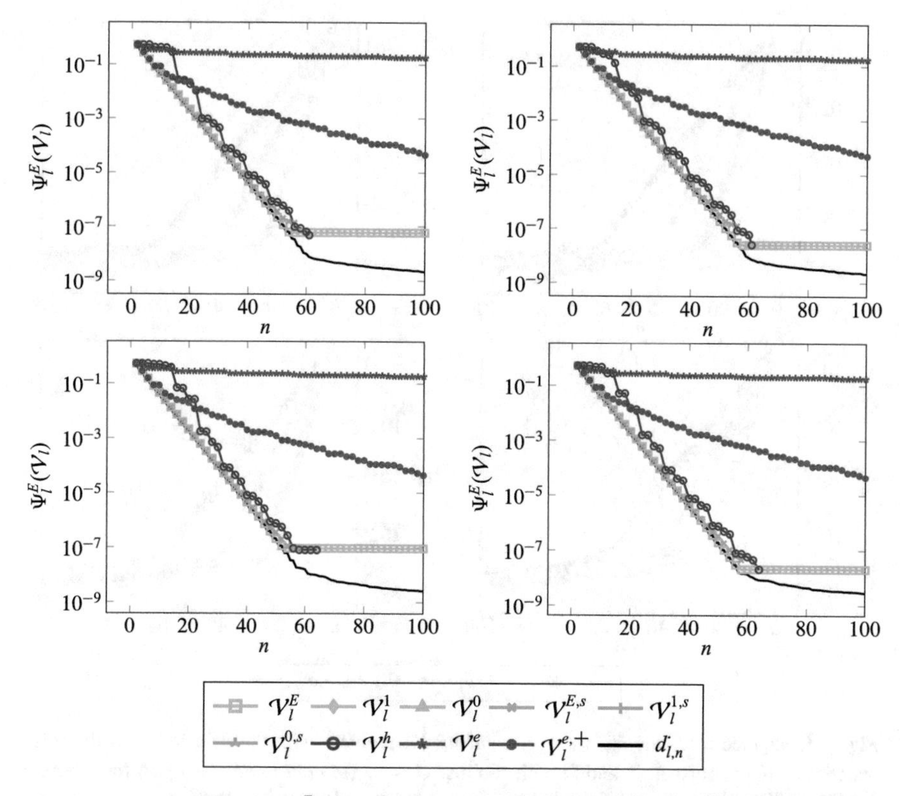

Fig. 6 Computed sup − infs Ψ_l^E for ω_4 and our scalar model problem with no inclusions and the three material configurations $\mathbb{A}$, $\mathbb{B}$ and $\mathbb{C}$ with contrast $C = \frac{1}{2}$ *(top left to bottom right)*. $d_{l,n}^E$ and $\Psi_l^E(\mathcal{V}_l)$ show similar behavior in all four plots, i.e. there is little dependence on the microstructure

6.4 Contrast Dependence of $\mathcal{V}_l$

Another important issues is the robustness of particular spaces $\mathcal{V}_l$ with respect to the contrast. In Fig. 9 we show plots of the computed sup − infs $\Psi_l^E(\mathcal{V}_l)$ for different contrasts obtained for the optimal space $\mathcal{V}_l^E$, the optimal space $\mathcal{V}_l^{E,s}$ using the cheaper construction with $\mathcal{H}_l^{+,s}$ instead of $\mathcal{H}_l^+$, the space of $\mathcal{L}$-harmonic polynomials $\mathcal{H}_l^h \subset \mathcal{H}_l^{+,s}$, and the space $\mathcal{V}_l^{e,+}$ of Efendiev et al. [11] with oversampling. From these plots we can clearly see that the spaces $\mathcal{V}_l^E, \mathcal{V}_l^{E,+}$ are rather robust with respect to varying contrast C whereas $\mathcal{V}_l^h$ shows slight

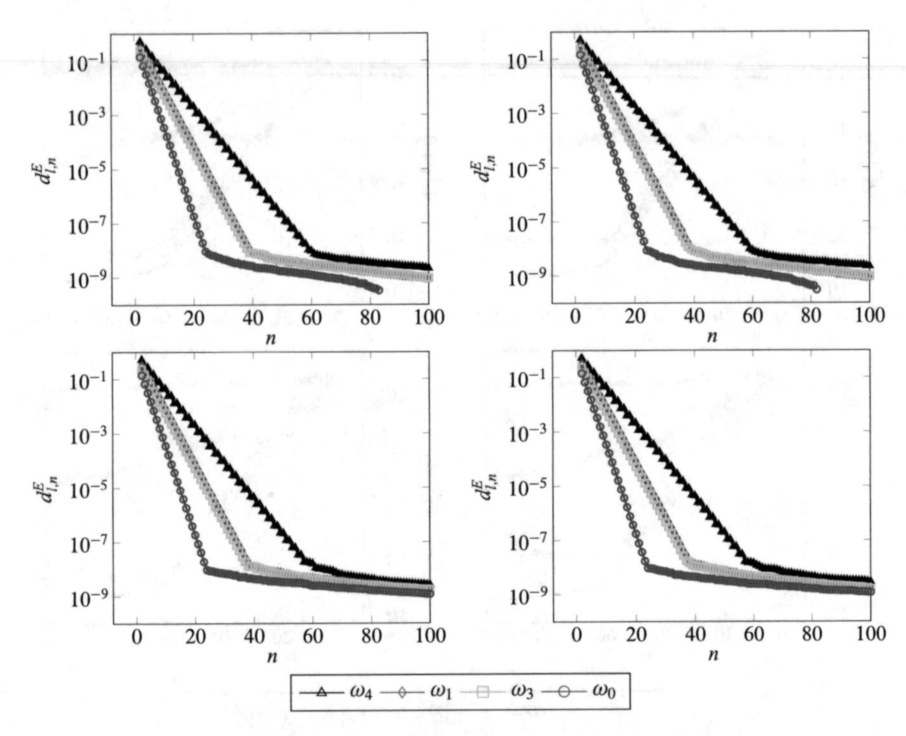

Fig. 7 Computed n-widths $d^E_{l,n}$ for our scalar model problem with no inclusions and the three material configurations $\mathbb{A}$, $\mathbb{B}$ and $\mathbb{C}$ with contrast $C = \frac{1}{2}$ (*top left to bottom right*) for different patches ω_l. The plots are essentially independent of the considered microstructure

fluctuations. This is probably due to the smoothing and stabilizing effect of the generalized eigenvalue problem that turns $\mathcal{V}^h_l \subset \mathcal{H}^{+,s}_l$ into $\mathcal{V}^{E,s}_l$ which is much more robust with regards to C. The space $\mathcal{V}^{e,+}_l$ shows more dependence on C which, however, seems to be related to results on the dependence of the eigenvalues $\eta^e_{l,k}, \eta^{e,+}_{l,k}$ on the contrast C and the number of inclusions, see [11].

6.5 Optimality and Costs of $\mathcal{V}_l$

From Fig. 10 we see that $\mathcal{V}^h_l$ is a very cost effective close to optimal space, which can be lightly improved on via a generalized eigenvalue problem yielding $\mathcal{V}^{\{E,1,0\},s}_l$. This, however, leads to fixed costs in the construction of $\mathcal{H}^{+,s}_l$. Moreover, $\mathcal{V}^{e,+}_l$ yields a cost effective space of shape functions, albeit, as seen in other plots, requiring more shape functions.

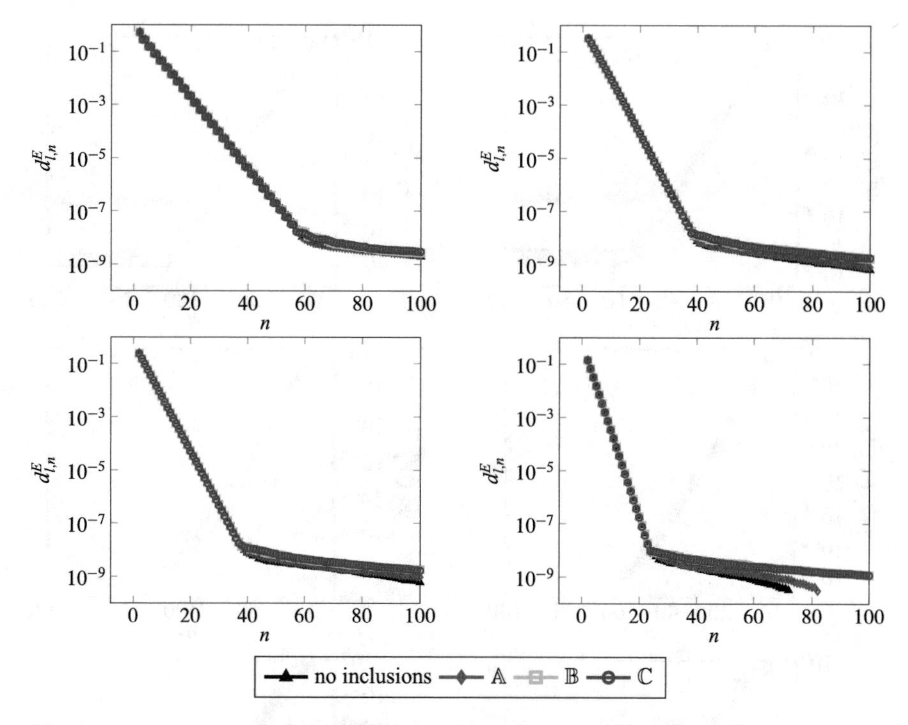

Fig. 8 Computed n-widths $d^E_{l,n}$ for our scalar model problem with contrast $C = \frac{1}{2}$ obtained for the patches $\omega_4, \omega_1, \omega_3, \omega_0$ (*top left to bottom right*) for different material configurations

From Fig. 11, where we give the optimality ratio Λ^E_l, we see that up to mesh precision we have optimality of $\mathcal{V}_l^{\{E,1,0\}}$ and $\mathcal{V}_l^{\{E,1,0\},s}$ and a very similar performance of $\mathcal{V}_l^h$ while $\mathcal{V}^e$ and $\mathcal{V}^{e,+}$ do not perform comparably well with respect to this measure.

6.6 Example of Global PUM Solution

Finally, we present some first results using the computed multiscale basis functions as enrichments in a PUM discretization of the global problem. Here, we consider our scalar model problem on material configuration B and consider two different types of refinements of the global PUM space. Obviously, we can keep the number of PUM patches ω_l fixed and consider an increasing number of enrichment functions, i.e. a p-type refinement. However, we may also consider a cheaper h-type refinement, where we increase the number of PUM patches and restrict the employed enrichments to the refined patches. The p-type refinement should of course yield a spectrally convergent method whereas the convergence behavior of the described

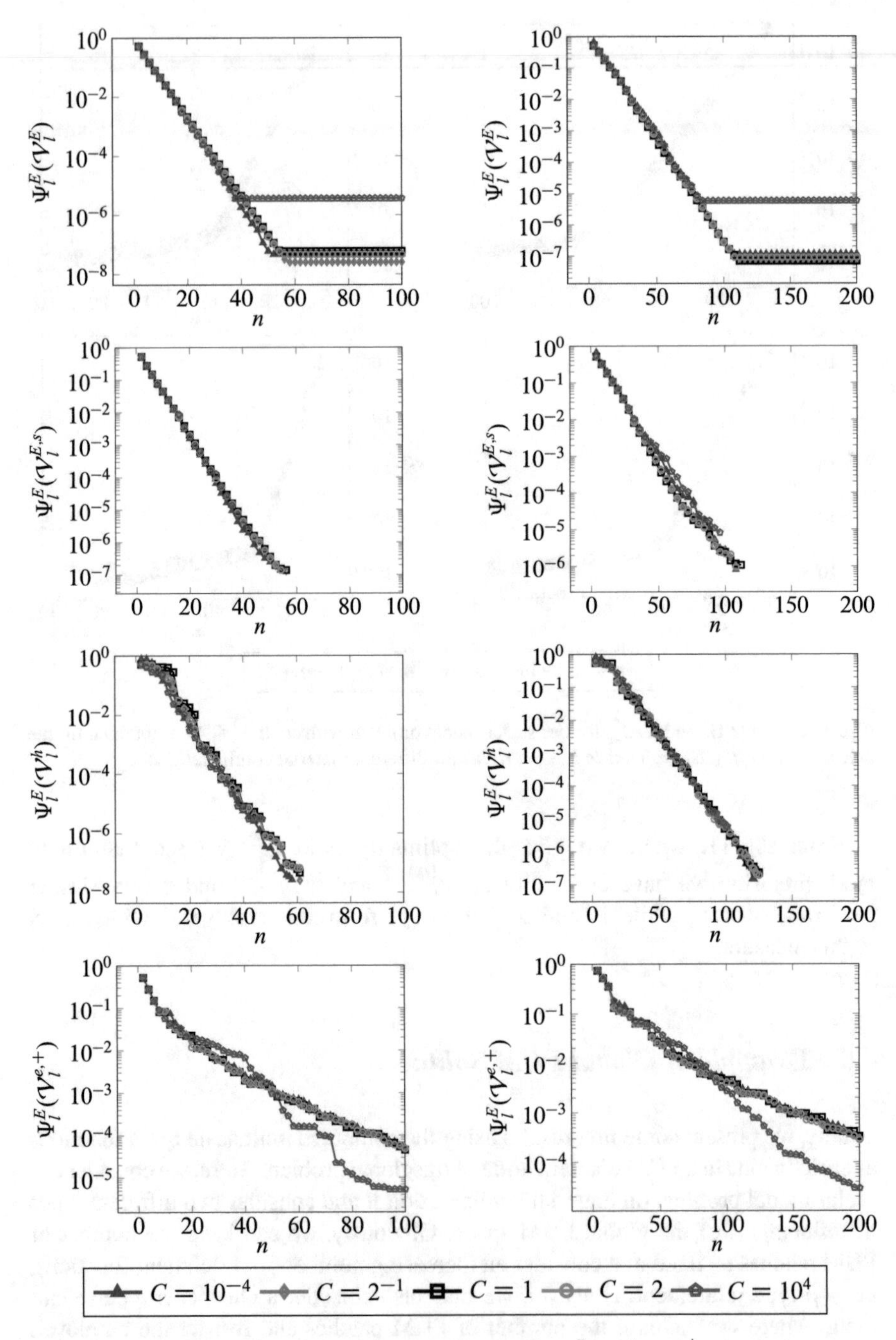

Fig. 9 Computed sup − infs $\Psi_l^E(\mathcal{V}_l)$ for the spaces $\mathcal{V}_l^E$, $\mathcal{V}_l^{E,s}$, $\mathcal{V}_l^h$, $\mathcal{V}_l^{e,+}$ (*top to bottom*) for our scalar model problem (*left*) and linear elasticity (*right*) on the center patch ω_4 and material configuration $\mathbb{A}$ with different contrasts C

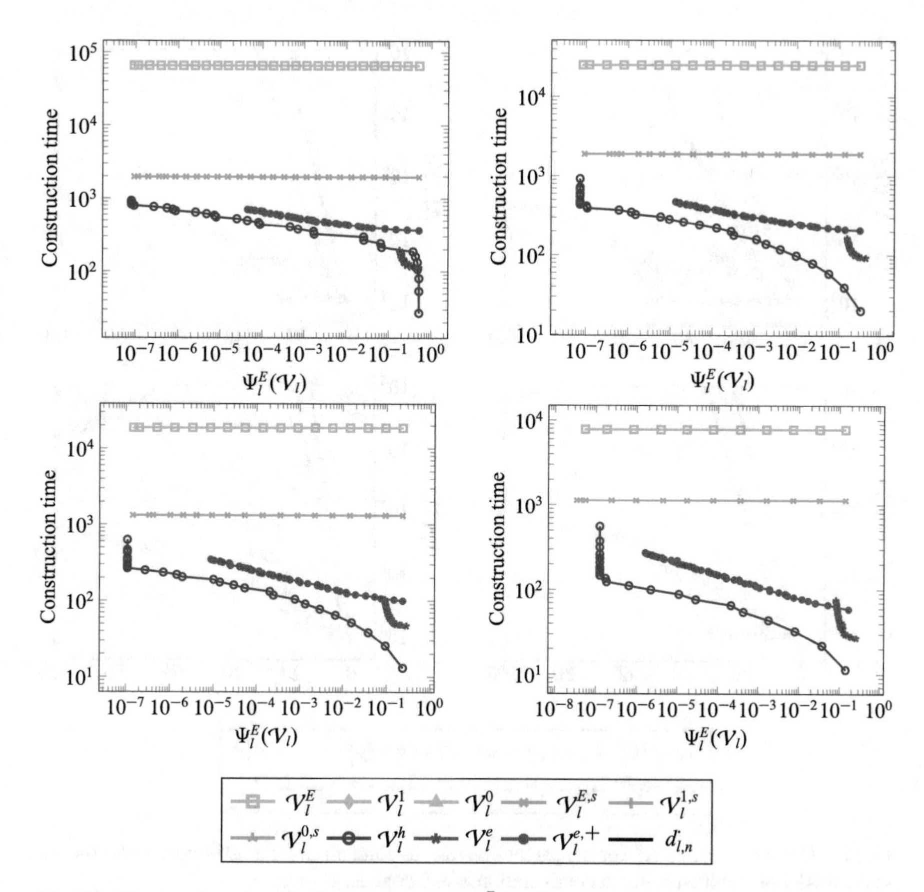

Fig. 10 Construction time of $\mathcal{V}_l$ plotted against Ψ_l^E for the patches ω_4, ω_1, ω_3, and ω_0 (*top left to bottom right*) for our scalar model problem on material configuration $\mathbb{B}$ and contrast $C = \frac{1}{2}$.

h-refinement should yield an algebraic rate. The plots depicted in Fig. 12 clearly shows the anticipated behavior of our p-type refined PUM with respect to the L^2-norm whereas for the H^1-norm the observed error reduction stalls. Similarly, we find that the h-refined PUM in the L^2-norm shows an algebraic convergence behavior as anticipated and essentially no error reduction in H^1. Note, however, that we have to use a numerically determined reference solution to compute the errors and thus may observe a numerical artifact only. Here, a more detailed study is necessary to identify the issue. From the contour plots depicted in Figs. 13 and 14 at least we find no obvious flaw in our PUM approximation.

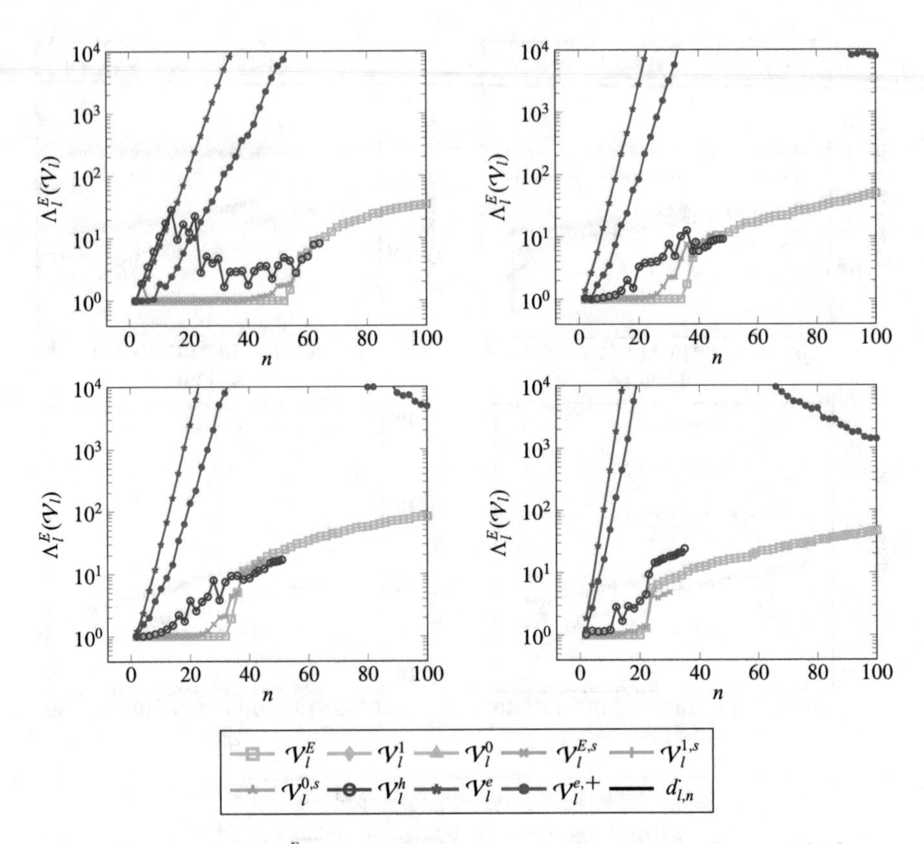

Fig. 11 Optimality ratios Λ_l^E for the patches ω_4, ω_1, ω_3, and ω_0 (*top left to bottom right*) for our scalar model problem on material configuration $\mathbb{B}$ and contrast $C = \frac{1}{2}$

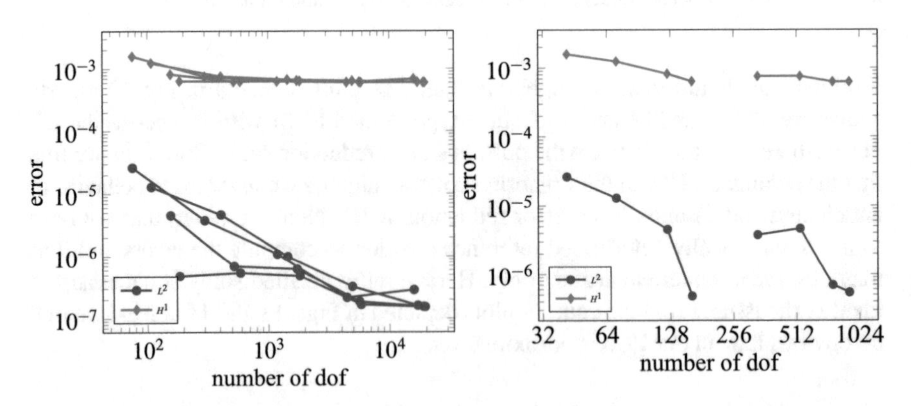

Fig. 12 Computed global errors for our scalar model problem on material configuration B with contrast $C = 10^4$ using an h-refined PUM (*left*) using a fixed number of local enrichments ($n_l = 4, 8, 12, 20$) and a p-refined PUM on two different covers with 9 and 36 patches

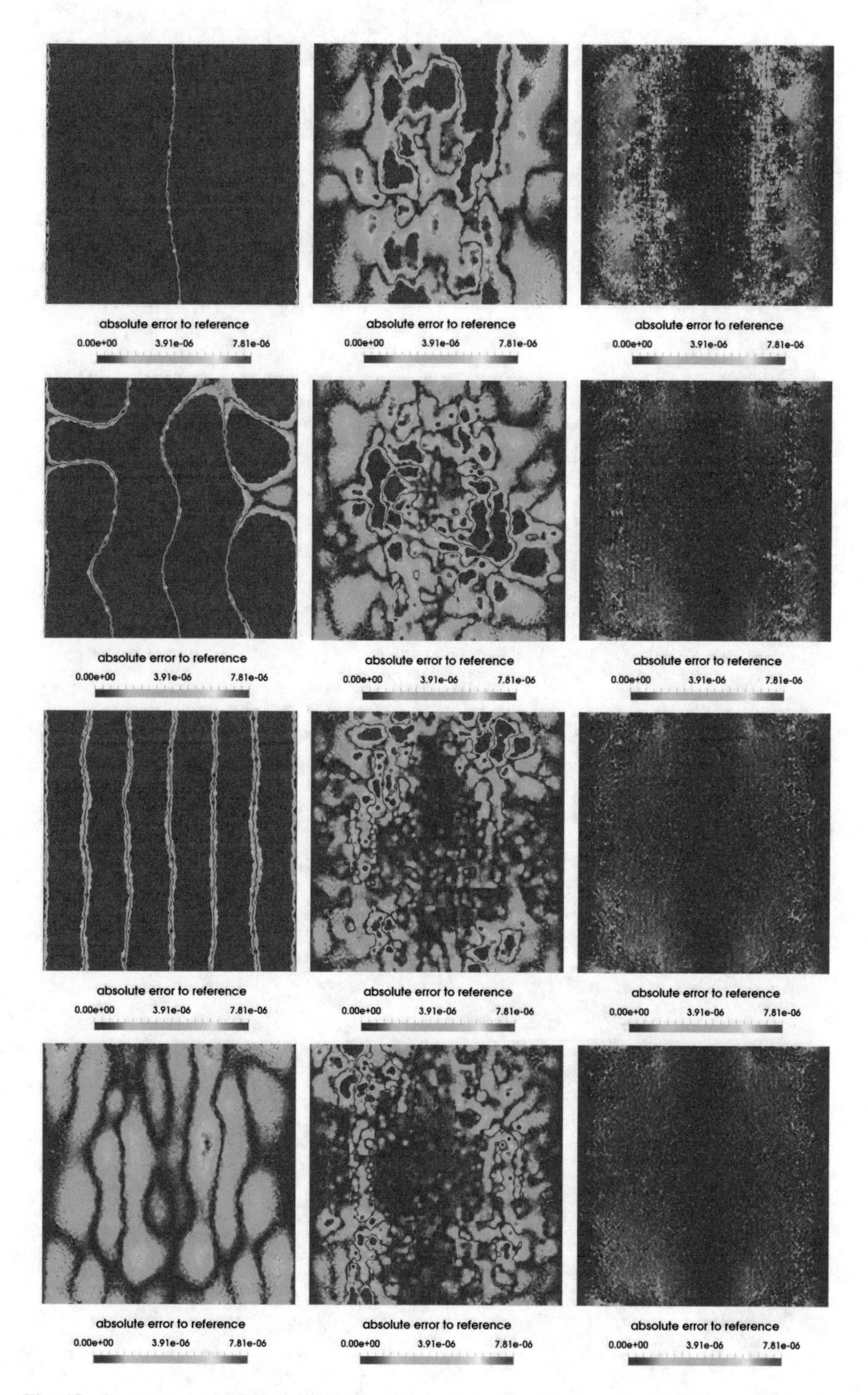

Fig. 13 Contour plots of the absolute error on three h-refinement levels (*left to right*) using $4, 8, 12, 20$ enrichments (*top to bottom*)

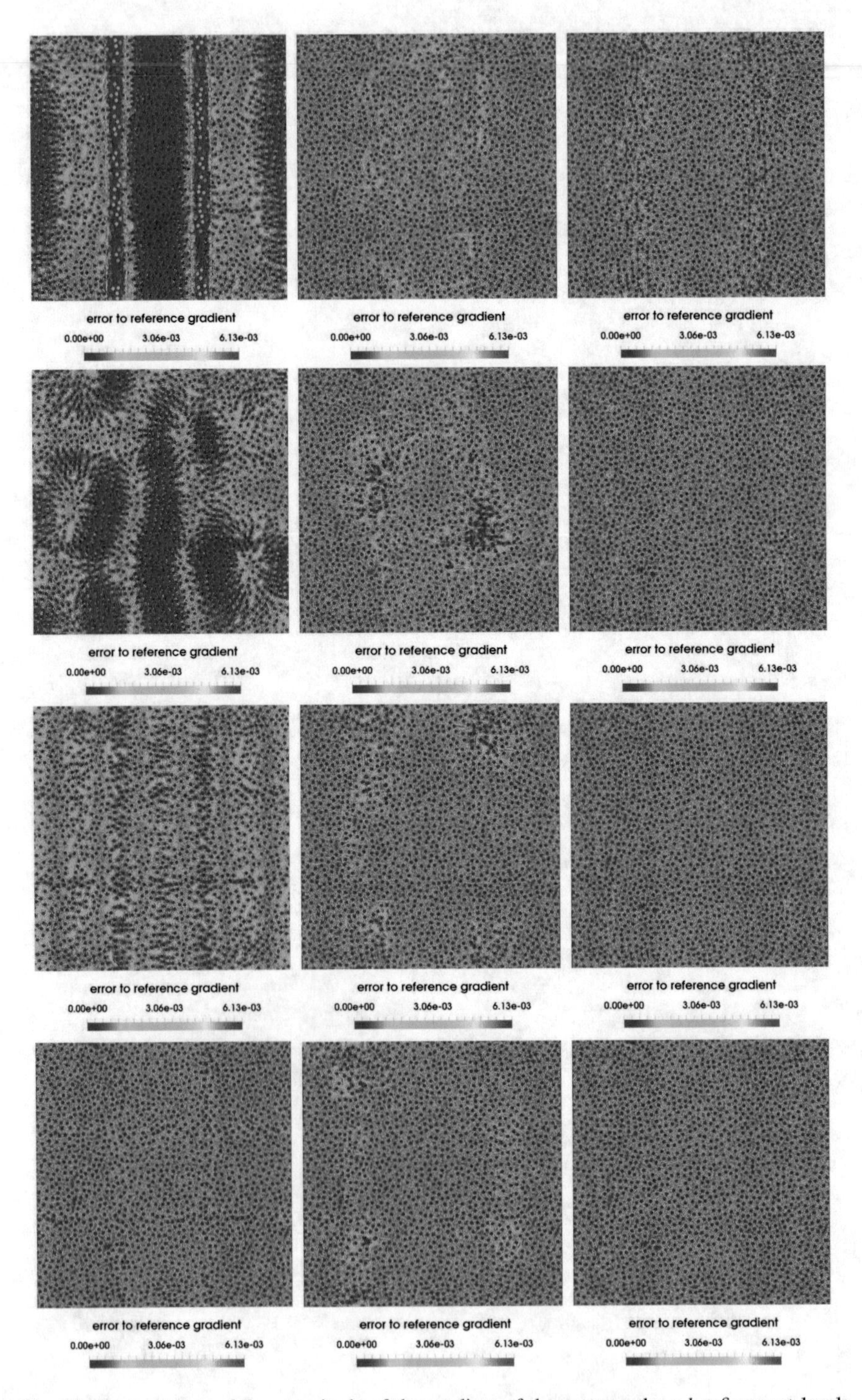

Fig. 14 Contour plots of the magnitude of the gradient of the error on three h-refinement levels (*left to right*) using $4, 8, 12, 20$ enrichments (*top to bottom*)

7 Concluding Remarks

We have presented a framework to compare the approximation power of local spaces $\mathcal{V}_l$ via the computation of the respective sup – infs $\Psi_l^{\cdot}(\mathcal{V}_l)$ and a comparison of $\Psi_l^{\cdot}(\mathcal{V}_l)$ to the optimal n-width $d_{l,n}^{\cdot} = \min \Psi_l^{\cdot}(\mathcal{V}_l)$. We focused on heterogeneous materials with the overall goal of determining high quality enrichments for a global PUM discretization via numerical homogenization techniques. Here, we considered microstructures induced by well separated circular inclusions only. The quality measures and local approximation spaces constructed in this study are both applicable to the scalar heat equation and vectorial linear elastic equation.

Our numerical results with respect to the n-widths indicate that $d_{l,n}^{\cdot}$ and the corresponding optimal spaces of eigenvectors $\mathcal{V}_l^{\cdot}$ can in fact be computed in a stable fashion provided the employed approximation space (in our case the FE mesh) is sufficiently refined. Moreover, our results show that the minimal number of shape functions n needed for $d_{l,n}^E < \epsilon$ is independent of the microstructure Ω_2 and the contrast C. Note, however, that this requires that all shape functions are computed for this particular value of the contrast C and the considered microstructure, i.e., are perfectly adapted to the problem.

The obtained results for the sup – infs showed that the optimal spaces $\mathcal{V}_l^{\{E,1,0\}}$ are all optimal for all three measures $\Psi_l^{\{E,1,0\}}(\mathcal{V}_l)$ but rather expensive to compute. Here, however, the major factor in the computation cost is the construction of $\mathcal{W}_l = \mathcal{H}_l^+$ and not the solution of the generalized eigenvalue problem. Moreover, we found that the construction of $\mathcal{W}_l \approx \mathcal{H}_l^{+,s}$ based on a_l^+-harmonic extensions of harmonic polynomials yields spaces $\mathcal{V}_l^{\{E,1,0\},+}$ that are of similar quality but much cheaper to construct. In fact, it appears to be the case that using the a_l^+-harmonic extensions of polynomials $\mathcal{V}_l^h \subset \mathcal{H}_l^{+,s}$ directly yields almost optimal performance even without the need to solve a generalized eigenvalue problem.

Finally, we found that the space $\mathcal{V}_l^{e,+}$ also appears to yield a cost-effective approximation which only requires the solution of a generalized eigenvalue problem that is readily obtained from the global stiffness matrix and a weighted mass matrix. Even though this approach requires roughly 2–3 times the minimal number of shape functions to achieve the same accuracy, the overall computational cost is far smaller than for the construction of optimal shape functions.

Acknowledgements This work was in part sponsored by the Sonderforschungsbereich 1060 of the Deutsche Forschungsgemeinschaft.

References

1. M.S. Alnæs, J. Blechta, J. Hake, A. Johansson, B. Kehlet, A. Logg, C. Richardson, J. Ring, M.E. Rognes, G.N. Wells, The fenics project version 1.5. Arch. Numer. Softw. **3**(100), 9–23 (2015)
2. I. Babuška, R. Lipton, Optimal local approximation spaces for generalized finite element methods with application to multiscale problems. Multiscale Model. Simul. **9**(1), 373–406 (2011)
3. I. Babuška, J.M. Melenk, The partition of unity finite element method: basic theory and applications. Comput. Methods Appl. Mech. Eng. **139**, 289–314 (1996). Special Issue on Meshless Methods
4. I. Babuška, J.M. Melenk, The partition of unity method. Int. J. Numer. Methods Eng. **40**, 727–758 (1997)
5. I. Babuška, G. Caloz, J.E. Osborn, Special finite element methods for a class of second order elliptic problems with rough coefficients. SIAM J. Numer. Anal. **31**, 945–981 (1994)
6. I. Babuška, U. Banerjee, J.E. Osborn, On principles for the selection of shape functions for the generalized finite element method. Comput. Methods Appl. Mech. Eng. **191**(49), 5595–5629 (2002)
7. I. Babuška, X. Huang, R. Lipton, Machine computation using the exponentially convergent multiscale spectral generalized finite element method. ESAIM: Math. Model. Numer. Anal. **48**, 493–515 (2014)
8. A. Buck, O. Iliev, H. Andrä, Multiscale finite element coarse spaces for the analysis of linear elastic composites. Technical Report 212, Fraunhofer ITWM (2012)
9. C.-C. Chu, I. Graham, T.-Y. Hou, A new multiscale finite element method for high-contrast elliptic interface problems. Math. Comput. **79**(272), 1915–1955 (2010)
10. P.G. Ciarlet, *Mathematical Elasticity, Volume I: Three Dimensional Elasticity*. Studies in Mathematics and its Applications, vol. 20 (Elsevier, Amsterdam, 1988)
11. Y. Efendiev, J. Galvis, X.-H. Wu, Multiscale finite element methods for high-contrast problems using local spectral basis functions. J. Comput. Phys. **230**(4), 937–955 (2011)
12. J.A. Evans, Y. Bazilevs, I. Babuška, T.J.R. Hughes, n-widths, sup–infs, and optimality ratios for the k-version of the isogeometric finite element method. Comput. Methods Appl. Mech. Eng. **198**(21), 1726–1741 (2009)
13. C.O. Horgan, Korn's inequalities and their applications in continuum mechanics. SIAM Rev. **37**(4), 491–511 (1995)
14. A. Kolmogoroff, Über die beste Annäherung von Funktionen einer gegebenen Funktionenklasse. Ann. Math. **37**(1), 107–110 (1936)
15. G. Leoni, *A First Course in Sobolev Spaces*. Graduate Studies in Mathematics, vol. 105 (American Mathematical Society, Providence, RI, 2009)
16. H. Owhadi, L. Zhang, Localized bases for finite-dimensional homogenization approximations with nonseparated scales and high contrast. Multiscale Model. Simul. **9**(4), 1373–1398 (2011)
17. A. Pinkus, *n-Widths in Approximation Theory*. Ergebnisse der Mathematik und ihrer Grenzgebiete, vol. 3. Folge/A series of Modern Surveys in Mathematics, vol. 7 (Springer, Berlin, 1985)
18. M.A. Schweitzer, Stable enrichment and local preconditioning in the particle–partition of unity method. Numer. Math. **118**(1), 137–170 (2011)
19. M.A. Schweitzer, Generalizations of the finite element method. Cent. Eur. J. Math, **10**, 3–24 (2012)
20. M.A. Schweitzer, Variational mass lumping in the partition of unity method. SIAM J. Sci. Comput. **35**(2), A1073–A1097 (2013)
21. M.A. Schweitzer, A. Ziegenhagel, in *Rapid Enriched Simulation Application Development with Puma*, ed. by M. Griebel, M.A. Schweitzer. Scientific Computing and Algorithms in Industrial Simulations (Springer, Berlin, 2017)

Embedding Enriched Partition of Unity Approximations in Finite Element Simulations

Marc Alexander Schweitzer and Albert Ziegenhagel

Abstract In this paper we present a general approach to embed arbitrary approximation spaces into classical finite element simulations in a non-intrusive fashion. To this end, we employ a global partition of unity method to splice the two independent approximation spaces together. The main goal of this research is to enable the timely evaluation of novel discretization approaches and meshfree techniques in an industrial context by embedding them into large scale finite element simulations. We present some numerical results showing the generality and effectiveness of our approach.

1 Introduction

Even though the classical finite element method (FEM) is the work horse of computational science and engineering today it has several limitations and drawbacks. Some of these issues were the initial starting points for research on meshfree methods (MM), generalized or extended finite element methods (GFEM/XFEM) and the partition of unity method (PUM) in the early 1990s. A number of developments in these fields by now found their way back into commercial finite element packages. However, the full spectrum of experimental and even well-established techniques in MM, GFEM/XFEM and PUM are not available in large scale commercial FEM packages which renders the evaluation of the developed techniques in industrial applications unfeasible at the moment.

M.A. Schweitzer
Institut für Numerische Simulation, Rheinische Friedrich-Wilhelms-Universität Bonn, Wegelerstr. 6, 53115 Bonn, Germany

Fraunhofer Institute for Algorithms and Scientific Computing SCAI Sankt Augustin, Germany
e-mail: schweitzer@ins.uni-bonn.de; marc.alexander.schweitzer@scai.fraunhofer.de

A. Ziegenhagel (✉)
Fraunhofer-Institut für Algorithmen und Wissenschaftliches Rechnen SCAI, Schloss Birlinghoven, 53754 Sankt Augustin, Germany
e-mail: albert.ziegenhagel@scai.fraunhofer.de

M. Griebel, M.A. Schweitzer (eds.), *Meshfree Methods for Partial Differential Equations VIII*, Lecture Notes in Computational Science and Engineering 115,
DOI 10.1007/978-3-319-51954-8_10

In this paper we present a very simple and non-intrusive strategy which in principle allows for the embedding of any approximation technique in any FEM simulation thereby enabling the timely evaluation of novel discretization techniques in an industrial context. Our approach is based on the PUM and can be viewed as a generalization of [5–7]. To document the flexibility and generality of our approach we present numerical results of a FEM simulation with an embedded enriched PUM approximation near re-entrant corners where we employ various enrichment schemes with singular and higher order polynomial enrichment functions. These results clearly show that we attain optimal convergence rates for the global FEM with embedded enriched approximation spaces without any adaptive mesh refinement or numerical artefacts.

2 Partition of Unity Method

The Partition of Unity Method (PUM) was introduced in [2, 3] and is based on [4]. The abstract ingredients which make up a PUM space

$$V^{\mathrm{PU}} := \sum_{i=1}^{N} \varphi_i V_i = \mathrm{span}\langle \varphi_i \vartheta_i^m \rangle; \tag{1}$$

are a partition of unity (PU) $\{\varphi_i : i = 1,\ldots,N\}$ and a collection of local approximation spaces $V_i := V_i(\omega_i) := \mathrm{span}\langle \vartheta_i^m \rangle_{m=1}^{d_{V_i}}$ defined on the patches $\omega_i := \mathrm{supp}(\varphi_i)$ for $i = 1,\ldots,N$. Thus, the shape functions of a PUM space are simply defined as the products of the PU functions φ_i and the local approximation functions ϑ_i^m. The PU functions provide the locality and global regularity of the product functions $\varphi_i \vartheta_i^m$ whereas the functions ϑ_i^m equip V^{PU} with its approximation power. Note that there are no constraints imposed on the choice of the local spaces V_i, i.e. they are completely independent of each other. Thus, the PUM approach allows to utilize a priori information about the sought solution locally by using so-called enrichment functions or physics-based basis functions in general [11]. Here, we usually employ local approximation spaces V_i of the form

$$V_i = \mathcal{P}_i + \mathcal{E}_i = \mathrm{span}\langle \psi_i^t \rangle + \mathrm{span}\langle \eta_i^s \rangle,$$

where $\mathcal{P}_i$ denotes a space of local polynomials and $\mathcal{E}_i$ accounts for non-smooth local features such as kinks, discontinuities and singularities of the solution on the patch ω_i. In our setting, however, we will not follow this local approach which focusses on approximation properties but we take a more global point of view which is in spirit closer to a domain decomposition line of thought, see e.g. [12]. To this end, let us consider a very simple cover of the domain Ω into just two overlapping patches Ω_0 and Ω_1 with respective PU functions Φ_0 and Φ_1, i.e. $\Phi_o + \Phi_1 \equiv 1$ on $\Omega \subset \Omega_0 \cup \Omega_1$. Then, let us choose the local approximation space V_0 on the patch Ω_0

to be a classical FE space defined on a respective mesh $\Omega_{0,h}$ which discretizes Ω_0. On the other patch Ω_1 we can choose any other approximation spaces V_1 since it is completely independent of V_0 by construction; for instance we could choose another FE space on a non-matching mesh $\Omega_{1,H}$ [5, 6]. Throughout this paper we will use a secondary enriched PUM as described above to demonstrate that any approximation space can be embedded in a classical FE simulation in a non-intrusive fashion by our approach, compare also [7].

3 Embedding PUM into Finite Element Simulations

Using this PU, the global approximation space V^{PU} on Ω, according to (1), is given by

$$V^{\mathrm{PU}} := \Phi_0 V_0 + \Phi_1 V_1 = \mathrm{span}\langle \Phi_0 \phi_j \rangle + \mathrm{span}\langle \Phi_1 \varphi_i \vartheta_i^m \rangle, \tag{2}$$

where ϕ_j denote the nodal FE basis functions given on the mesh $\Omega_{0,h}$ and the $\varphi_i \vartheta_i^m$ denote the basis functions of the local PUM space V_1 on the patch Ω_1. At first sight, it seems that we actually need to access each FE basis function ϕ_j to multiply it by the PU function Φ_0 which obviously would not be possible when using a commercial FE package. Fortunately, this is not the case when we employ a so-called flat-top PU [8, 9], i.e. we choose $\{\Phi_0, \Phi_1\}$ such that there exist

$$\begin{aligned} &\tilde{\Omega}_0 \subset \Omega_0 \text{ such that } \Phi_0|_{\tilde{\Omega}_0} \equiv 1, \quad \tilde{\Omega}_0 \cap \Omega_1 = \emptyset, \\ &\tilde{\Omega}_1 \subset \Omega_1 \text{ such that } \Phi_1|_{\tilde{\Omega}_1} \equiv 1, \quad \tilde{\Omega}_1 \cap \Omega_0 = \emptyset, \end{aligned}$$

and to this end we define

$$\hat{\Omega} := \Omega_0 \cap \Omega_1, \quad \tilde{\Omega}_0 := \Omega_0 \setminus \hat{\Omega}, \quad \text{and} \quad \tilde{\Omega}_1 := \Omega_1 \setminus \hat{\Omega}.$$

For simplicity, let us furthermore assume that $\tilde{\Omega}_0$ is already resolved by the mesh $\Omega_{0,h}$, i.e. $\tilde{\Omega}_0$ is discretized by a subset $\tilde{\Omega}_{0,h}$ of elements of the original mesh $\Omega_{0,h}$ on Ω_0. Then, we can rewrite our approximation space (2) as

$$V^{\mathrm{PU}} := \mathrm{span}\langle \phi_{\tilde{j}} \rangle + \mathrm{span}\langle \Phi_0 \phi_j \rangle + \mathrm{span}\langle \varphi_{\tilde{i}} \vartheta_{\tilde{i}}^m \rangle + \mathrm{span}\langle \Phi_1 \varphi_i \vartheta_i^m \rangle, \tag{3}$$

where the first term now involves only those FE basis functions with $\mathrm{supp}\, \phi_{\tilde{j}} \subset \tilde{\Omega}_0$ whereas in the second term we consider the FE basis functions with $\mathrm{supp}\, \phi_j \cap \hat{\Omega} \neq \emptyset$. Analogously, the third term in (3) accounts for $\varphi_{\tilde{i}} \vartheta_{\tilde{i}}^m$ with $\mathrm{supp}\, \varphi_{\tilde{i}} \subset \tilde{\Omega}_1$ and the last term for $\varphi_i \vartheta_i^m$ which satisfy $\mathrm{supp}\, \varphi_i \cap \hat{\Omega} \neq \emptyset$. Thus, the first and third terms in (3), i.e.,

$$V_0^{\mathrm{FE}} := \mathrm{span}_{\mathrm{supp}\, \phi_{\tilde{j}} \subset \tilde{\Omega}_0} \langle \phi_{\tilde{j}} \rangle \quad \text{and} \quad V_1^{\mathrm{PU}} := \mathrm{span}_{\mathrm{supp}\, \varphi_{\tilde{i}} \vartheta_{\tilde{i}}^m \subset \tilde{\Omega}_1} \langle \phi_{\tilde{i}} \rangle, \tag{4}$$

are completely agnostic to the construction described above and thus we can employ any FE package, commercial or academic, (or any other method) to discretize our model on $\tilde{\Omega}_0$ and $\tilde{\Omega}_1$ without any intrusion into the respective code or access to its implementation. We must essentially be concerned with the implementation of the discretization by

$$\hat{V} := \text{span}_{\text{supp}\,\phi_j \cap \hat{\Omega} \neq \emptyset} \langle \Phi_0 \phi_j \rangle + \text{span}_{\text{supp}\,\varphi_i \vartheta_i^m \cap \hat{\Omega} \neq \emptyset} \langle \Phi_1 \varphi_i \vartheta_i^m \rangle \tag{5}$$

on the overlapping region $\hat{\Omega} = \Omega_0 \cap \Omega_1$ only. Here, we need to have access to the employed FE mesh $\Omega_{0,h}$ and the respective element types as well as to the implementation of the method we wish to embed into the FE simulation, i.e. V_1 on Ω_1. To this end, our implementation of the proposed embedding scheme provides a number of classical finite elements so that we can use an arbitrary FE package on $\tilde{\Omega}_0$, our original PUM implementation on $\tilde{\Omega}_1$ and this new interface implementation on the overlap $\hat{\Omega}$ for the discretization. Then, we can assemble the overall stiffness matrix K and the load-vector f

$$K := \begin{pmatrix} K_0^{\text{FE}} & \hat{K}_0 & 0 \\ \hat{K}_0^T & \hat{K} & \hat{K}_1 \\ 0 & \hat{K}_1^T & K_1^{\text{PU}} \end{pmatrix}, \quad f := \begin{pmatrix} f_0^{\text{FE}} \\ \hat{f} \\ f_1^{\text{PU}} \end{pmatrix},$$

where K_0^{FE}, f_0^{FE} and K_1^{PU}, f_1^{PU} are the stiffness matrices and load-vectors obtained directly from the two non-overlapping spaces (4) and the other terms $\hat{K}_0$, $\hat{K}$, $\hat{K}_1, \hat{f}$ involve the space (5) on the overlap $\hat{\Omega}$.

For the iterative solution of the resulting global linear system we can obviously employ available solvers S_0^{FE} and S_1^{PU} for the two blocks K_0^{FE} and K_1^{PU} to define a simple preconditioner for the global stiffness K via

$$P := \begin{pmatrix} S_0^{\text{FE}} & & \\ & \hat{S} & \\ & & S_1^{\text{PU}} \end{pmatrix},$$

where we may use a direct solver to construct $\hat{S}$ for $\hat{K}$ (since the overlap $\hat{\Omega}$ is typically small compared with $\tilde{\Omega}_0$ and $\tilde{\Omega}_1$) or any other approximate inverse $\hat{S}$.

4 Numerical Results

In this section we present some results of our numerical experiments using the embedded enriched PUM within a classical FE simulation as discussed above. To this end, we introduce some shorthand notation for various norms of the error

$u - u^{\text{PU}}$, i.e., we define

$$e_{L^2} := \frac{\|u - u^{\text{PU}}\|_{L^2}}{\|u\|_{L^2}}, \quad e_{H^1} := \frac{\|\nabla(u - u^{\text{PU}})\|_{L^2}}{\|\nabla u\|_{L^2}}. \tag{6}$$

For each of these error norms we compute the respective algebraic convergence rate ρ by considering the error norms of two consecutive refinement levels $l-1$ and l

$$\rho := -\frac{\log\left(\frac{\|u-u_l^{\text{PU}}\|}{\|u-u_{l-1}^{\text{PU}}\|}\right)}{\log\left(\frac{\text{DOF}(l)}{\text{DOF}(l-1)}\right)}, \quad \text{where } \text{DOF}(k) := \text{DOF}_0(k) + \text{DOF}_1(k), \tag{7}$$

$\text{DOF}_0(k) := \dim V_{0,k}$ and $\text{DOF}_1(k) = \dim V_{1,k}$, which corresponds to the classical h^γ notation for uniform h-refinement where

$$\gamma := \frac{\log\left(\frac{\|u-u_l^{\text{PU}}\|}{\|u-u_{l-1}^{\text{PU}}\|}\right)}{\log(\frac{1}{2})} \tag{8}$$

such that $\gamma = \rho d$ in $\mathbb{R}^d$, i.e. $\gamma = 2\rho$ in two space dimensions.

The model problem considered in the following is the simple Poisson problem

$$\begin{aligned} -\Delta u &= f \text{ in } \Omega \subset \mathbb{R}^2, \\ u &= g \text{ on } \Gamma_D \subset \partial\Omega, \\ \frac{\partial u}{\partial n} &= h \text{ on } \Gamma_N = \partial\Omega \setminus \Gamma_D, \end{aligned} \tag{9}$$

on an L-shaped domain $\Omega := [-1, 1]^2 \setminus [0, 1]^2$, compare Fig. 1. Prescribing vanishing Dirichlet boundary conditions at the re-entrant corner, i.e. on $\Gamma_D :=$

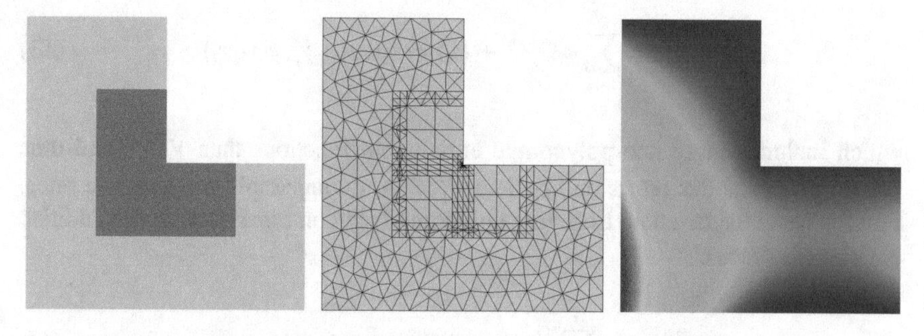

Fig. 1 Sketch of domain Ω and the sub-domains $\tilde{\Omega}_0$ and Ω_1 (*left*). A coarse mesh $\Omega_{0,h}$ defined on Ω_0 and the overlapping patches ω_i defined on Ω_1 (*center*). A contour plot of the solution (11)

$\partial\Omega \cap [0,1]^2$, the solutions to (9) exhibit the singular behavior

$$u_s(r,\theta) := r^{\frac{2}{3}} \sin\left(\frac{2\theta - \pi}{3}\right) \tag{10}$$

near the origin. In our experiment, we choose the data f, g and h such that the solution to (9) is given by

$$u = (x^2 + xy - y^2 + 1)u_s \tag{11}$$

which would usually require an adaptive mesh refinement (AMR) near the re-entrant corner when using classical FEM to obtain an optimal convergence rate of $\rho_{L^2} = 1$ and $\rho_{H^1} = 1/2$. Instead of employing AMR near the re-entrant corner, we will embed an enriched PUM approximation at the origin, compare Fig. 1, to account for the singular behavior (10). Thus, we anticipate that a uniform refinement of the FE mesh $\Omega_{0,h}$ will already yield these optimal convergence rates if the enriched PUM space V_1 resolves the singular solution (11) with sufficiently accuracy.

We consider three different enriched PUM spaces V_1 in our experiments to demonstrate that our approach is truely suitable to embed an arbitrary approximation scheme into a FE simulation. First, we choose as local approximation spaces V_i on the patches ω_i linear Legendre polynomials $\mathcal{P}_i$ and the one-dimensional enrichment space $\mathcal{E}_i = \operatorname{span}\langle \eta_i \rangle$ with $\eta_i := u_s|_{\omega_i}$ to define the space

$$V_1^{1+u_s} := \sum_{i=1}^{N} \varphi_i(\mathcal{P}_i^1 + \mathcal{E}_i) = \operatorname{span}\langle \varphi_i \psi_i^t, \varphi_i \eta_i \rangle \tag{12}$$

which is known to yield optimal approximation to solutions of the form (11) by uniform refinement of the patches ω_i, see e.g. [10]. Then, we consider the somewhat larger space

$$V_1^{1\times u_s} := \sum_{i=1}^{N} \varphi_i \mathcal{P}_i^1(1 + \mathcal{E}_i) = \operatorname{span}\langle \varphi_i \psi_i^t, \varphi_i \psi_i^t \eta_i \rangle \tag{13}$$

which includes more non-polynomial enrichment functions than $V_1^{1+u_s}$ and thus should yield smaller errors on Ω_1 than $V_1^{1+u_s}$ yet comparable convergence rates. Finally, we consider cubic Legendre polynomials in combination with the additive enrichment space $\mathcal{E}_i$

$$V_1^{3+u_s} := \sum_{i=1}^{N} \varphi_i(\mathcal{P}_i^3 + \mathcal{E}_i) = \operatorname{span}\langle \varphi_i \psi_i^t, \varphi_i \eta_i \rangle \tag{14}$$

which is also larger than $V_1^{1+u_s}$ but contains additional higher order polynomials and should also attain smaller errors on Ω_1 than $V_1^{1+u_s}$.

To control the refinement of the two independent local approximation spaces V_0 and V_1 we compute norms of the true errors $u - u^{\mathrm{PU}}$ on the subdomains $\tilde{\Omega}_0$, $\hat{\Omega}$ and $\tilde{\Omega}_1$, i.e. we compute

$$\hat{e} := \|u - u^{\mathrm{PU}}\|_{\hat{\Omega}}, \quad \tilde{e}^0 := \|u - u^{\mathrm{PU}}\|_{\tilde{\Omega}_0}, \quad \tilde{e}^1 := \|u - u^{\mathrm{PU}}\|_{\tilde{\Omega}_1}. \tag{15}$$

Our simple refinement scheme is based on balancing the error in the subdomains, i.e. we aim at attaining

$$\tilde{e}^0 \approx 3\tilde{e}^1 \text{ and } \hat{e} \leq \tilde{e}^0. \tag{16}$$

Since we anticipate that all enriched spaces V_1 provide high-quality approximations to (11) with at least the convergence rates $\rho_{L^2} = 1$ and $\rho_{H^1} = 1/2$, we refine only the classical FE space V_0 on Ω_0 by uniform mesh refinement if (16) is satisfied. If (16) is not satisfied we refine both V_0 and V_1 by uniform refinement.

The results of our numerical experiments are given in Tables 1, 2 and 3, compare also Fig. 2. From the computed convergence rates (7) it is obvious that all three embedded schemes attain show an optimal error reduction, i.e. the global convergence rate is limited by the classical FE part. With respect to the performance of the three different enriched spaces V_1 we find that $V_1^{3+u_s}$ requires the least amount of refinement in Ω_1. Just 132 degrees of freedom in $V_1^{3+u_s}$ are sufficient to obtain

Table 1 Measured errors (6) and (15) and the respective convergence rates (7) using $V_1^{1+u_s}$ (12)

l	DOF	DOF$_0$	DOF$_1$	e_{L^2}	ρ_{L^2}	e_{H^1}	ρ_{H^1}	$e^0_{L^2}$	$e^0_{H^1}$	$\hat{e}_{L^2}$	$\hat{e}_{H^1}$	$e^1_{L^2}$	$e^1_{H^1}$
1	158	110	48	1.673_{-2}	2.5	1.489_{-1}	1.2	2.13_{-2}	4.20_{-1}	3.50_{-3}	6.10_{-2}	3.89_{-3}	8.29_{-2}
2	335	287	48	5.267_{-3}	1.5	7.738_{-2}	0.9	4.99_{-3}	2.05_{-1}	2.83_{-3}	4.05_{-2}	3.87_{-3}	8.26_{-2}
3	1,204	1,015	189	1.143_{-3}	1.2	3.702_{-2}	0.6	1.20_{-3}	9.93_{-2}	4.63_{-4}	1.71_{-2}	7.69_{-4}	3.75_{-2}
4	4,487	3,737	750	3.313_{-4}	0.9	1.887_{-2}	0.5	3.36_{-4}	5.03_{-2}	1.19_{-4}	8.99_{-3}	2.49_{-4}	1.99_{-2}
5	17,447	14,489	2,958	8.892_{-5}	1.0	9.550_{-3}	0.5	8.77_{-5}	2.53_{-2}	3.11_{-5}	4.57_{-3}	7.06_{-5}	1.04_{-2}
6	68,772	57,039	11,733	2.253_{-5}	1.0	4.765_{-3}	0.5	2.20_{-5}	1.26_{-2}	7.70_{-6}	2.27_{-3}	1.82_{-5}	5.23_{-3}
7	273,048	226,360	46,688	5.666_{-6}	1.0	2.379_{-3}	0.5	5.48_{-6}	6.28_{-3}	1.95_{-6}	1.14_{-3}	4.64_{-6}	2.65_{-3}
8	1,085,230	899,123	186,107	1.421_{-6}	1.0	1.191_{-3}	0.5	1.37_{-6}	3.14_{-3}	4.87_{-7}	5.76_{-4}	1.17_{-6}	1.33_{-3}

Table 2 Measured errors (6) and (15) and the respective convergence rates (7) using $V_1^{1\times u_s}$ (13)

l	DOF	DOF$_0$	DOF$_1$	e_{L^2}	ρ_{L^2}	e_{H^1}	ρ_{H^1}	$e^0_{L^2}$	$e^0_{H^1}$	$\hat{e}_{L^2}$	$\hat{e}_{H^1}$	$e^1_{L^2}$	$e^1_{H^1}$
1	389	110	279	1.811_{-2}	0.9	1.458_{-1}	0.5	2.37_{-2}	4.21_{-1}	2.09_{-3}	4.97_{-2}	1.03_{-3}	1.08_{-2}
2	566	287	279	4.397_{-3}	3.8	7.095_{-2}	1.9	5.74_{-3}	2.05_{-1}	5.33_{-4}	2.07_{-2}	2.91_{-4}	7.05_{-3}
3	1,294	1,015	279	1.013_{-3}	1.8	3.446_{-2}	0.9	1.32_{-3}	9.93_{-2}	1.35_{-4}	1.10_{-2}	1.08_{-4}	6.20_{-3}
4	4,016	3,737	279	2.726_{-4}	1.2	1.756_{-2}	0.6	3.45_{-4}	5.03_{-2}	4.54_{-5}	6.11_{-3}	8.32_{-5}	5.88_{-3}
5	15,587	14,489	1,098	6.713_{-5}	1.0	8.817_{-3}	0.5	8.64_{-5}	2.53_{-2}	1.23_{-5}	3.40_{-3}	1.29_{-5}	1.75_{-3}
6	61,305	57,039	4,266	1.888_{-5}	0.9	4.448_{-3}	0.5	2.16_{-5}	1.26_{-2}	8.40_{-6}	2.10_{-3}	8.83_{-6}	1.86_{-3}
7	243,091	226,360	16,731	4.966_{-6}	1.0	2.230_{-3}	0.5	5.37_{-6}	6.28_{-3}	2.37_{-6}	1.12_{-3}	2.84_{-6}	1.12_{-3}
8	965,059	899,123	65,936	1.364_{-6}	0.9	1.125_{-3}	0.5	1.35_{-6}	3.14_{-3}	7.03_{-7}	5.69_{-4}	9.39_{-7}	6.96_{-4}

Table 3 Measured errors (6) and (15) and the respective convergence rates (7) using $V_1^{3+u_s}$ (14)

l	DOF	DOF$_0$	DOF$_1$	e_{L^2}	ρ_{L^2}	e_{H^1}	ρ_{H^1}	$e^0_{L^2}$	$e^0_{H^1}$	$\hat{e}_{L^2}$	$\hat{e}_{H^1}$	$e^1_{L^2}$	$e^1_{H^1}$
1	242	110	132	1.807_{-2}	4.0	1.458_{-1}	2.1	2.4_{-2}	4.2_{-1}	2.1_{-3}	5.1_{-2}	1.0_{-3}	1.2_{-2}
2	419	287	132	4.375_{-3}	2.6	7.101_{-2}	1.3	5.7_{-3}	2.1_{-1}	5.4_{-4}	2.2_{-2}	3.1_{-4}	7.3_{-3}
3	1,147	1,015	132	9.906_{-4}	1.5	3.446_{-2}	0.7	1.3_{-3}	9.9_{-2}	1.4_{-4}	1.2_{-2}	1.4_{-4}	4.7_{-3}
4	3,869	3,737	132	2.531_{-4}	1.1	1.749_{-2}	0.6	3.2_{-4}	5.0_{-2}	4.4_{-5}	6.7_{-3}	9.3_{-5}	2.8_{-3}
5	14,621	14,489	132	8.109_{-5}	0.9	8.838_{-3}	0.5	6.8_{-5}	2.5_{-2}	1.9_{-5}	3.7_{-3}	8.0_{-5}	2.1_{-3}
6	57,564	57,039	525	1.609_{-5}	1.2	4.369_{-3}	0.5	2.0_{-5}	1.3_{-2}	2.0_{-6}	1.4_{-3}	6.4_{-6}	4.1_{-4}
7	228,398	226,360	2,038	4.030_{-6}	1.0	2.177_{-3}	0.5	5.2_{-6}	6.3_{-3}	5.1_{-7}	7.0_{-4}	8.5_{-7}	1.7_{-4}

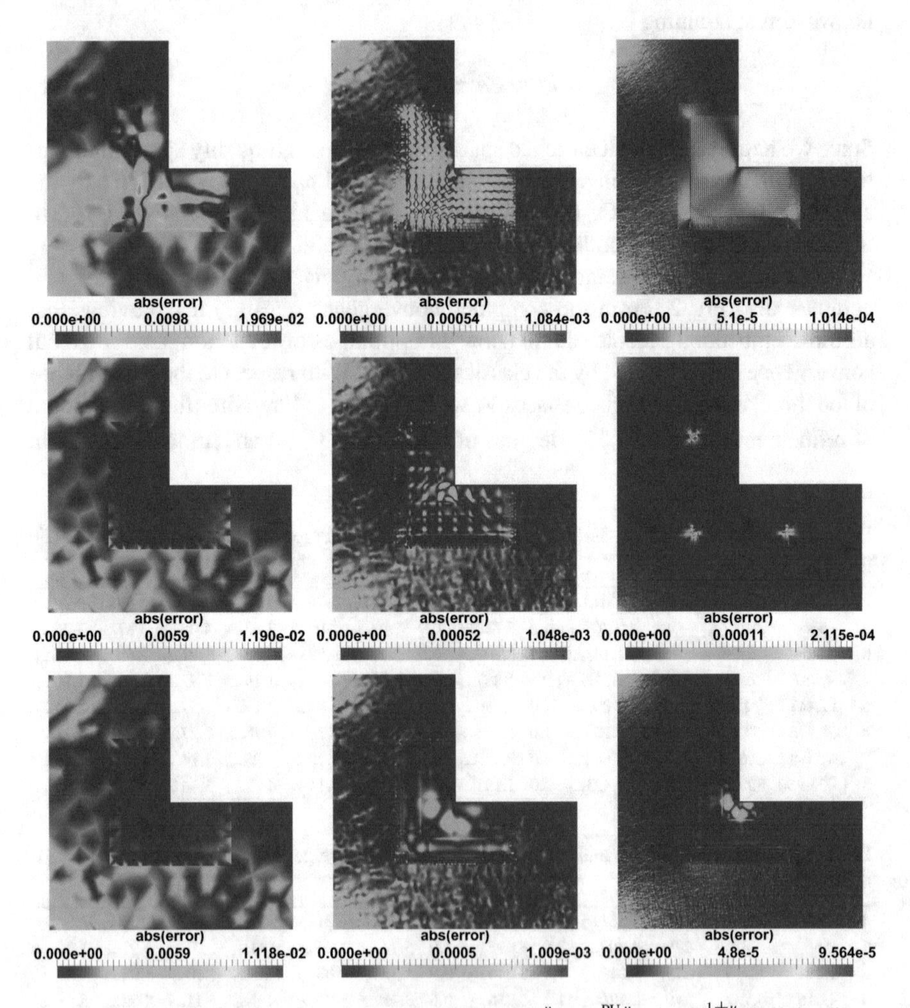

Fig. 2 Contour plots of the attained approximation errors $\|u - u_l^{\mathrm{PU}}\|$ using $V_1^{1+u_s}$ (12), i.e. linear polynomials with additive enrichment (*top*), using $V_1^{1\times u_s}$ (13) with multiplicative enrichment (*center*) and $V_1^{3+u_s}$ (14), i.e. cubic polynomials with additive enrichment (*bottom*), on the respective refinement levels $l = 2, 4, 6$ (*left to right*)

a global error of less than 1%, whereas $V^{1\times u_s}$ needs DOF $=$ 1098 and the most commonly employed space $V_1^{1+u_s}$ must employ DOF $=$ 2958. Note however that the classical FE space V_0 must be a lot finer with DOF $=$ 14489 even though the solution in Ω_0 is smooth to yield the global accuracy of less than 1%. Moreover, a global FE approach would require AMR and a much larger number of degrees of freedom near the origin to resolve the singularity of the solution (11). Overall these results clearly show the effectiveness of using an embedding of physics-based basis functions into a FE simulation to reduce the total number of degrees of freedom substantially and that the presented approach is able to embed an arbitrary approximation space into a FE simulation. From the contour plots depicted in Fig. 2 we can also see that our embedding approach is free from any artefacts in the overlap $\hat{\Omega}$. The distribution of the errors in $\hat{\Omega}$ is essentially determined by the choice of V_1 (since V_0 is fixed). Recall that $V^{1\times u_s}$ contains more non-polynomial enrichment functions which should provide better approximation near the origin whereas the space $V_1^{3+u_s}$ is more effective further away from the origin. This anticipated behavior can clearly be observed from the contour plots of the respective errors depicted in Fig. 2.

5 Concluding Remarks

In the paper we presented a simple technique which enables the timely evaluation of novel discretization techniques by embedding them into classical FE simulations. Our approach is non-intrusive and thus can be employed also in conjunction with large scale commercial FEM packages so that novel discretization techniques may be evaluated in industrial applications. So far we have used the presented technique successfully to embed enriched PUM spaces into the academic code FEniCS [1] and the commercial package Abaqus (http://www.3ds.com/products-services/simulia/products/abaqus/).

References

1. M.S. Alnæs, J. Blechta, J. Hake, A. Johansson, B. Kehlet, A. Logg, C. Richardson, J. Ring, M.E. Rognes, G.N. Wells, The fenics project version 1.5. Arch. Numer. Softw. **3**(100), 9–23 (2015)
2. I. Babuška, J.M. Melenk, The partition of unity finite element method: basic theory and applications. Comput. Methods Appl. Mech. Eng. **139**, 289–314 (1996). Special Issue on Meshless Methods
3. I. Babuška, J.M. Melenk, The partition of unity method. Int. J. Numer. Methods Eng. **40**, 727–758 (1997)
4. I. Babuška, G. Caloz, J.E. Osborn, Special finite element methods for a class of second order elliptic problems with rough coefficients. SIAM J. Numer. Anal. **31**, 945–981 (1994)
5. C. Bacuta, J. Xu, Partition of unity for the Stokes problem on nonmatching grids, in *Proceedings of the 2003 Copper Mountain Conference on Multigrid*, 2003

6. C. Bakuta, J. Chen, Y. Huang, J. Xu, L. Zikatanov, Partition of unity method on nonmatching grids for the Stokes problem. J. Numer. Math. **13**(3), 157–169 (2005)
7. P. Gupta, J.P. Pereira, D.-J. Kim, C.A. Duarte, T. Eason, Analysis of three-dimensional fracture mechanics problems: a non-intrusive approach using a generalized finite element method. Eng. Fract. Mech. **90**, 41–64 (2012)
8. M.A. Schweitzer, *A Parallel Multilevel Partition of Unity Method for Elliptic Partial Differential Equations*. Lecture Notes in Computational Science and Engineering, vol. 29 (Springer, Berlin, 2003)
9. M.A. Schweitzer, *Meshfree and Generalized Finite Element Methods* (Habilitation, Institute for Numerical Simulation, University of Bonn, 2008)
10. M.A. Schweitzer, Stable enrichment and local preconditioning in the particle–partition of unity method. Numer. Math. **118**(1),137–170 (2011)
11. M.A. Schweitzer, Generalizations of the finite element method. Central Eur. J. Math. **10**, 3–24 (2012)
12. B.F. Smith, P.E. Bjørstad, W.D. Gropp, *Domain Decomposition: Parallel Multilevel Methods for Elliptic Partial Differential Equations* (Cambridge University Press, Cambridge, 1996)

Building a Numerical Framework to Model Gas-Liquid-Solid Interactions Using Meshfree Interpolation Methods

Chu Wang and Lucy T. Zhang

Abstract In this work, we present a numerical framework that can model and simulate gas-liquid-solid three-phase interactions. A non-boundary-fitted approach is developed to simultaneously accommodate the moving gas-liquid interfaces and deforming solid. The connectivity-free front tracking method (CFFT) is adopted to track the gas-liquid interface, where an approximation-correction step is used to construct an indicator field without requiring the connectivity of the interfacial points. Therefore, topological change such as free surfaces with bubble breaking up and coalescing can be handled more easily and robustly. The fluid-solid interactions are modeled using the modified immersed finite element method (mIFEM). A more realistic and accurate solid movement and deformation are achieved by solving the solid dynamics, rather than been imposed as in the original IFEM. The coupling of the two algorithms is achieved using a meshfree interpolation function, the reproducing kernel particle method. The concept of constructing the indicator function to distinguish gas from liquid and fluid from solid naturally combines the CFFT and mIFEM algorithms together, and simulate the complex 3-phase physical system in a cohesive manner.

1 Introduction

The interactions among gas, liquid, and solid is an important physical phenomenon that appears in nature and many engineering applications. Due to the complex entities involved in the multiphase flows, namely, gas bubbles, liquid drops and solid particles, accurate analysis using either experimental or analytical techniques maybe hindered. Despite numerous efforts in the past decades, researchers are still actively working on developing numerical algorithms that can be efficient, accurate, and

C. Wang
Convergent Science, Madison, WI, USA
e-mail: chu.wang@convergecfd.com

L.T. Zhang (✉)
Rensselaer Polytechnic Institute, Troy, NY, USA
e-mail: zhanglucy@rpi.edu

M. Griebel, M.A. Schweitzer (eds.), *Meshfree Methods for Partial Differential Equations VIII*, Lecture Notes in Computational Science and Engineering 115,
DOI 10.1007/978-3-319-51954-8_11

easily implementable. One of the most notable challenges is the interface tracking, where a robust numerical scheme that can handle large interface deformation and topological changes is desired.

Various numerical schemes have been developed to treat gas-liquid multiphase flows. The most commonly used computational approaches are: the front capturing method and the front tracking method. Volume of fluid (VOF) method [3, 8] and level-set method [7, 38] root from the front capturing approach. The interface is *implicitly* modeled as an isosurface of a scalar function such as the indicator (color) function in VOF or signed distance function in level-set. They are relatively easy to implement and the interface topology change is naturally handled. However, the VOF method can form discontinuities at interfaces between grid boundaries when the interface is highly deformed in an un-resolved grid, whereas the level-set method may result in unphysical total mass change in a long simulation [28]. The front tracking method [23, 31, 32, 37], on the other hand, models the interface as markers or points that are connected using Lagrangian interface elements, therefore the interface is *explicitly* represented. This approach results in good total volume conservation and accurate moving interface capturing. However, the required logical connectivities of the interfacial points bring numerical difficulties in treating interface topology changes, especially for three dimensional cases. Therefore, it would be of a great improvement if a numerical algorithm can combine the merit of both front tracking and front capturing approaches. The connectivity-free front tracking method (CFFT) [36] combines the merits of both the front tracking and front capturing techniques where the interface is explicitly represented by interfacial points to achieve better volume conservation and the topology change of the interface is handled without considering the logical connectivity of the interfacial points. A curvature projection scheme is adopted to minimized the effect of spurious currents coming from the irregular curvature calculations.

Besides gas-liquid multiphase flows, modeling the fluid-structure interactions is also a fascinating topic in three-phase interaction problems. Since the fluid may involve two phases (gas and liquid), their interactions with the solid phase can raise many numerical difficulties and complexity. Many numerical approaches have been developed to deal with fluid-structure interactions. In general, they can be grouped into two categories: boundary-fitted and non-boundary-fitted approaches. One of the commonly used boundary fitted approach is the Arbitrary Lagrangian Eulerian (ALE) [9, 11, 12, 15, 16, 41] method. The fluid-solid interface is explicitly tracked using a boundary-fitted mesh. This approach can capture the interface and interfacial solution relatively accurately However, re-meshing or mesh-updating processes when mesh is severely distorted can be computationally expensive. To tackle this problem, the non-boundary-fitted approach can be used to avoid the re-meshing process. A widely used method following this approach is called the immersed boundary (IB) method, which was initially proposed by Peskin to study the blood flow around heart valves [22, 24–26]. The problem with the IB approach is the lack of more realistic representations of the solid, which hinders the accurate assessment of the material behavior and its deformation, or even the surrounding fluid solutions. The immersed finite element method [6, 14, 20, 40, 42] resolves this issue by allowing the solid domain to be constructed independently with a Lagrangian mesh and be described using a more detailed constitutive model such as linear elastic, hyperelastic and viscoelastic materials. However, the original IFEM method relies on the 'soft material' assumption that the solid follows the movement of the fluid, where the dynamics of the solid is not been solved, but rather, imposed based on the dynamics of the fluid. Therefore, an improvement of the method is needed when the solid behaves very different from the fluid, or in the cases that the movement of the solid dominates and changes the fluid flow. This leads to the development of the modified IFEM method (mIFEM) [35], which yields more accurate solid solution by solving the dynamics of the solid.

In this study, we are presenting a fully coupled numerical algorithm that can model gas-liquid-solid multiphase flows using the non-boundary-fitted approach by meticulously integrating the aforementioned CFFT and mIFEM methods. The indicator function naturally unites the two algorithms together without defining simultaneously different levels as it would be required in level-sets. The indicator function is constructed using the interpolation function of the reproducing kernel particle method (RKPM) for its high order and non-connectivity features.

The outline of this paper is listed as follows. In Sect. 2, the numerical framework for 3-phase modeling that couples CFFT and modified IFEM are presented. This framework is not specific to CFFT and IFEM, rather it is applicable to any combination of existing multiphase and fluid-structure interaction solvers. In Sect. 2.3, several 2-D and 3-D numerical examples are studied to validate the numerical approach. Finally, conclusions are drawn in Sect. 3.

2 Numerical Framework

In this section, we first briefly review the mathematical formulations for connectivity-free front tracking method and the modified immersed finite element method, respectively. The coupling scheme of the two methods through the indicator setup and the interpolation function is then introduced.

2.1 Multi-Fluid Interface Tracking Using the Connectivity-Free Front Tracking Method

The governing equations for an isothermal multiphase flow can be described using a single set of Navier-Stokes equations with fluid properties varying across the interface. The multi-fluid is treated as 'one-fluid' without the need to handle the jump condition across the interface [10]. The surface tension force can be treated as a singular source term which is included in the momentum equation.

Together, the continuity and momentum equations for incompressible fluid can be expressed as follows:

$$\nabla \cdot \mathbf{u} = 0, \tag{1}$$

$$\rho \frac{\partial \mathbf{u}}{\partial t} + \rho(\mathbf{u} \cdot \nabla)\mathbf{u} = -\nabla p + \mu \nabla^2 \mathbf{u} + \rho \mathbf{g} + \mathbf{F}_\sigma, \tag{2}$$

where $\mathbf{u}(\mathbf{x}, t)$ is the velocity, $p(\mathbf{x}, t)$ is the pressure, ρ and μ are the fluid density and viscosity, respectively, $\mathbf{g}$ is gravity, and $\mathbf{F}_\sigma$ is the surface tension force. To evaluate the surface tension, interfacial properties such as unit normal, curvature, and surface area of each interfacial point are required. The traditional front tracking method relies on the connectivity of the interfacial points to obtain these properties. However, when the interface undergoes topological changes such as bubble coalescing and breaking up reconstructing this interface connectivity is also required. This leads to the development of our recent CFFT algorithm [36] which does not require the interface connectivity by constructing the indicator field with a correction step.

2.1.1 Indicator Construction

An indicator function I is defined as 1 and 0, respectively, for each fluid,

$$I = \begin{cases} 1 & \text{for fluid 1,} \\ 0 & \text{for fluid 2.} \end{cases} \tag{3}$$

Fluid parameters such as the density and viscosity for a given location with coordinate $\mathbf{x}$ can be expressed using the indicator function accordingly such that:

$$\rho(\mathbf{x}) = I(\mathbf{x}) \cdot \rho_1 + (1 - I(\mathbf{x})) \cdot \rho_2, \tag{4}$$

$$\mu(\mathbf{x}) = I(\mathbf{x}) \cdot \mu_1 + (1 - I(\mathbf{x})) \cdot \mu_2. \tag{5}$$

In the front tracking method with connectivity [32], the indicator is obtained by solving Poisson's equation $\nabla^2 I(\mathbf{x}) = \nabla \cdot \int_\Gamma \mathbf{n}\Phi(\mathbf{x} - \mathbf{x}_\Gamma) d\Gamma$, where $\mathbf{n}$ is the unit normal of the interface, Γ is the interface between the two phases, x_Γ are the points on the interface, and Φ is the interpolation function. This step requires the unit normal of the interfacial points to be computed a priori. However, without connectivity the unit normal, and thus the indicator field has to be constructed using an alternative way.

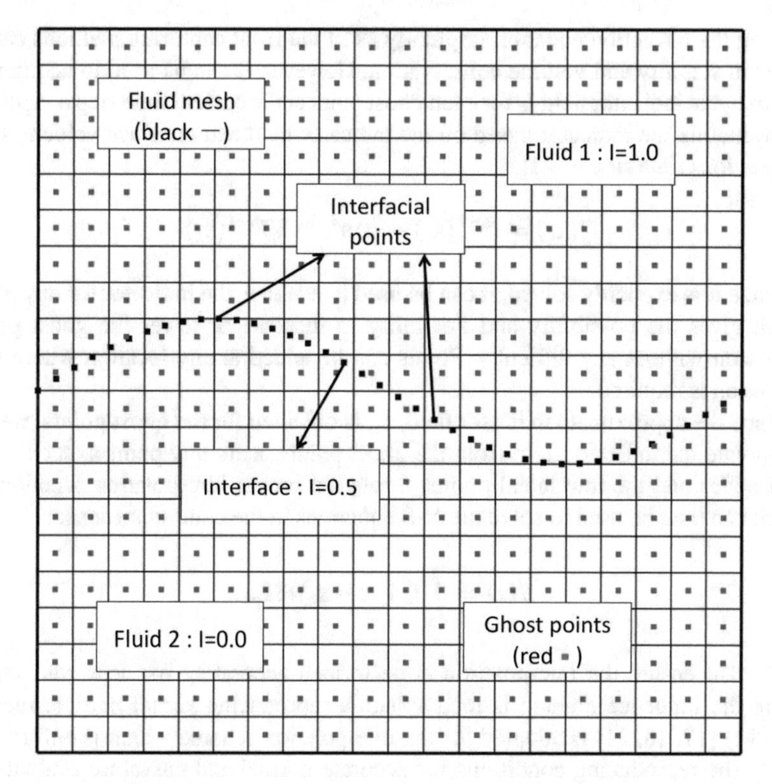

Fig. 1 Schematics for fluid mesh, ghost points, and indicator function set up

An approximation-correction procedure is developed to construct the indicator field: (1) obtain an approximate indicator field based on ghost points; (2) correct this indicator field based on the current updated position of the interfacial points. To achieve this, we first define a set of ghost points ($\mathbf{x}_g$) that are placed in each background fluid element. Here, we choose the center of each fluid element or cell as the ghost points, see Fig. 1. One can use more points within each fluid element to achieve better resolution.

The initial position of the interface is given based on the initial configuration of the problem setup, the approximate indicator for the ghost points is, therefore, known (either 1 or 0). As time marches forward, the indicator or signed distance function, is advected by solving an advection equation at each current time step:

$$\frac{dI}{dt} + \mathbf{u} \cdot \nabla I = 0. \tag{6}$$

Solving the advection equation implicitly as in the front capturing methods causes issues in stability and volume conservation. However, it can be used to acquire the approximate indicator field I_a for each ghost point at the current time step n *explicitly* by evaluating the indicator based on the indicator field and interface velocity from the previous time step $n-1$:

$$I_a^n(\mathbf{x}_g) = I^{n-1}(\mathbf{x}_g) - \Delta t \mathbf{u}^{n-1} \cdot \nabla I^{n-1}(\mathbf{x}_g). \tag{7}$$

Since it is explicitly solved, it can be used to evaluate the indicator for any point, which gives the possibility and flexibility to increase or lower the ghost points resolution without any difficulty. Points can be added at any location where high resolution is required.

Once the approximate indicator field, I_a, is obtained for the ghost points, we can interpolate the indicator field from the ghost points, $\mathbf{x}_g$ to any point $\mathbf{x}$, such as the fluid nodes and the interfacial points, through a proper interpolation function, Φ, which can then be used to solve the N-S equations in the continuum sense:

$$I(\mathbf{x}) = \int_g I_a \Phi(\mathbf{x} - \mathbf{x}_g) d\Omega_g. \tag{8}$$

To also ensure the interpolation is performed accurately for near wall region where the influence domain is incomplete, a reproducing kernel particle method (RKPM) [17, 18, 21] is adopted for the interpolation scheme which is enforced to satisfy the reproducing conditions for accurate normal and curvature evaluations. For detailed implementation, please refer to [36].

Adopting the basic idea of the level-set method that the level contour represents the interface, we set the interface to have a constant indicator value. To accomplish this, a correction term δI should be added to the indicator function, I, to ensure that the interface coincides exactly with a constant indicator contour. Combining the interpolated approximate indicator, Eq. (8), with the correction term, δI, the indicator function for any points can be written as:

$$I(\mathbf{x}) = \sum_{p=1}^{N_p} \delta I_p \Phi(\mathbf{x} - \mathbf{x}_p) + \int_g I_a \Phi(\mathbf{x} - \mathbf{x}_g) d\Omega_g, \tag{9}$$

where the subscript p denotes the interfacial points, N_p is the number of interfacial points. The correction term needs to be solved for every interfacial point, p. If the indicator of all the interfacial points is set to be 0.5, i.e. $I(\mathbf{x}_p) = 0.5$, then

$$\sum_{p'=1}^{N_p} \delta I_{p'} \Phi(\mathbf{x}_p - \mathbf{x}_{p'}) = 0.5 - \int_g I_a \Phi(\mathbf{x}_p - \mathbf{x}_g) d\Omega_g. \tag{10}$$

Here the subscript p' is used to differentiate from p in the summation. Repeating Eq. (10) for every interfacial point would generate a linear system with $RANK = N_p$. Solving the linear system yields the correction value δI_p for each interfacial point. Typically, the number of interfacial points is much less than the number of fluid nodes, therefore the computational cost of Eq. (10) is negligible. Once the correction of the indicator function is solved, the final indicator field for the fixed fluid mesh can be obtained using Eq. (9).

2.1.2 Surface Tension Force Without Connectivity

Once the indicator field is known, the unit normal, **n**, and curvature, k can be calculated by differentiating the indicator function:

$$\mathbf{n} = -\frac{\nabla I}{|\nabla I|}, \tag{11}$$

$$k = \nabla \cdot \mathbf{n}. \tag{12}$$

In order to solve the N-S equation, the surface tension force which is a singular source term in the momentum equation, Eq. (2), should be distributed to the fixed fluid grid properly. Here we compute the singular surface tension force by converting a point force into a volume force using a continuum surface tension force (CSF) approach [2]:

$$\mathbf{F}_{sv}(\mathbf{x}) = \sigma k(\mathbf{x}) \frac{\nabla I(\mathbf{x})}{[I]} \frac{\rho(\mathbf{x})}{<\rho>}, \tag{13}$$

where $[I]$ is the jump of the indicator function, $\rho(\mathbf{x})$ is the weighted density, and $<\rho>$ is the average density at the interface. The advantage of using this approach is to avoid the calculation of arc length or surface area, which is difficult to obtain without connectivity.

One issue in the CSF approach is the spurious currents (irregular velocity field) generated by the inaccurate curvature calculation near the interface [29]. The CSF converts the surface force into a volume force by evaluating Eq. (13). Directly calculating curvature at fluid nodes instead of the interfacial points can only get a first order accuracy at best [7]. Also, when the curvature is calculated at a fluid node with the indicator approaches to 1 or 0, the curvature is more likely to be inaccurate due to the noise in the indicator. Therefore, we adopt another approach, the curvature projection scheme presented in [7], to evaluate the curvature at the fluid node that any surface variable (such as curvature k) can be projected to the whole fluid domain by solving:

$$\nabla k \cdot \nabla I = 0. \tag{14}$$

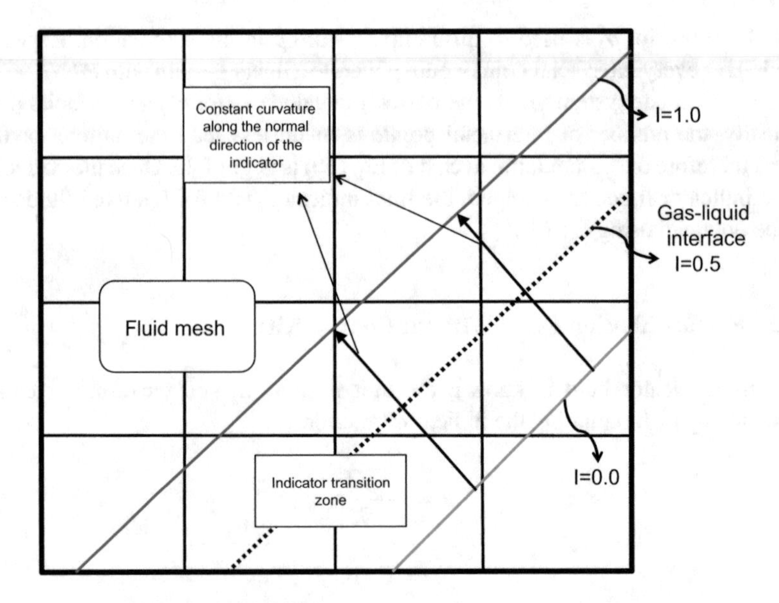

Fig. 2 Schematics for curvature projection

Figure 2 shows the schematics for the curvature projection scheme. The first step is to construct an indicator transition zone, which is a narrow band spanning several fluid elements across the interface. The results of Eq. (14) let the curvature on the interface which is more accurately evaluated to be projected along the normal direction of the indicator within this indicator transition zone. By imposing the same constant curvature in the transition zone, it eliminates the inaccurate curvature evaluation caused by the noise in the indicator. As the mesh resolution increases near the interface which results in the bandwidth of the transition zone to approach zero, the curvature on the interface can be accurately reproduced. Since the interfacial points are not coinciding with the fluid nodes, Eq. (14) cannot be solved directly on the fluid domain. The detailed implementation can be found in [33]. To overcome this issue, we develop the following numerical procedure: (1) identify all the fluid elements that contain the interfacial points; (2) project the nodal points of these elements to the interface using the projection scheme (which will be discussed in Sect. 2.2.2) along the direction of ∇I to find out the projected points; (3) evaluate the curvature of the projected points, and set the corresponding nodal points to have the same curvature; (4) set the curvature of the nodal points as a boundary condition and solve Eq. (14) to project the curvature to the fluid domain.

Here, we demonstrate the spurious currents effect yielded from both with and without the curvature projection scheme. Dimensionless units are used for all parameters. The computational domain is a 6×6 rectangular box, which is discretized with 4096 uniform quadrilateral elements. The box is filled with gas. A circular liquid drop with a radius of 1.0 is placed in the center of the domain. Both the gas and liquid have the same density, $\rho_g = \rho_l = 4.0$, and viscosity

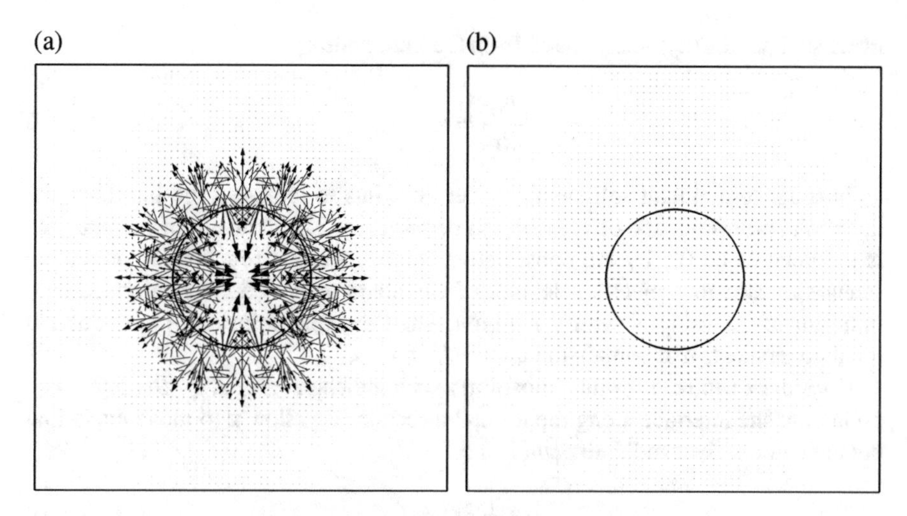

Fig. 3 Spurious currents for with and without curvature projection scheme. (**a**) Without curvature projection scheme. (**b**) With curvature projection scheme

$\mu_g = \mu_l = 1.0$. The surface tension is set to be $\sigma = 0.357$. The box has no-slip boundary on all sides. The simulation is carried out for 200 time steps with a time step size of $\Delta t = 10^{-4}$. Since the initial velocity field is zero, the exact solution of the velocity should be zero at all time.

Figure 3 shows the spurious currents for both with and without curvature projection scheme calculations. The scales of the velocity vectors for both cases are the same. It is evident that with curvature projection scheme, the spurious currents are significantly decreased. In fact, the value of the spurious currents are expected to be in the order of $0.01\sigma/\mu = 3.57 \times 10^{-3}$ [4, 5, 13, 27]. In our study, the ℓ_∞, which is defined as $\ell_\infty = (max||\mathbf{u}||)$, for without and with curvature projection scheme are 2.22×10^{-3} and 7.60×10^{-5}, respectively. Both solutions behave well in terms of the spurious currents magnitude (less than 3.57×10^{-3}). However, our improved curvature method reduces the spurious currents by two orders of magnitude, which dramatically increases the accuracy of the calculation.

2.1.3 Interface Advection and Points Regeneration Scheme

After properly incorporating the surface tension force, the N-S equations (1) and (2) can be solved to obtain the velocity and pressure fields using stabilized finite element method by adding streamline-upwind/Petrov-Galerkin (SUPG) and pressure-stablizing/Petrov-Galerkin (PSPG) stabilization terms [1, 42]. The velocity for each interfacial point is interpolated from the fluid mesh through the interpolation function Φ. The interface position, $\mathbf{x}_p$, is then advected using the velocity of the interfacial point $\mathbf{u}(\mathbf{x}_p)$ interpolated from the fluid nodes:

$$\frac{d\mathbf{x}_p}{dt} = \mathbf{u}(\mathbf{x}_p). \tag{15}$$

In order to maintain sufficient number of points to represent the interface, the interfacial points need to be regenerated periodically. The indicator of the interface is enforced to be 0.5 by the construction of the indicator function. A projection scheme is used to project a selection of candidate points near the interface onto indicator of 0.5 using Newton's iteration. The candidate points are pre-defined in the fluid elements that contain the interfacial points.

Consider a candidate point at position $\mathbf{x}$ with indicator value $I(\mathbf{x})$. To project this point onto the interface along the normal direction, we allow $\mathbf{x}$ to move $\delta\mathbf{x}$ so that the indicator of the candidate point is 0.5.

$$I(\mathbf{x} + \delta\mathbf{x}) = 0.5. \tag{16}$$

If we perform a first order Taylor expansion, we obtain:

$$I(\mathbf{x}) + \delta\mathbf{x} \cdot \nabla I(\mathbf{x}) \approx 0.5. \tag{17}$$

Since the projection is along the normal direction $\mathbf{n} = \frac{\nabla I(\mathbf{x})}{|\nabla I(\mathbf{x})|}$, we have

$$\delta\mathbf{x} \times \nabla I(\mathbf{x}) = 0. \tag{18}$$

Based on Eqs. (17) and (18), $\delta\mathbf{x}$ is solved and the candidate point is projected to a new position $\mathbf{x}' = \mathbf{x} + \delta\mathbf{x}$. This scheme is performed several times until $|I(\mathbf{x}') - 0.5| < \epsilon$, where ϵ is a set tolerance, then a new interfacial point is regenerated. However, if the distance between the newly regenerated point and those have already been regenerated is less than a set value, then the newly regenerated point would not be used. This procedure is to avoid too many unnecessary interfacial points within a short interface segment.

To treat the interface topology change such as deleting points when two interface is merging, the candidate points must be carefully selected. Torres et al. [30] suggests to pre-select the candidate points from either side of the interface, say $I(\mathbf{x}) > 0.5 + \epsilon$ or $I(\mathbf{x}) < 0.5 - \epsilon$. Since the points near the contacting surface do not belong to this range, these points as well as the contact surface are deleted. An indicator field can be constructed based on the newly regenerated points.

2.2 *Fluid-Solid Interactions Using the Modified Immersed Finite Element Method*

The kinematics of the computational domain that contains both fluid and solid is shown in Fig. 4. Let us consider a computational domain, Ω, which is consisted of gas domain (Ω^{g}), liquid domain (Ω^{l}) and solid domain (Ω^{s}), such that $\Omega \equiv \Omega^{\mathrm{l}} \cup \Omega^{\mathrm{g}} \cup \Omega^{\mathrm{s}}$. The gas and liquid domains together can be treated as the fluid domain such that $\Omega^{\mathrm{f}} \equiv \Omega^{\mathrm{g}} \cup \Omega^{\mathrm{l}}$. The gas-liquid interface is denoted as Γ^{gl}. The solid and fluid domains are intersected by a common interface Γ^{s}.

Several assumptions are made as follows:

1. *The solid structure is fully immersed in the entire domain.* Following this assumption, the solid cannot move out of the fluid domain. The fluid-solid interface needs to be within the computational domain in order to evaluate the interactions correctly.
2. *The fluid exists everywhere in the domain,* Ω. This assumption allows us to generate independent mesh for the fluid and the solid, which avoids frequent mesh updating schemes. This approach is considered as non-boundary fitted technique, as mentioned earlier. Since the solid is immersed in the fluid and the fluid exists in the whole domain Ω, an overlapping region forms. This overlapping region is 'artificial' because it physically does not exists. Therefore its effects (viscous and inertial) must be eliminated.
3. *The fluid-solid interface abides the no-slip and/or traction boundary conditions.* These two boundary conditions can co-exist, but cannot overlap. The solid material can be described using various solid constitutive laws, which provides feasibility to model different materials.

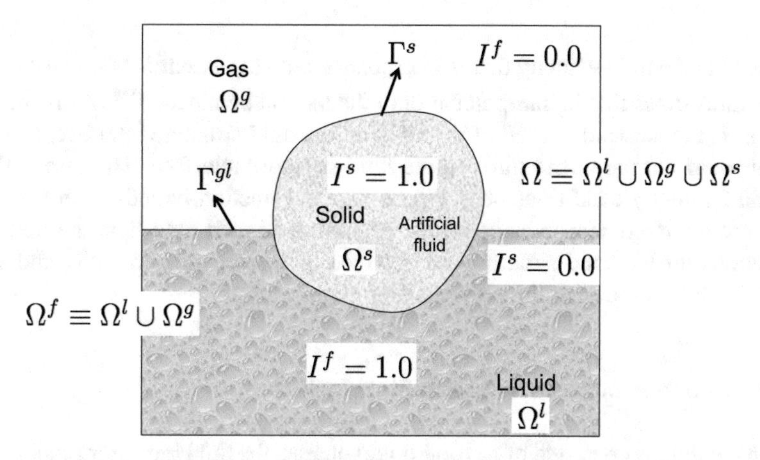

Fig. 4 Kinematics of the computational domain

2.2.1 Solid and Fluid Domains

The original IFEM let the solid follow the movement of the artificial fluid by applying the no-slip condition within the entire solid domain which includes the interface and the interior. The solid governing equation is only used to evaluate the interaction force that influences the solution of the fluid. In this modified IFEM (mIFEM) technique, we let the solid dynamics to be solved by applying boundary conditions that are evaluated based on the fluid velocity and stress on the fluid-solid interface. This algorithm takes into account the dynamics of the solid where the solid interior deformation is been solved rather than been imposed. This is more reasonable approach since the artificial fluid is not real anyway, we can manipulate it to behave more like a solid.

In solid domain Ω^s, the solid governing equation can be described as

$$\rho^s \frac{Dv_i^s}{Dt} = \rho^s u_{i,tt}^s = \sigma_{ij,j}^s + \rho^s g_i, \tag{19}$$

where $\mathbf{u}^s$ is the solid displacement, $\mathbf{v}^s$ is the solid velocity, the total derivative operator D is used to evaluate the acceleration of the solid, ρ^s is the solid density, $\mathbf{g}$ is gravity, the solid stress $\sigma_{kl}^s = c_{ijkl}\epsilon_{ij}^s + \eta_{ijkl}\epsilon_{ij,t}^s$, $\epsilon_{ij}^s = \frac{1}{2}(u_{i,j}^s + u_{j,i}^s)$. By choosing different combinations of c_{ijkl} and η_{ijkl}, the solid can be modeled as linear elastic, visco-linear elastic, hyper-elastic and etc.

The Dirichlet and Neumann boundary conditions for the solid can be defined as follows:

$$u_i^s = q_i = v_i^f \Delta t \quad \text{on } \Gamma^{sq}, \tag{20}$$

$$\sigma_{ij}^s n_j^s = h_i = -\sigma_{ij}^f n_j^f \quad \text{on } \Gamma^{sh}. \tag{21}$$

where v^f is the fluid velocity that is interpolated onto the essential boundary Γ^{sq}, σ_{ij}^f is the fluid stress that is interpolated onto the natural boundary Γ^{sh}, Δt is the time size, n^s is the outward normal of the solid on the fluid-structure interface, and n^f is the outward normal of the fluid on the fluid-structure interface. The essential and natural boundary conditions obey the no-slip and traction boundary on the fluid-structure interface, respectively. At the beginning of each time step, the boundary conditions for the solid structure are obtained by interpolating velocity and stress from the fluid solution of the previous time step.

2.2.2 Coupling Between Solid and Fluid

When the fluid is consisted of gas and liquid phases, the fluid indicator function I^f is used to differentiate the gas and liquid ($I^f = 1.0$ in liquid domain Ω^l and $I^f = 0.0$ in gas domain Ω^g), the real fluid can be treated as 'one fluid' with properties varying across the gas-liquid interface. The gas-liquid indicator function can be obtained following the CFFT method as shown in Sect. 2.1. The density and viscosity of the real fluid are calculated as follows:

$$\rho^{\mathrm{f}} = \rho^{\mathrm{l}} I^{\mathrm{f}} + \rho^{\mathrm{g}}(1 - I^{\mathrm{f}}), \quad \mu^{\mathrm{f}} = \mu^{\mathrm{l}} I^{\mathrm{f}} + \mu^{\mathrm{g}}(1 - I^{\mathrm{f}}). \tag{22}$$

Similarly, when treating the fluid-solid interactions, a new independent set of solid indicator function I^{s} is defined where $I^{\mathrm{s}} = 1.0$ in solid domain Ω^{s} and $I^{\mathrm{s}} = 0.0$ in fluid domain Ω^{f}. Since the normal of the solid boundary can be easily calculated from the solid mesh, a smooth indicator field is achieved by solving Poisson's equation near the fluid-solid interface following the front tracking method.

The adoption of fluid and solid indicator makes it natural to couple the multi-fluid and fluid-structure treatments together. The gas, the liquid, and the artificial fluid which follows the dynamics of the solid, are treated as one single fluid with properties described as follows:

$$\bar{\rho} = \rho^{\mathrm{s}} I^{\mathrm{s}} + \rho^{\mathrm{f}}(1 - I^{\mathrm{s}}), \quad \bar{\mu} = \mu^{\mathrm{s}} I^{\mathrm{s}} + \mu^{\mathrm{f}}(1 - I^{\mathrm{s}}). \tag{23}$$

The coupling of the fluid and the solid is easier to be explained using the concept of virtual work (or weak form). With a test function $\delta \mathbf{v}$, the weak form for momentum equation of the entire computational domain is,

$$\int_{\Omega^{\mathrm{f}}} \delta v_i \left(\rho^{\mathrm{f}} \frac{D v_i^{\mathrm{f}}}{Dt} - \sigma_{ij,j}^{\mathrm{f}} - \rho^{\mathrm{f}} g_i - F_i^{\sigma} \right) d\Omega + \int_{\Omega^{\mathrm{s}}} \delta v_i \left(\rho^{\mathrm{s}} \frac{D v_i^{\mathrm{s}}}{Dt} - \sigma_{ij,j}^{\mathrm{s}} - \rho^{\mathrm{s}} g_i \right) d\Omega = 0. \tag{24}$$

The first integral is the virtual work done by the real fluid, whereas the second integral is the virtual work done by the solid.

Since the fluid is filled in the entire computational domain (based on Assumption 2), Eq. (24) needs to be re-arranged to account for that, but in the meantime the effects of the artificial fluid must be eliminated.

Re-arranging Eq. (24) yields,

$$\int_{\Omega^{\mathrm{f}}} \delta v_i \left(\rho^{\mathrm{f}} \frac{D v_i^{\mathrm{f}}}{Dt} - \sigma_{ij,j}^{\mathrm{f}} - \rho^{\mathrm{f}} g_i - F_i^{\sigma} \right) d\Omega + \int_{\Omega^{\mathrm{s}}} \delta v_i \left(\rho^{\mathrm{s}} \frac{D v_i^{\mathrm{f}}}{Dt} - \sigma_{ij,j}^{\mathrm{f}} - \rho^{\mathrm{s}} g_i \right) d\Omega + \int_{\Omega^{\mathrm{s}}} \delta v_i \left(\rho^{\mathrm{s}} \left(\frac{D v_i^{\mathrm{s}}}{Dt} - \frac{D v_i^{\mathrm{f}}}{Dt} \right) - (\sigma_{ij,j}^{\mathrm{s}} - \sigma_{ij,j}^{\mathrm{f}}) \right) d\Omega = 0. \tag{25}$$

The second integral is the virtual work done by the artificial fluid in the solid domain. Re-arranging Eq. (25) is to include the artificial fluid terms in the solid domain without contradicting the equilibrium. Combining the first two integral terms then yields the total work done by the entire computational domain that includes both real and artificial fluid. The work done by the artificial fluid is then subtracted in the third term from the work done by the solid.

We define the integrand in the third integral, the balance of the forces in the artificial fluid and the solid, as the fluid-structure interaction force on the solid nodes:

$$F_i^{\mathrm{FSI}} = \left(\sigma_{ij,j}^{\mathrm{s}} - \sigma_{ij,j}^{\mathrm{f}}\right) - \rho^{\mathrm{s}}\left(\frac{Dv_i^{\mathrm{s}}}{Dt} - \frac{Dv_i^{\mathrm{f}}}{Dt}\right) \quad \text{in } \Omega^{\mathrm{s}}. \tag{26}$$

The interaction force F_i^{FSI} defined in Eq. (26) involves the calculation of the solid acceleration $\frac{Dv_i^{\mathrm{s}}}{Dt}$, the artificial fluid acceleration $\frac{Dv_i^{\mathrm{f}}}{Dt}$, and the derivatives of the solid and fluid stresses, $\sigma_{ij,j}^{\mathrm{s}}$ and $\sigma_{ij,j}^{\mathrm{f}}$, under current solid configuration. These four terms act as an external force on top of the solid domain to control the movement of artificial fluid. Once the evaluation of the fluid-structure interaction force is completed, it is used as an external force in the fluid to 'feel' the existence of the solid. *The artificial fluid is driven by the solid motion, which means the velocity of the artificial fluid should be as close to the solid ($\mathbf{v}^{\mathrm{f}} = \mathbf{v}^{\mathrm{s}}$ in Ω^{s}) as possible*. To enforce the artificial fluid to be very similar to the solid, we can further assume the artificial fluid is pseudo-compressible with same bulk modulus κ^{s} as the solid.

Together with the density with indicator function defined in Eq. (23), Eq. (25) becomes:

$$\int_{\Omega} \delta v_i \left(\bar{\rho}\frac{Dv_i^{\mathrm{f}}}{Dt} - \sigma_{ij,j}^{\mathrm{f}} - \bar{\rho}g_i - F_i^{\mathrm{FSI}} - F_i^{\sigma}\right) d\Omega = 0. \tag{27}$$

Its strong form is,

$$\bar{\rho}\frac{Dv_i^{\mathrm{f}}}{Dt} - \sigma_{ij,j}^{\mathrm{f}} - \bar{\rho}g_i - F_i^{\mathrm{FSI}} - F_i^{\sigma} = 0 \quad \text{in } \Omega \equiv \Omega^{\mathrm{f}} \cup \Omega^{\mathrm{s}}. \tag{28}$$

The interpolation functions are important in coupling the three phases. In this algorithm, the interpolation functions are used in three places.

1. When using connectivity-free gas-liquid interfacial points, as we do in CFFT, the approximate indicator values will be first interpolated from the ghost points onto the fluid grid, Eq. (8), as explained previously in Sect. 2.1.
2. Interpolating the fluid velocity and the fluid stress onto the solid domain when evaluating the fluid-solid interaction force $\mathbf{f}^{\mathrm{FSI}}$ as described in Eq. (26) in Sect. 2.2:

$$v_i^{\mathrm{s}}(\mathbf{x}^{\mathrm{s}}, t) = \int_{\Omega^{\mathrm{s}}} v_i^{\mathrm{f}}(\mathbf{x}^{\mathrm{f}}, t)\Phi(\mathbf{x}^{\mathrm{f}} - \mathbf{x}^{\mathrm{s}})d\Omega, \tag{29}$$

$$\sigma_{ij}^{\mathrm{s}}(\mathbf{x}^{\mathrm{s}}, t) = \int_{\Omega^{\mathrm{s}}} \sigma_{ij}^{\mathrm{f}}(\mathbf{x}^{\mathrm{f}}, t)\Phi(\mathbf{x}^{\mathrm{f}} - \mathbf{x}^{\mathrm{s}})d\Omega. \tag{30}$$

The interpolated values are also used for the solid boundary condition. Boundary conditions for the solid are interpolated from the fluid velocity and fluid stress onto the solid boundary Γ, as described in Eqs. (20) and (21) in Sect. 2.2.

3. Finally, the fluid-solid interaction force evaluated on the solid nodes is distributed back to the fluid domain:

$$F_i^{\mathrm{FSI,f}}(\mathbf{x}^{\mathrm{f}}, t) = \int_{\Omega} F_i^{\mathrm{FSI,s}}(\mathbf{x}^{\mathrm{s}}, t)\Phi(\mathbf{x}^{\mathrm{f}} - \mathbf{x}^{\mathrm{s}})d\Omega. \quad (31)$$

This two-way interpolation between the fluid and the solid is also done in the IB and IFEM methods. To obtain accurate interpolations, a high-order interpolation function, Φ is adopted, which is acquired through the RKPM procedure [18]. There are different choices for the interpolation functions [34]. Here we choose the RKPM interpolation for its capability in dealing with gas-liquid interfaces near a boundary for free surfaces and dealing with non-uniform background fluid grid.

The following lists the numerical framework for simulating the three-phase interactions. The CFFT multiphase gas-liquid solver is wrapped inside the mIFEM fluid-structure interaction solver. The overall solution procedure can be summarized as follows:

1. Solve solid dynamic governing equation, given solid boundary conditions interpolated from the velocity and stress from the previous time step.
2. Construct the fluid and solid indicator field.
3. Evaluate fluid-structure interaction forces on solid nodes.
4. Distribute the fluid-structure interaction force to the background fluid.
5. Solve multiphase gas-liquid using CFFT:
   (a) Construct the fluid indicator based on the gas-liquid interface.
   (b) Calculate the surface tension force using continuum surface tension force approach.
   (c) Evaluate the weighted density and viscosity using the indicator function.
   (d) Solve Navier-Stokes equations to obtain the velocity and pressure solutions.
   (e) Advect the gas-liquid interface based on the fluid velocity.
6. Interpolate the fluid velocity and stress onto the solid domain, proceed to next time step.

2.3 Numerical Examples

2.3.1 Rayleigh-Taylor Instability Test

The first validation case is the Rayleigh-Taylor instability study to verify the accuracy and the CFFT implementation. The Rayleigh-Taylor instability is to examine the instability of an interface between two fluid where the heavier fluid lies upon the lighter fluid with gravity. If the interface is disturbed initially by a small perturbation, the instability will drive the interface to grow exponentially with time. The linear analysis shows that the growth rate n can be predicted as [2, 19, 39]:

$$n^2 = Kg\left(A - \frac{K^2\sigma}{g(\rho_1 + \rho_2)}\right), \tag{32}$$

where K is the wavenumber of the perturbation, constant $A = (\rho_2 - \rho_1)/(\rho_1 + \rho_2)$, g is the gravity acceleration which is perpendicular to the interface, σ is the surface tension force, and ρ_1 and ρ_2 are the densities of lighter and heavier fluid, respectively.

The simulation is carried out in a 6×1 rectangular domain with 50,000 uniform quadrilateral elements. The simulation parameters are chosen as: $\rho_1 = 1.225$, $\rho_2 = 0.1694$, viscosities are the same $\mu_1 = \mu_2 = 0.00313$, $\sigma = 0.05$ and $g = 9.8$. The initial interface is placed in the center of the domain and is perturbed as:

$$y = 0.001\cos(2\pi x). \tag{33}$$

In this example, the wave number is $K = 2\pi$, and the analytical growth rate calculated from Eq. (32) is found to be $n = 6.14$. Ideally, the interface should grow following $h = 0.001e^{nt}$, where h is defined as the maximum height of the interface with respect to the initial interface position.

Figure 5 shows the interface shapes obtained from the simulation at different times. Figure 6 compares the interface development, which tracks the maximum height of the interface at different times, between the linearized analytical solution and our numerical results. It can be clearly seen that for most of the simulation time ($t < 1$) the numerical result matches the analytical solution very well. For later stage ($t > 1$), because the interface undergoes large deformation, the linear assumption from the analytical solution is no longer valid. Therefore, the simulation result starts to deviate from the analytical solution.

2.3.2 3-D Floating Object on a Breaking Dam

In this example, a 3-D simulation of a dam breaking problem with a floating deformable solid object placed on the gas-liquid interface (free-surface) is carried out to show the capability of the coupled algorithm to solve complex gas-liquid-solid 3-phase flows. The geometrical setup is shown in Fig. 7. The simulation is performed in a 2 cm × 0.5 cm × 1 cm computational domain with a rigid column of size 0.2 cm × 0.2 cm × 1 cm placed in the center of the domain as an obstacle to add into the complexity. The domain is discretized with 119,500 hexahedral elements. All the boundaries are slip walls. The liquid column is initially placed at the left lower corner, with all sides to be all 0.5 cm that has a fillet radius of 0.1 cm. The density and viscosity of the liquid are 1.0 g/cm^3 and 0.01 P; they are 0.001 g/cm^3 and 0.0001 P for the gas phase, respectively. A solid sphere with a diameter of

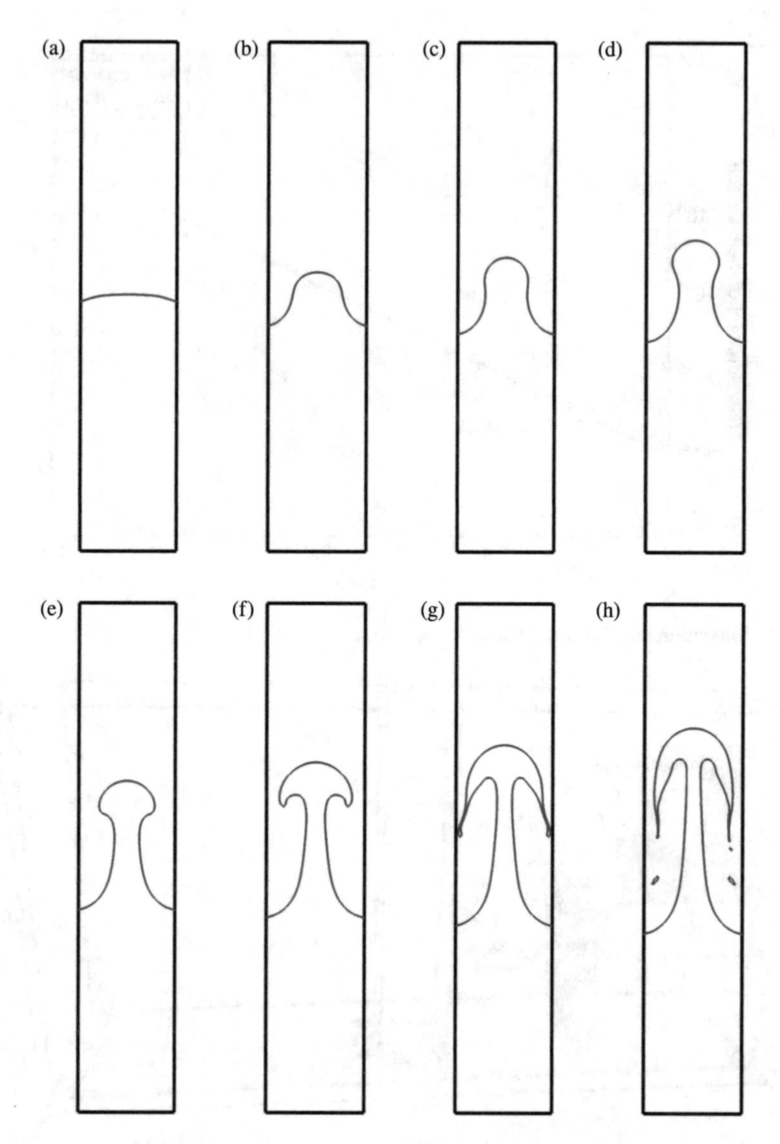

Fig. 5 Interface shapes at different times for Rayleigh-Taylor instability. (**a**) t=0.5. (**b**) t=0.9. (**c**) t=1.0. (**d**) t=1.1. (**e**) t=1.2. (**f**) t=1.3. (**g**) t=1.4. (**h**) t=1.5

0.2 cm is placed on top of the liquid column and is discretized with 2768 hexahedral elements. The distance between the center of the sphere and the left wall is 0.25 cm, and 0.35 cm from the front wall. The density of the solid is 0.5 g/cm^3, which is half of the liquid, making it stay afloat. The sphere is deformable described with linear elastic material that has Young's modulus of 10^4 dyn/cm^2 and Poisson's ratio of 0.3. The damping factor is 100 P. The gravity is set as 5.0 cm/s^2 acting in the negative

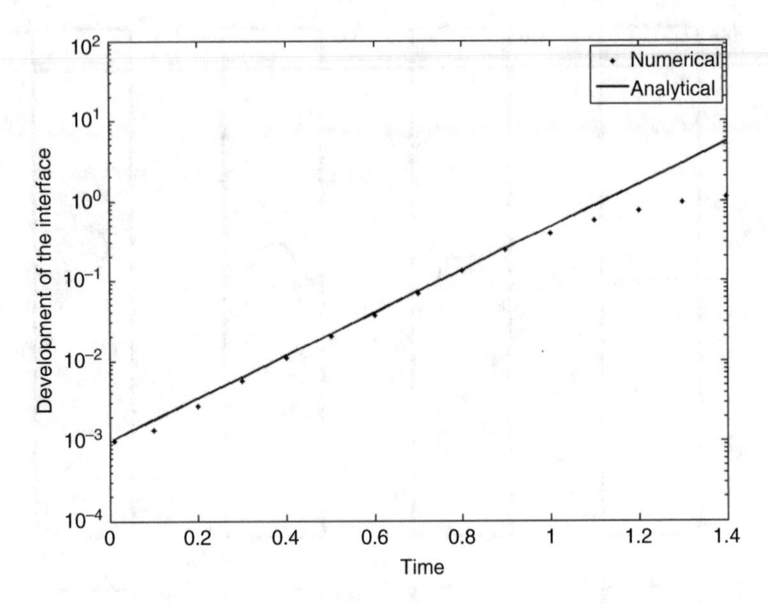

Fig. 6 Maximum interface height at different times

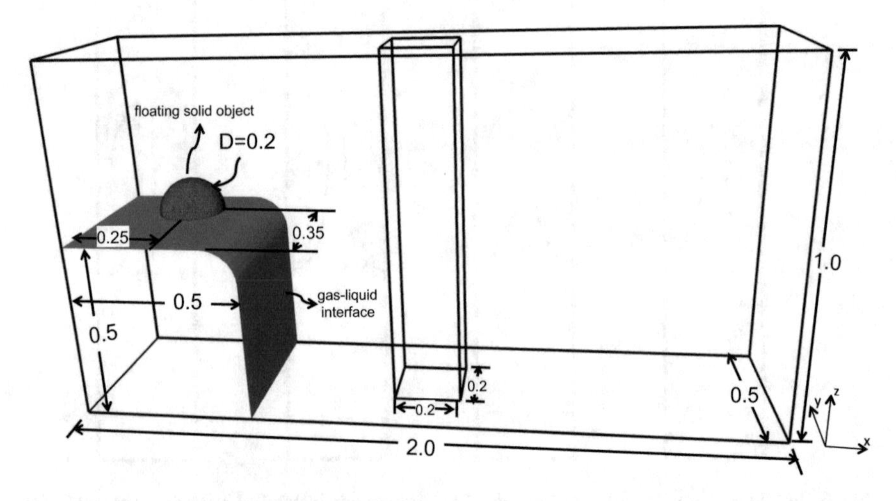

Fig. 7 Schematics of a floating solid object on a breaking dam

vertical or z-direction. The surface tension is neglected here since we consider a large scale problem where the surface tension force is negligible. The time step size is 2×10^{-3} s. The simulation is carried out for 2500 time steps which provides sufficient amount of time for the liquid column to hit the right wall and reflect back.

Figures 8 and 9 show the snapshots of the sphere floating along a breaking liquid column at different time steps from different view points. The collapsing liquid column falls down and moves forward due to gravity, and bifurcates as it hits the

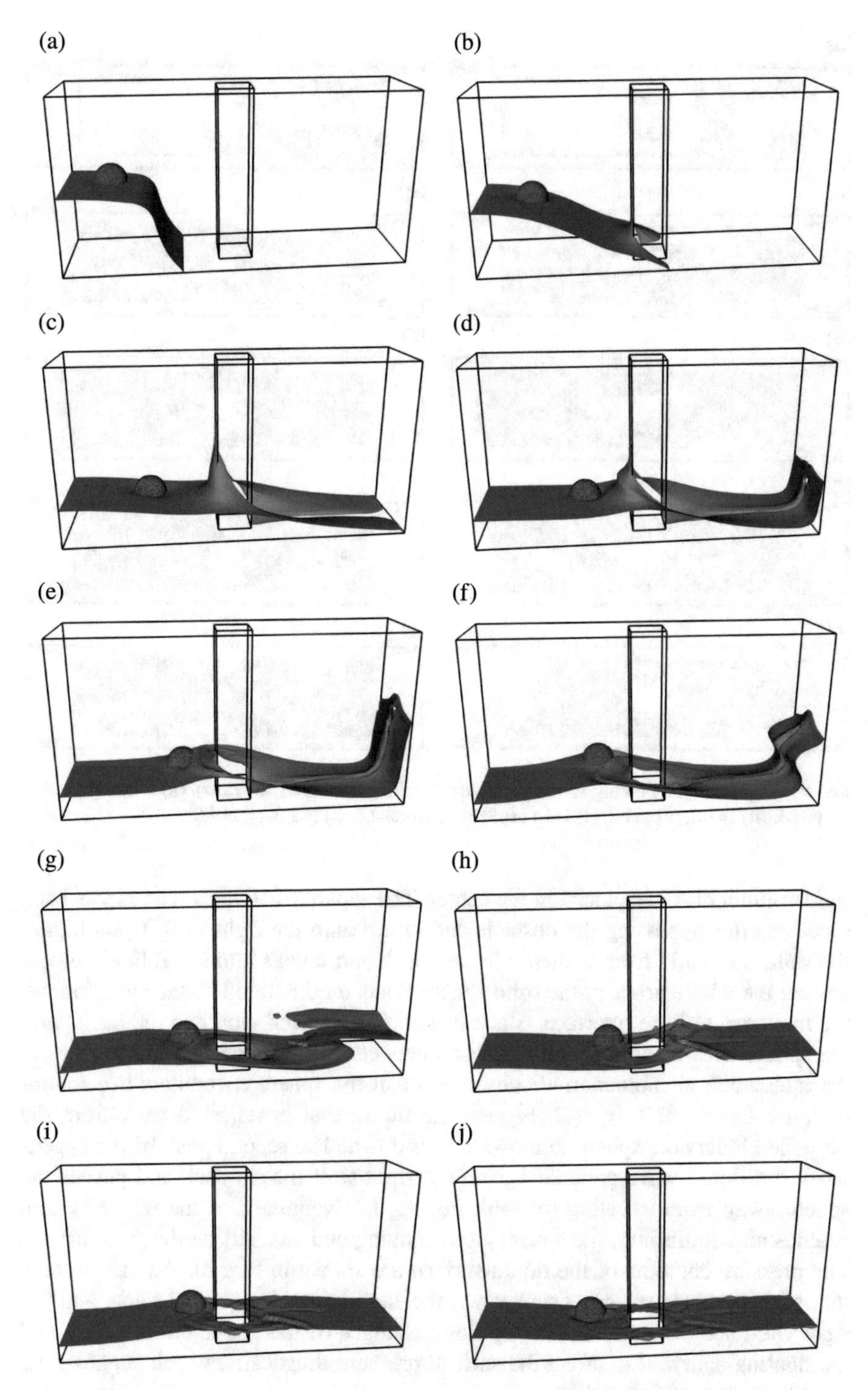

Fig. 8 Snapshots of a floating solid sphere on a breaking dam (horizontal view). (**a**) t=0.2. (**b**) t=0.4. (**c**) t=0.8. (**d**) t=1.0. (**e**) t=1.2. (**f**) t=1.6. (**g**) t=2.0. (**h**) t=2.8. (**i**) t=3.6. (**j**) t=4.8

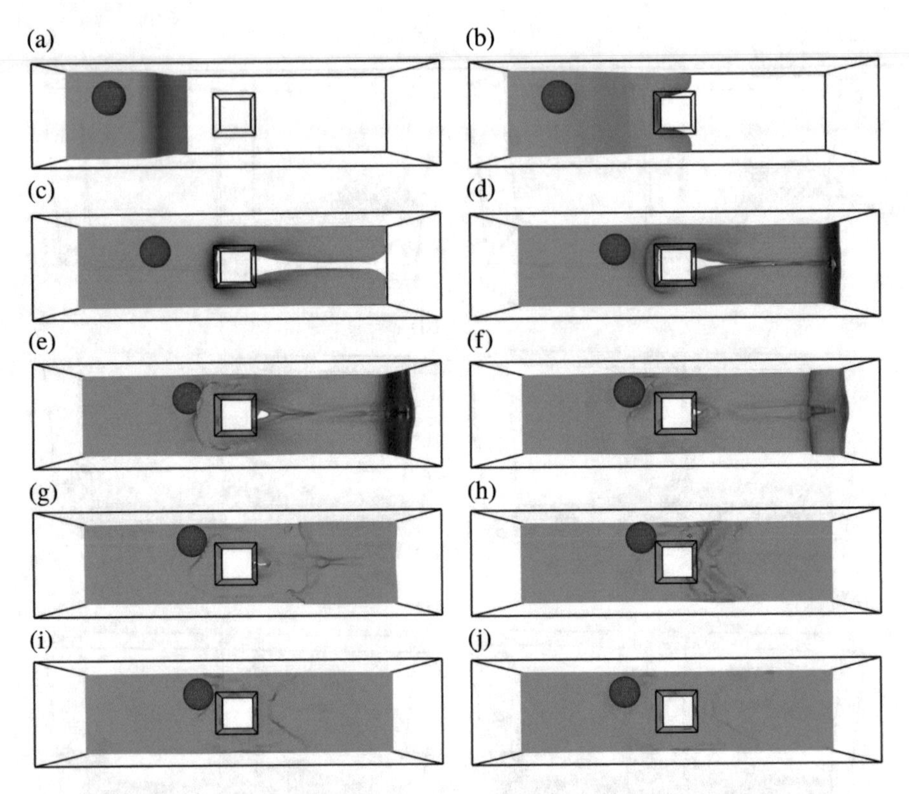

Fig. 9 Snapshots of a floating solid sphere on a breaking dam (vertical view). (**a**) t=0.2. (**b**) t=0.4. (**c**) t=0.8. (**d**) t=1.0. (**e**) t=1.2. (**f**) t=1.6. (**g**) t=2.0. (**h**) t=2.8. (**i**) t=3.6. (**j**) t=4.8

rigid column obstacle placed in the center. The separated streams then merge back together after bypassing the obstacle and climb onto the right wall. Upon hitting the wall, the liquid front is then reflected back and breaks into small liquid drops. During the whole process, the solid sphere floats on the liquid surface and follows the movement of the interface. Since it is initially placed closer to the back wall, the sphere tries to go through the space between the column and the back wall. An interesting phenomenon we observe is that the sphere encounters two rounds of 'push-back'. At $t = 1.2$, Fig. 9e, the liquid that is reflected back from the obstacle hinders the sphere to move further down. The second push-back happens when the liquid wave generated from the right wall travels back and pushes the sphere away from traveling forward, Fig. 9g, h. Eventually, as the whole system reaches an equilibrium, the sphere stops moving and sits still on the free surface. The pressure contours of the liquid surface are shown in Fig. 10. We can observe that high pressure regions occur when the liquid front hits the obstacle and the right wall, see Fig. 10b, d. The topology changes of the gas-liquid interface and the floating sphere also affect the surface pressure drastically, which suggests the complex nature of the problem.

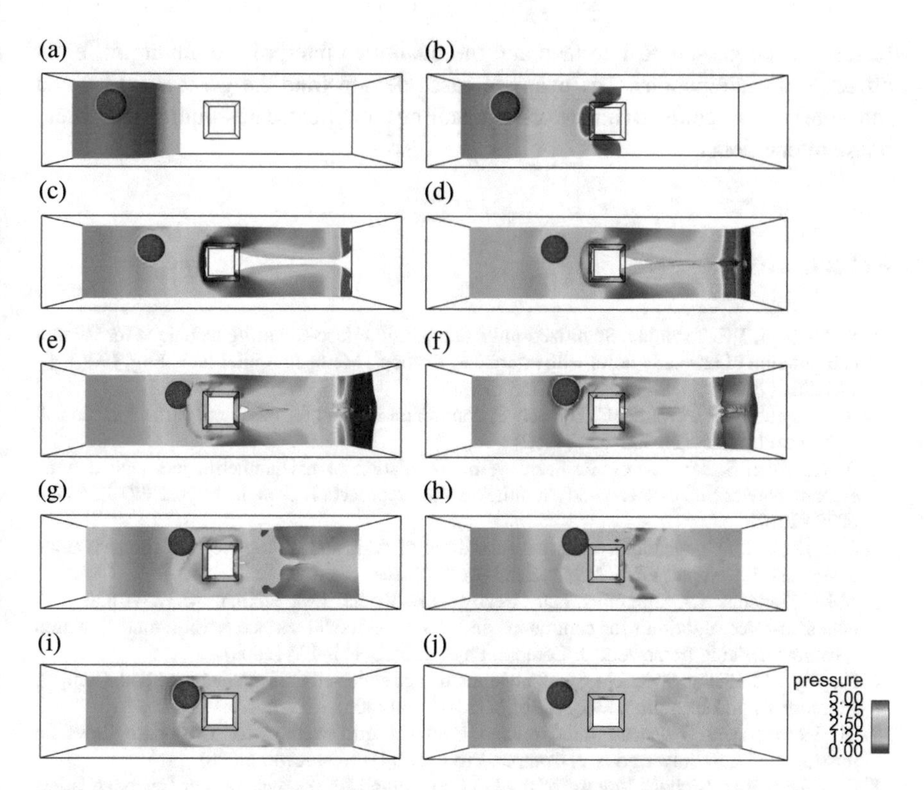

Fig. 10 Snapshots of a floating solid sphere on a breaking dam (pressure field on the liquid surface). (**a**) t=0.2. (**b**) t=0.4. (**c**) t=0.8. (**d**) t=1.0. (**e**) t=1.2. (**f**) t=1.6. (**g**) t=2.0. (**h**) t=2.8. (**i**) t=3.6. (**j**) t=4.8

3 Conclusions

In this work, we have presented a robust framework that can model the complex gas-liquid-solid three-phase interactions. The connectivity-free front tracking method (CFFT) that models gas-liquid multiphase flows is integrated into the fluid-structure interaction algorithm, the mIFEM, seamlessly through the construction of an indictor field. The CFFT adopts an approximation-correction step to build the indicator field without the connectivity of the interfacial points. This strategy significantly reduces the complexity in reconstructing the interface when topology changes occur. A curvature projection scheme helps minimize the spurious currents that are generated from the irregular curvature calculation which is commonly seen in other algorithms. The fluid-solid interactions are handled using the modified immersed finite element method, which provides more realistic solid movement and deformation by solving the solid dynamics. The non-boundary-fitted approach that is used in the algorithm avoids the re-meshing procedure. RKPM interpolation functions are used throughout the framework to accommodate the connectivity-free feature of the gas-liquid interface and the quantities interpolated among different phases from interactions. The example cases demonstrate the great accuracy and robustness of the coupled framework in handling complicated gas-liquid-solid three-phase interactions.

References

1. S. Aliabadi, T.E. Tezduyar, Stabilized-finite-element/interface-capturing technique for parallel computation of unsteady flows with interfaces. Comput. Methods Appl. Mech. Eng. **190**(3–4), 243–261 (2000)
2. J.U. Brackbill, D.B. Kothe, C. Zemach, A continuum method for modeling surface tension. J. Comput. Phys. **100**(2), 335–354 (1992)
3. A. Cervone, S. Manservisi, R. Scardovelli, Simulation of axisymmetric jets with a finite element Navier-Stokes solver and a multilevel VOF approach. J. Comput. Phys. **229**(19), 6853–6873 (2010)
4. J.-B. Dupont, D. Legendre, Numerical simulation of static and sliding drop with contact angle hysteresis. J. Comput. Phys. **229**(7), 2453–2478 (2010)
5. M.M. Francois, S.J. Cummins, E.D. Dendy, D.B. Kothe, J.M. Sicilian, M.W. Williams, A balanced-force algorithm for continuous and sharp interfacial surface tension models within a volume tracking framework. J. Comput. Phys. **213**(1), 141–173 (2006)
6. M. Gay, L.T. Zhang, W.K. Liu, Stent modeling using immersed finite element method. Comput. Methods Appl. Mech. Eng. **195**(33–36), 4358–4370 (2006)
7. M. Herrmann, A balanced force refined level set grid method for two-phase flows on unstructured flow solver grids. J. Comput. Phys. **227**(4), 2674–2706 (2008)
8. C.W. Hirt, B.D. Nichols, Volume of fluid (VOF) method for the dynamics of free boundaries. J. Comput. Phys. **39**(1), 201–225 (1981)
9. H.H. Hu, N.A. Patankar, M.Y. Zhu, Direct numerical simulations of fluid-solid systems using the arbitrary Lagrangian-Eulerian technique. J. Comput. Phys. **169**(2), 427–462 (2001)
10. J. Hua, J.F. Stene, P. Lin, Numerical simulation of 3D bubbles rising in viscous liquids using a front tracking method. J. Comput. Phys. **227**(6), 3358–3382 (2008)
11. A. Huerta, W.K. Liu, Viscous flow with large free surface motion. Comput. Methods Appl. Mech. Eng. **69**(3), 277–324 (1988)
12. T.J.R. Hughes, W.K. Liu, T.K. Zimmermann, Lagrangian-Eulerian finite element formulation for incompressible viscous flows. Comput. Methods Appl. Mech. Eng. **29**(3), 329–349 (1981)
13. B. Lafaurie, C. Nardone, R. Scardovelli, S. Zaleski, G. Zanetti, Modelling merging and fragmentation in multiphase flows with SURFER. J. Comput. Phys. **113**(1), 134–147 (1994)
14. Y. Liu, W.K. Liu, Rheology of red blood cell aggregation by computer simulation. J. Comput. Phys. **220**(1), 139–154 (2006)
15. W.K. Liu, D.C. Ma, Computer implementation aspects for fluid-structure interaction problems. Comput. Methods Appl. Mech. Eng. **31**(2), 129–148 (1982)
16. W.K. Liu, H. Chang, J.-S. Chen, T. Belytschko, Arbitrary Lagrangian-Eulerian Petrov-Galerkin finite elements for nonlinear continua. Comput. Methods Appl. Mech. Eng. **68**(3), 259–310 (1988)
17. W.K. Liu, S. Jun, S. Li, J. Adee, B. Belytschko, Reproducing kernel particle methods for structural dynamics. Int. J. Numer. Methods Eng. **38**(10), 1655–1679 (1995)
18. W.K. Liu, S. Jun, Y.F. Zhang, Reproducing kernel particle methods. Int. J. Numer. Methods Fluids **20**(8–9), 1081–1106 (1995)
19. J. Liu, S. Koshizuka, Y. Oka, A hybrid particle-mesh method for viscous, incompressible, multiphase flows. J. Comput. Phys. **202**(1), 65–93 (2005)

20. W.K. Liu, Y. Liu, D. Farrell, L.T. Zhang, X.S. Wang, Y. Fukui, N. Patankar, Y. Zhang, C. Bajaj, J. Lee, J. Hong, X. Chen, H. Hsu, Immersed finite element method and its applications to biological systems. Comput. Methods Appl. Mech. Eng. **195**(13–16), 1722–1749 (2006)
21. W.K. Liu, D.W. Kim, S. Tang, Mathematical foundations of the immersed finite element method. Comput. Mech. **39**(3), 211–222 (2007) (English)
22. M.F. McCracken, C.S. Peskin, A vortex method for blood flow through heart valves. J. Comput. Phys. **35**(2), 183–205 (1980)
23. M. Muradoglu, G. Tryggvason, A front-tracking method for computation of interfacial flows with soluble surfactants. J. Comput. Phys. **227**(4), 2238–2262 (2008)
24. C.S. Peskin, Numerical analysis of blood flow in the heart. J. Comput. Phys. **25**(3), 220–252 (1977)
25. C.S. Peskin, D.M. McQueen, A three-dimensional computational method for blood flow in the heart I. Immersed elastic fibers in a viscous incompressible fluid. J. Comput. Phys. **81**(2), 372–405 (1989)
26. C.S. Peskin, D.M. McQueen, Mechanical equilibrium determines the fractal fiber architecture of aortic heart valve leaflets. Am. J. Phys. **266**(1), 319–328 (1994)
27. Y. Renardy, M. Renardy, PROST: a parabolic reconstruction of surface tension for the volume-of-fluid method. J. Comput. Phys. **183**(2), 400–421 (2002)
28. S. Shin, I. Yoon, D. Juric, The local front reconstruction method for direct simulation of two- and three-dimensional multiphase flows. J. Comput. Phys. **230**(17), 6605–6646 (2011)
29. E. Shirani, N. Ashgriz, J. Mostaghimi, Interface pressure calculation based on conservation of momentum for front capturing methods. J. Comput. Phys. **203**(1), 154–175 (2005)
30. D.J. Torres, J.U. Brackbill, The point-set method: front-tracking without connectivity. J. Comput. Phys. **165**(2), 620–644 (2000)
31. G. Tryggvason, B. Bunner, A. Esmaeeli, D. Juric, N. Al-Rawahi, W. Tauber, J. Han, S. Nas, Y.-J. Jan, A front-tracking method for the computations of multiphase flow. J. Comput. Phys. **169**(2), 708–759 (2001)
32. S.O. Unverdi, G. Tryggvason, A front-tracking method for viscous, incompressible, multi-fluid flows. J. Comput. Phys. **100**(1), 25–37 (1992)
33. C. Wang, Numerical modeling of three-phase (gas-liquid-solid) flows with connectivity-free multi-fluid interface treatment and non-boundary-fitted techniques for fluid-structure interactions. Ph.D. thesis, Rensselaer Polytechnic Institute (2013)
34. X. Wang, L.T. Zhang, Interpolation functions in the immersed boundary and finite element methods. Comput. Mech. **45**, 321–334 (2010)
35. C. Wang, L.T. Zhang, Numerical modeling of gas-liquid-solid interactions: gas-liquid free surfaces interacting with deformable solids. Comput. Methods Appl. Mech. Eng. **286**, 123–146 (2013)
36. C. Wang, X. Wang, L. Zhang, Connectivity-free front tracking method for multiphase flows with free surfaces. J. Comput. Phys. **241**, 58–75 (2013)
37. J.A.S. Witteveen, B. Koren, P.G. Bakker, An improved front tracking method for the Euler equations. J. Comput. Phys. **224**(2), 712–728 (2007)
38. J.-J. Xu, Z. Li, J. Lowengrub, H. Zhao, A level-set method for interfacial flows with surfactant. J. Comput. Phys. **212**(2), 590–616 (2006)
39. K. Yokoi, Efficient implementation of THINC scheme: a simple and practical smoothed VOF algorithm. J. Comput. Phys. **226**(2), 1985–2002 (2007)
40. L.T. Zhang, M. Gay, Immersed finite element method for fluid-structure interactions. J. Fluids Struct. **23**(6), 839–857 (2007)
41. L.T. Zhang, G.J. Wagner, W.K. Liu, Modeling and simulation of fluid structure interaction by meshfree and FEM. Commun. Numer. Methods Eng. **19**(8), 615–621 (2003)
42. L.T. Zhang, A. Gerstenberger, X. Wang, W.K. Liu, Immersed finite element method. Comput. Methods Appl. Mech. Eng. **193**(21–22), 2051–2067 (2004)

Editorial Policy

1. Volumes in the following three categories will be published in LNCSE:

i) Research monographs
ii) Tutorials
iii) Conference proceedings

Those considering a book which might be suitable for the series are strongly advised to contact the publisher or the series editors at an early stage.

2. Categories i) and ii). Tutorials are lecture notes typically arising via summer schools or similar events, which are used to teach graduate students. These categories will be emphasized by Lecture Notes in Computational Science and Engineering. **Submissions by interdisciplinary teams of authors are encouraged.** The goal is to report new developments – quickly, informally, and in a way that will make them accessible to non-specialists. In the evaluation of submissions timeliness of the work is an important criterion. Texts should be well-rounded, well-written and reasonably self-contained. In most cases the work will contain results of others as well as those of the author(s). In each case the author(s) should provide sufficient motivation, examples, and applications. In this respect, Ph.D. theses will usually be deemed unsuitable for the Lecture Notes series. Proposals for volumes in these categories should be submitted either to one of the series editors or to Springer-Verlag, Heidelberg, and will be refereed. A provisional judgement on the acceptability of a project can be based on partial information about the work: a detailed outline describing the contents of each chapter, the estimated length, a bibliography, and one or two sample chapters – or a first draft. A final decision whether to accept will rest on an evaluation of the completed work which should include

- at least 100 pages of text;
- a table of contents;
- an informative introduction perhaps with some historical remarks which should be accessible to readers unfamiliar with the topic treated;
- a subject index.

3. Category iii). Conference proceedings will be considered for publication provided that they are both of exceptional interest and devoted to a single topic. One (or more) expert participants will act as the scientific editor(s) of the volume. They select the papers which are suitable for inclusion and have them individually refereed as for a journal. Papers not closely related to the central topic are to be excluded. Organizers should contact the Editor for CSE at Springer at the planning stage, see *Addresses* below.

In exceptional cases some other multi-author-volumes may be considered in this category.

4. Only works in English will be considered. For evaluation purposes, manuscripts may be submitted in print or electronic form, in the latter case, preferably as pdf- or zipped ps-files. Authors are requested to use the LaTeX style files available from Springer at http://www.springer.com/gp/authors-editors/book-authors-editors/manuscript-preparation/5636 (Click on LaTeX Template → monographs or contributed books).

For categories ii) and iii) we strongly recommend that all contributions in a volume be written in the same LaTeX version, preferably LaTeX2e. Electronic material can be included if appropriate. Please contact the publisher.

Careful preparation of the manuscripts will help keep production time short besides ensuring satisfactory appearance of the finished book in print and online.

5. The following terms and conditions hold. Categories i), ii) and iii):

Authors receive 50 free copies of their book. No royalty is paid.
Volume editors receive a total of 50 free copies of their volume to be shared with authors, but no royalties.

Authors and volume editors are entitled to a discount of 33.3 % on the price of Springer books purchased for their personal use, if ordering directly from Springer.

6. Springer secures the copyright for each volume.

Addresses:

Timothy J. Barth
NASA Ames Research Center
NAS Division
Moffett Field, CA 94035, USA
barth@nas.nasa.gov

Michael Griebel
Institut für Numerische Simulation
der Universität Bonn
Wegelerstr. 6
53115 Bonn, Germany
griebel@ins.uni-bonn.de

David E. Keyes
Mathematical and Computer Sciences
and Engineering
King Abdullah University of Science
and Technology
P.O. Box 55455
Jeddah 21534, Saudi Arabia
david.keyes@kaust.edu.sa

and

Department of Applied Physics
and Applied Mathematics
Columbia University
500 W. 120 th Street
New York, NY 10027, USA
kd2112@columbia.edu

Risto M. Nieminen
Department of Applied Physics
Aalto University School of Science
and Technology
00076 Aalto, Finland
risto.nieminen@aalto.fi

Dirk Roose
Department of Computer Science
Katholieke Universiteit Leuven
Celestijnenlaan 200A
3001 Leuven-Heverlee, Belgium
dirk.roose@cs.kuleuven.be

Tamar Schlick
Department of Chemistry
and Courant Institute
of Mathematical Sciences
New York University
251 Mercer Street
New York, NY 10012, USA
schlick@nyu.edu

Editor for Computational Science
and Engineering at Springer:
Martin Peters
Springer-Verlag
Mathematics Editorial IV
Tiergartenstrasse 17
69121 Heidelberg, Germany
martin.peters@springer.com

Lecture Notes in Computational Science and Engineering

1. D. Funaro, *Spectral Elements for Transport-Dominated Equations.*
2. H.P. Langtangen, *Computational Partial Differential Equations.* Numerical Methods and Diffpack Programming.
3. W. Hackbusch, G. Wittum (eds.), *Multigrid Methods V.*
4. P. Deuflhard, J. Hermans, B. Leimkuhler, A.E. Mark, S. Reich, R.D. Skeel (eds.), *Computational Molecular Dynamics: Challenges, Methods, Ideas.*
5. D. Kröner, M. Ohlberger, C. Rohde (eds.), *An Introduction to Recent Developments in Theory and Numerics for Conservation Laws.*
6. S. Turek, *Efficient Solvers for Incompressible Flow Problems.* An Algorithmic and Computational Approach.
7. R. von Schwerin, ***M**ulti **B**ody **S**ystem **SIM**ulation.* Numerical Methods, Algorithms, and Software.
8. H.-J. Bungartz, F. Durst, C. Zenger (eds.), *High Performance Scientific and Engineering Computing.*
9. T.J. Barth, H. Deconinck (eds.), *High-Order Methods for Computational Physics.*
10. H.P. Langtangen, A.M. Bruaset, E. Quak (eds.), *Advances in Software Tools for Scientific Computing.*
11. B. Cockburn, G.E. Karniadakis, C.-W. Shu (eds.), *Discontinuous Galerkin Methods.* Theory, Computation and Applications.
12. U. van Rienen, *Numerical Methods in Computational Electrodynamics.* Linear Systems in Practical Applications.
13. B. Engquist, L. Johnsson, M. Hammill, F. Short (eds.), *Simulation and Visualization on the Grid.*
14. E. Dick, K. Riemslagh, J. Vierendeels (eds.), *Multigrid Methods VI.*
15. A. Frommer, T. Lippert, B. Medeke, K. Schilling (eds.), *Numerical Challenges in Lattice Quantum Chromodynamics.*
16. J. Lang, *Adaptive Multilevel Solution of Nonlinear Parabolic PDE Systems.* Theory, Algorithm, and Applications.
17. B.I. Wohlmuth, *Discretization Methods and Iterative Solvers Based on Domain Decomposition.*
18. U. van Rienen, M. Günther, D. Hecht (eds.), *Scientific Computing in Electrical Engineering.*
19. I. Babuška, P.G. Ciarlet, T. Miyoshi (eds.), *Mathematical Modeling and Numerical Simulation in Continuum Mechanics.*
20. T.J. Barth, T. Chan, R. Haimes (eds.), *Multiscale and Multiresolution Methods.* Theory and Applications.
21. M. Breuer, F. Durst, C. Zenger (eds.), *High Performance Scientific and Engineering Computing.*
22. K. Urban, *Wavelets in Numerical Simulation.* Problem Adapted Construction and Applications.
23. L.F. Pavarino, A. Toselli (eds.), *Recent Developments in Domain Decomposition Methods.*

24. T. Schlick, H.H. Gan (eds.), *Computational Methods for Macromolecules: Challenges and Applications.*
25. T.J. Barth, H. Deconinck (eds.), *Error Estimation and Adaptive Discretization Methods in Computational Fluid Dynamics.*
26. M. Griebel, M.A. Schweitzer (eds.), *Meshfree Methods for Partial Differential Equations.*
27. S. Müller, *Adaptive Multiscale Schemes for Conservation Laws.*
28. C. Carstensen, S. Funken, W. Hackbusch, R.H.W. Hoppe, P. Monk (eds.), *Computational Electromagnetics.*
29. M.A. Schweitzer, *A Parallel Multilevel Partition of Unity Method for Elliptic Partial Differential Equations.*
30. T. Biegler, O. Ghattas, M. Heinkenschloss, B. van Bloemen Waanders (eds.), *Large-Scale PDE-Constrained Optimization.*
31. M. Ainsworth, P. Davies, D. Duncan, P. Martin, B. Rynne (eds.), *Topics in Computational Wave Propagation.* Direct and Inverse Problems.
32. H. Emmerich, B. Nestler, M. Schreckenberg (eds.), *Interface and Transport Dynamics.* Computational Modelling.
33. H.P. Langtangen, A. Tveito (eds.), *Advanced Topics in Computational Partial Differential Equations.* Numerical Methods and Diffpack Programming.
34. V. John, *Large Eddy Simulation of Turbulent Incompressible Flows.* Analytical and Numerical Results for a Class of LES Models.
35. E. Bänsch (ed.), *Challenges in Scientific Computing - CISC 2002.*
36. B.N. Khoromskij, G. Wittum, *Numerical Solution of Elliptic Differential Equations by Reduction to the Interface.*
37. A. Iske, *Multiresolution Methods in Scattered Data Modelling.*
38. S.-I. Niculescu, K. Gu (eds.), *Advances in Time-Delay Systems.*
39. S. Attinger, P. Koumoutsakos (eds.), *Multiscale Modelling and Simulation.*
40. R. Kornhuber, R. Hoppe, J. Périaux, O. Pironneau, O. Wildlund, J. Xu (eds.), *Domain Decomposition Methods in Science and Engineering.*
41. T. Plewa, T. Linde, V.G. Weirs (eds.), *Adaptive Mesh Refinement – Theory and Applications.*
42. A. Schmidt, K.G. Siebert, *Design of Adaptive Finite Element Software.* The Finite Element Toolbox ALBERTA.
43. M. Griebel, M.A. Schweitzer (eds.), *Meshfree Methods for Partial Differential Equations II.*
44. B. Engquist, P. Lötstedt, O. Runborg (eds.), *Multiscale Methods in Science and Engineering.*
45. P. Benner, V. Mehrmann, D.C. Sorensen (eds.), *Dimension Reduction of Large-Scale Systems.*
46. D. Kressner, *Numerical Methods for General and Structured Eigenvalue Problems.*
47. A. Boriçi, A. Frommer, B. Joó, A. Kennedy, B. Pendleton (eds.), *QCD and Numerical Analysis III.*
48. F. Graziani (ed.), *Computational Methods in Transport.*
49. B. Leimkuhler, C. Chipot, R. Elber, A. Laaksonen, A. Mark, T. Schlick, C. Schütte, R. Skeel (eds.), *New Algorithms for Macromolecular Simulation.*

50. M. Bücker, G. Corliss, P. Hovland, U. Naumann, B. Norris (eds.), *Automatic Differentiation: Applications, Theory, and Implementations.*

51. A.M. Bruaset, A. Tveito (eds.), *Numerical Solution of Partial Differential Equations on Parallel Computers.*

52. K.H. Hoffmann, A. Meyer (eds.), *Parallel Algorithms and Cluster Computing.*

53. H.-J. Bungartz, M. Schäfer (eds.), *Fluid-Structure Interaction.*

54. J. Behrens, *Adaptive Atmospheric Modeling.*

55. O. Widlund, D. Keyes (eds.), *Domain Decomposition Methods in Science and Engineering XVI.*

56. S. Kassinos, C. Langer, G. Iaccarino, P. Moin (eds.), *Complex Effects in Large Eddy Simulations.*

57. M. Griebel, M.A Schweitzer (eds.), *Meshfree Methods for Partial Differential Equations III.*

58. A.N. Gorban, B. Kégl, D.C. Wunsch, A. Zinovyev (eds.), *Principal Manifolds for Data Visualization and Dimension Reduction.*

59. H. Ammari (ed.), *Modeling and Computations in Electromagnetics: A Volume Dedicated to Jean-Claude Nédélec.*

60. U. Langer, M. Discacciati, D. Keyes, O. Widlund, W. Zulehner (eds.), *Domain Decomposition Methods in Science and Engineering XVII.*

61. T. Mathew, *Domain Decomposition Methods for the Numerical Solution of Partial Differential Equations.*

62. F. Graziani (ed.), *Computational Methods in Transport: Verification and Validation.*

63. M. Bebendorf, *Hierarchical Matrices.* A Means to Efficiently Solve Elliptic Boundary Value Problems.

64. C.H. Bischof, H.M. Bücker, P. Hovland, U. Naumann, J. Utke (eds.), *Advances in Automatic Differentiation.*

65. M. Griebel, M.A. Schweitzer (eds.), *Meshfree Methods for Partial Differential Equations IV.*

66. B. Engquist, P. Lötstedt, O. Runborg (eds.), *Multiscale Modeling and Simulation in Science.*

67. I.H. Tuncer, Ü. Gülcat, D.R. Emerson, K. Matsuno (eds.), *Parallel Computational Fluid Dynamics 2007.*

68. S. Yip, T. Diaz de la Rubia (eds.), *Scientific Modeling and Simulations.*

69. A. Hegarty, N. Kopteva, E. O'Riordan, M. Stynes (eds.), *BAIL* 2008 – *Boundary and Interior Layers.*

70. M. Bercovier, M.J. Gander, R. Kornhuber, O. Widlund (eds.), *Domain Decomposition Methods in Science and Engineering XVIII.*

71. B. Koren, C. Vuik (eds.), *Advanced Computational Methods in Science and Engineering.*

72. M. Peters (ed.), *Computational Fluid Dynamics for Sport Simulation.*

73. H.-J. Bungartz, M. Mehl, M. Schäfer (eds.), *Fluid Structure Interaction II - Modelling, Simulation, Optimization.*

74. D. Tromeur-Dervout, G. Brenner, D.R. Emerson, J. Erhel (eds.), *Parallel Computational Fluid Dynamics 2008.*

75. A.N. Gorban, D. Roose (eds.), *Coping with Complexity: Model Reduction and Data Analysis.*

76. J.S. Hesthaven, E.M. Rønquist (eds.), *Spectral and High Order Methods for Partial Differential Equations.*

77. M. Holtz, *Sparse Grid Quadrature in High Dimensions with Applications in Finance and Insurance.*

78. Y. Huang, R. Kornhuber, O.Widlund, J. Xu (eds.), *Domain Decomposition Methods in Science and Engineering XIX.*

79. M. Griebel, M.A. Schweitzer (eds.), *Meshfree Methods for Partial Differential Equations V.*

80. P.H. Lauritzen, C. Jablonowski, M.A. Taylor, R.D. Nair (eds.), *Numerical Techniques for Global Atmospheric Models.*

81. C. Clavero, J.L. Gracia, F.J. Lisbona (eds.), *BAIL 2010 – Boundary and Interior Layers, Computational and Asymptotic Methods.*

82. B. Engquist, O. Runborg, Y.R. Tsai (eds.), *Numerical Analysis and Multiscale Computations.*

83. I.G. Graham, T.Y. Hou, O. Lakkis, R. Scheichl (eds.), *Numerical Analysis of Multiscale Problems.*

84. A. Logg, K.-A. Mardal, G. Wells (eds.), *Automated Solution of Differential Equations by the Finite Element Method.*

85. J. Blowey, M. Jensen (eds.), *Frontiers in Numerical Analysis - Durham 2010.*

86. O. Kolditz, U.-J. Gorke, H. Shao, W. Wang (eds.), *Thermo-Hydro-Mechanical-Chemical Processes in Fractured Porous Media - Benchmarks and Examples.*

87. S. Forth, P. Hovland, E. Phipps, J. Utke, A. Walther (eds.), *Recent Advances in Algorithmic Differentiation.*

88. J. Garcke, M. Griebel (eds.), *Sparse Grids and Applications.*

89. M. Griebel, M.A. Schweitzer (eds.), *Meshfree Methods for Partial Differential Equations VI.*

90. C. Pechstein, *Finite and Boundary Element Tearing and Interconnecting Solvers for Multiscale Problems.*

91. R. Bank, M. Holst, O. Widlund, J. Xu (eds.), *Domain Decomposition Methods in Science and Engineering XX.*

92. H. Bijl, D. Lucor, S. Mishra, C. Schwab (eds.), *Uncertainty Quantification in Computational Fluid Dynamics.*

93. M. Bader, H.-J. Bungartz, T. Weinzierl (eds.), *Advanced Computing.*

94. M. Ehrhardt, T. Koprucki (eds.), *Advanced Mathematical Models and Numerical Techniques for Multi-Band Effective Mass Approximations.*

95. M. Azaïez, H. El Fekih, J.S. Hesthaven (eds.), *Spectral and High Order Methods for Partial Differential Equations ICOSAHOM 2012.*

96. F. Graziani, M.P. Desjarlais, R. Redmer, S.B. Trickey (eds.), *Frontiers and Challenges in Warm Dense Matter.*

97. J. Garcke, D. Pflüger (eds.), *Sparse Grids and Applications – Munich 2012.*

98. J. Erhel, M. Gander, L. Halpern, G. Pichot, T. Sassi, O. Widlund (eds.), *Domain Decomposition Methods in Science and Engineering XXI.*

99. R. Abgrall, H. Beaugendre, P.M. Congedo, C. Dobrzynski, V. Perrier, M. Ricchiuto (eds.), *High Order Nonlinear Numerical Methods for Evolutionary PDEs - HONOM 2013.*

100. M. Griebel, M.A. Schweitzer (eds.), *Meshfree Methods for Partial Differential Equations VII.*

101. R. Hoppe (ed.), *Optimization with PDE Constraints - OPTPDE 2014.*

102. S. Dahlke, W. Dahmen, M. Griebel, W. Hackbusch, K. Ritter, R. Schneider, C. Schwab, H. Yserentant (eds.), *Extraction of Quantifiable Information from Complex Systems.*

103. A. Abdulle, S. Deparis, D. Kressner, F. Nobile, M. Picasso (eds.), *Numerical Mathematics and Advanced Applications - ENUMATH 2013.*

104. T. Dickopf, M.J. Gander, L. Halpern, R. Krause, L.F. Pavarino (eds.), *Domain Decomposition Methods in Science and Engineering XXII.*

105. M. Mehl, M. Bischoff, M. Schäfer (eds.), *Recent Trends in Computational Engineering - CE2014.* Optimization, Uncertainty, Parallel Algorithms, Coupled and Complex Problems.

106. R.M. Kirby, M. Berzins, J.S. Hesthaven (eds.), *Spectral and High Order Methods for Partial Differential Equations - ICOSAHOM'14.*

107. B. Jüttler, B. Simeon (eds.), *Isogeometric Analysis and Applications 2014.*

108. P. Knobloch (ed.), *Boundary and Interior Layers, Computational and Asymptotic Methods – BAIL 2014.*

109. J. Garcke, D. Pflüger (eds.), *Sparse Grids and Applications – Stuttgart 2014.*

110. H. P. Langtangen, *Finite Difference Computing with Exponential Decay Models.*

111. A. Tveito, G.T. Lines, *Computing Characterizations of Drugs for Ion Channels and Receptors Using Markov Models.*

112. B. Karazösen, M. Manguoğlu, M. Tezer-Sezgin, S. Göktepe, Ö. Uğur (eds.), *Numerical Mathematics and Advanced Applications - ENUMATH 2015.*

113. H.-J. Bungartz, P. Neumann, W.E. Nagel (eds.), *Software for Exascale Computing - SPPEXA 2013-2015.*

114. G.R. Barrenechea, F. Brezzi, A. Cangiani, E.H. Georgoulis (eds.), *Building Bridges: Connections and Challenges in Modern Approaches to Numerical Partial Differential Equations.*

115. M. Griebel, M.A. Schweitzer (eds.), *Meshfree Methods for Partial Differential Equations VIII.*

116. C.-O. Lee, X.-C. Cai, D.E. Keyes, H.H. Kim, A. Klawonn, E.-J. Park, O.B. Widlund (eds.), *Domain Decomposition Methods in Science and Engineering XXIII.*

For further information on these books please have a look at our mathematics catalogue at the following URL: `www.springer.com/series/3527`

Monographs in Computational Science and Engineering

1. J. Sundnes, G.T. Lines, X. Cai, B.F. Nielsen, K.-A. Mardal, A. Tveito, *Computing the Electrical Activity in the Heart.*

For further information on this book, please have a look at our mathematics catalogue at the following URL: www.springer.com/series/7417

Texts in Computational Science and Engineering

1. H. P. Langtangen, *Computational Partial Differential Equations.* Numerical Methods and Diffpack Programming. 2nd Edition
2. A. Quarteroni, F. Saleri, P. Gervasio, *Scientific Computing with MATLAB and Octave.* 4th Edition
3. H. P. Langtangen, *Python Scripting for Computational Science.* 3rd Edition
4. H. Gardner, G. Manduchi, *Design Patterns for e-Science.*
5. M. Griebel, S. Knapek, G. Zumbusch, *Numerical Simulation in Molecular Dynamics.*
6. H. P. Langtangen, *A Primer on Scientific Programming with Python.* 5th Edition
7. A. Tveito, H. P. Langtangen, B. F. Nielsen, X. Cai, *Elements of Scientific Computing.*
8. B. Gustafsson, *Fundamentals of Scientific Computing.*
9. M. Bader, *Space-Filling Curves.*
10. M. Larson, F. Bengzon, *The Finite Element Method: Theory, Implementation and Applications.*
11. W. Gander, M. Gander, F. Kwok, *Scientific Computing: An Introduction using Maple and MATLAB.*
12. P. Deuflhard, S. Röblitz, *A Guide to Numerical Modelling in Systems Biology.*
13. M. H. Holmes, *Introduction to Scientific Computing and Data Analysis.*
14. S. Linge, H. P. Langtangen, *Programming for Computations* - A Gentle Introduction to Numerical Simulations with MATLAB/Octave.
15. S. Linge, H. P. Langtangen, *Programming for Computations* - A Gentle Introduction to Numerical Simulations with Python.
16. H.P. Langtangen, S. Linge, *Finite Difference Computing with PDEs* - A Modern Software Approach

For further information on these books please have a look at our mathematics catalogue at the following URL: www.springer.com/series/5151

Printed in the United States
By Bookmasters

Printed in the United States
By Bookmasters